Silicon Micromachining

This comprehensive book provides an overview of the key techniques used in the fabrication of micron-scale structures in silicon. Recent advances in these techniques have made it possible to create a new generation of microsystem devices, such as microsensors, accelerometers, micropumps, and miniature robots.

The authors underpin the discussion of each technique with a brief review of the fundamental physical and chemical principles involved. They pay particular attention to methods such as isotropic and anisotropic wet chemical etching, wafer bonding, reactive ion etching, and surface micromachining. There is a special section on bulk micromachining, and the authors also discuss release mechanisms for movable microstructures.

The book is a blend of detailed experimental and theoretical material, and it will be of great interest to graduate students and researchers in electrical engineering and materials science whose work involves the study of micro-electromechanical systems (MEMS).

Miko Elwenspoek, a professor of transducer technology at the University of Twente, studied physics at the Free University of West Berlin. He joined the University of Twente in 1987, where he heads the micromechanics group at the MESA Research Institute.

Henrie Jansen received his Ph.D. from the University of Twente on the subject of plasma etching in microsystem technology. After having worked for one year as a plasma researcher at the micromechanical group, he joined CSEM in Neuchâtel, Switzerland. There he spent one year as a specialist in plasma engineering. Currently, he is a plasma engineer at the University of Twente.

**Cambridge Studies in Semiconductor Physics
and Microelectronic Engineering: 7**

TITLES IN THIS SERIES

SILICON MICROMACHINING

M. ELWENSPOEK

and

H. V. JANSEN

CAMBRIDGE
UNIVERSITY PRESS

PUBLISHED BY THE PRESS SYNDICATE OF THE UNIVERSITY OF CAMBRIDGE
The Pitt Building, Trumpington Street, Cambridge, United Kingdom

CAMBRIDGE UNIVERSITY PRESS
The Edinburgh Building, Cambridge CB2 2RU, UK
40 West 20th Street, New York NY 10011–4211, USA
477 Williamstown Road, Port Melbourne, VIC 3207, Australia
Ruiz de Alarcón 13, 28014 Madrid, Spain
Dock House, The Waterfront, Cape Town 8001, South Africa

http://www.cambridge.org

© Cambridge University Press 1998

First published 1998
First paperback edition 2004

Typset in Times Roman 10/12.5 pt. in LAT$_E$X 2$_\varepsilon$ [TB]

A catalogue record for this book is available from the British Library

Library of Congress Cataloguing-in-Publication Data
Elwenspoek, M. (Miko), 1948–
Silicon micromachining / M. Elwenspoek, H. Jansen.
p.   cm. – (Cambridge studies in semiconductor physics and
microelectronic engineering; 7)
ISBN 0 521 59054 X (hardback)
1. Silicon.   2. Micromachining.   3. Etching.   I. Jansen, H.
(Henri)   II. Title.   III. Series.
TK7871.15.S55E49   1998
621.3815 – dc21   97-43732
                                                CIP

ISBN 0 521 59054 X hardback
ISBN 0 521 60767 1 paperback

Contents

Preface

This book on silicon micromachining has evolved from notes for a course "Etching Technology," which was developed in the framework of the European program COMETT. The organization of the course development was done by FSRM in Neuchâtel, Switzerland, and it was (and still is) part of an educational program in microsystems technology, named UETP MEMS. Part of the material has been used for lecture notes for the course "Micromechanical Devices and Systems: Etching of Silicon" at the University of Twente.

The book consists of two large blocks, the first one (Chapters 1–7) on micromachining using more "traditional" wet etching processes and wafer bonding, and a second block in which we concentrate on dry etching of silicon, in particular reactive ion etching (RIE). H. V. J. is the main author of the latter, and M. E. is the main author of the first block. Especially the second block has changed a lot as compared to the first versions of the material. In these early versions, the author of the dry etch part was Meint de Boer.

A number of colleagues actively contributed to the book by suggestions and supplying material. These are in the first place Rob Legtenberg, Han Gardeniers, Tonny Sonnenberg, Bas Deheij, Meint de Boer, and Erwin Berenschot, all of the Micromechanics group at MESA. The whole however would have been impossible without the work of our colleagues at the micromechanics group of MESA: Frans Blom, Siebe Bouwstra, Hans-Elias de Bree, Johannes Burger, Gert-Jan Burger, Gui Chengun, Job Elders, Twan Korthorst, Joost Van Kuijk, Stein Kuijpers, Theo Lammerink, Peter Leussink, Cees van Mullem, Cristina Neagu, Jasper Nijdam, Wietse Nijdam, Edwin Oosterbroek, Frans van de Pol, Albert Prak, Cees Van Rijn, Stefan Sanches, Edwin Smulders, Vincent Spiering, Jaap van Suchtelen, Niels Tas, Harrie Tilmans, Willem Tjerkstra, Erik van Veenendaal, Theo Veenstra, Henk Wensink, Remco Wiegerink, and Robert Zwijze. The contribution of all these people in one or another form is gratefully acknowledged. Special thanks are for Jan Fluitman as the inspiration of our group.

M. E. wishes also to express his thanks to the Uppsala (Sweden) Micromachining Group (in particular to Bertil Hök, Ylva Bäcklund, Leif Smidt, Lars Rosengren, and Karin Hermansson) for a wonderful time in their lab. He learned a lot of micromachining in Uppsala.

The ideas presented in Chapter 3 on chemical physics of wet etching go back to M. E.'s time as a post doc in Piet Bennema's group at the University of Nijmegen, The Netherlands.

1

Introduction

Etching and bonding of silicon are basic technologies in micro systems technology (MST). MST is currently developing very fast, partially due to the large financial backing which is put into its development by governments and companies.

The promises of MST are formidable. The main promise stems from the fact that MST is based on batch fabrication derived from integrated circuit technology. It is anticipated that MST will undergo an evolution similar to integrated circuit (IC) technology, with new inexpensive products that will in part revolutionize our lives. The ideas on the potential of MST range from devices which are nowadays found in nearly all automobiles (pressure sensors and accelerometers) to sensors for flow, temperature, force, position, magnetic fields, chemicals, light, IR-radiation, etc.; either inexpensive or with unmet performance. In our view, the next generation of devices will use micropumps, flow sensors, micromixers, microsieves, microreactors, etc. for dosing in medicine, biology, chemical, and biochemical analysis systems with applications in environmental monitoring, medicine, process control, and chemical analysis in general.

Looking further into the future, we will have microrobots in a vast number of applications. They will be found in charge coupled device (CCD) cameras for active optimizing of the position of the CCD chip, and transporting and positioning the optic components in CD setups. There will be robots that help physicians in microsurgery, robots that repair and maintain other microsystems, and microrobots that perform fabrication and inspection tasks in microclean rooms. Most of the latter are a few decades away, and – very likely – main future applications of MST are lacking in the list, being as unpredictable as the application of computers for games was twenty years ago.

At present we have to solve a great deal of problems. These problems are associated with the control of the technology; we have to learn how to design microsystems, and we have to develop tools for design, modelling, and simulation of microsystems. We are encountering new aspects of physics and chemistry when scaling down known systems. We have to translate three-dimensional principles, design, and fabrication to planar, say 2.5-dimensional, principles, design, and fabrication.

The microworld is different from the macroworld. The microworld is governed by surface forces such as surface tension and friction. Inertia and weight have meaning in microsystems only in exceptional cases, such as the accelerometer or the resonant sensor, and electrostatic forces and fluid shear forces are large and heavily deform mechanical constructions. Streaming is at extremely small Reynolds numbers, so all fluids appear quite viscous. Surfaces appear very rough, and mechanical constructions are very stiff and strong. In systems of μm size the temperature of the surroundings would manifest itself by violent Brownian motion.

To give an illustrative example, ants are unable to wash their face: they are not strong enough to break through the surface tension of a water droplet. Once inside a water droplet they would drown because they couldn't leave it. They could never maintain a fire, because the flame is too large. They could never read a book: the surface tension of the micropages would forbid ants to open the book. If we were to shrink by using a fantastic machine, as in the movie "The Fantastic Voyage," we would certainly be unable to survive a fight against an ant, just because we are wrongly designed for this scale.

The physics of the microworld and the resulting possibilities for microsystems were anticipated by R. P. Feynman in 1959. The article ("There is plenty room at the bottom") has been reprinted in the *Journal of Microelectromechanical Systems* [1]. Feynman anticipated the important role of photolithography, thin film techniques, high density recording, and electrostatic motors. Twenty years later, Feynman gave a second lecture on the subject, a written version of which can be found in [2].

When designing microsystems, we have to learn how this microworld feels. The designer has to be trained for a new intuition. If one looks at scaling from a phenomenological point of view, one finds some interesting trends. In Fig. 1.1 we show how the price per mass for machines and mechanical components depends on the scale. It is quite evident that small things are relatively more expensive than large things.

Figure 1.2 demonstrates a striking scale effect. The Reynolds number, as defined in the figure, lies on a single line for animals over the whole range between the smallest animals (bacteria) and the largest animals (whales).

The technology for microsystems is rooted in two quite different traditions. One derives from fine mechanics, and the other is based on photolithographic techniques. The latter allows fabrication of structures with finest details in the order of 1 μm and a relative precision of not more than say 1/10,000. The limit of fine mechanics is close to 10 μm; however, the relative precision is one or two orders larger. The most important difference between these

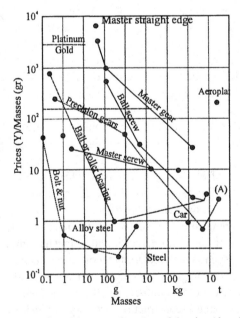

Fig. 1.1. Price per mass of machines as a function of the size (data from Hayashi [3]).

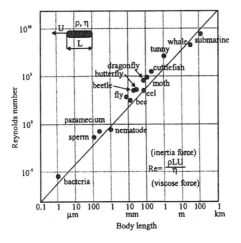

Fig. 1.2. Reynolds number as a function of size (from Hayashi [3]).

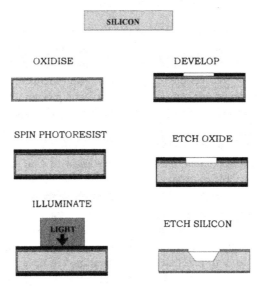

Fig. 1.3. Basic photolithographic process.

techniques is that the latter technology is suitable for batch processes but is restricted to projections of a two-dimensional structure in three dimensions.

This book concentrates on the latter. In Fig. 1.3 we show the basic photolithographic process. We see how "three-dimensional" structures are realized by using a two-dimensional structure. This "three-dimensional" structure is not truly three dimensional: we will always see that the structures derive from a pattern in a plane. In some way or other (Fig. 1.3) we only get projections. We cannot make screws and nuts, but they would not be of much help: we have no systems to assemble microsystems.

Here is one of the challenges for the microengineer: to design microsystems that need not be assembled part by part, but that come out of the production process assembled. We cannot make microcomponents and thereafter assemble them: we have to design microsystems. Microtechnology does not only restrict the design possibilities, but also offers new design

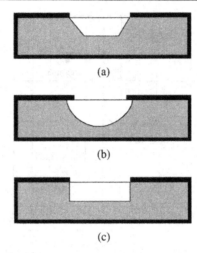

Fig. 1.4. Possibilities for etching. (a) and (c) anisotropic etching, (b) isotropic etching.

principles. We can fabricate microbridges with a thickness of 1μm and a length of 1 cm. In a macroworld this would correspond to a bridge of 1 km length and 10 cm thick!

In Fig. 1.4 the various possibilities for etching are indicated. All shapes shown can be realized by both dry and wet etching. However, wet chemical anisotropic etching is restricted to crystallographic orientations, while dry etching is not. Dry etching is therefore more flexible than wet etching, but the process is more difficult to control and the equipment is much more expensive.

Engineers who do their work in MST are faced with the fact that MST is a multidisciplinary field. In the optimal situation, the engineer must be trained in basic physics and chemistry in order to be able to understand design and modelling of the microdevices and systems. He needs to know quite a lot about mechanical engineering, in particular, strength of materials, theory of elasticity, and tribology. He needs the typical skills of an electrical engineer with respect to systems thinking, and he must be able to use the various simulation tools, such as finite elements and SPICE. He must be aquainted with the fundamentals of optics and magnetics. He will find that each application will have its own difficulties. For example, an application in the medical field demands that the engineer be able to communicate with the physician; an application in space travel again demands a totally different knowledge and background.

All this cannot be combined in a single person. Research in MST is only possible in teams in which physicists, chemists, and electrical and mechanical engineers co-operate. We believe however that the high degree of interdisciplinary demands in the long term that engineers are trained not in a single discipline, but start from one discipline to gain knowledge and skills from other disciplines as well. Therefore, it is necessary to design new curriculum in MST.

References

[1] R. P. Feynman, J.MEMS **1**, 60 (1992).

[2] S. D. Senturia, J.MEMS **4**, 309 (1995).

[3] T. Hayashi, Proc. MEMS'94, Oiso, Japan, January 25–28, p. 39, 1994.

2

Anisotropic wet chemical etching

2.1 Introduction

Wet chemical anisotropic etching of silicon is one of the key technologies of silicon micromachining. As opposed to "surface micromachining," where structures are fabricated from thin films that are released by films of a material that can be removed without attacking the structural material, wet chemical anisotropic etching is also referred to as "bulk micromachining," because in this technology the body of the silicon wafer is etched away.

The anisotropy of the etching stems from the crystal structure of silicon, and the shapes that can be realized by bulk micromachining are restricted to those that are bounded at least in part by slowly etching planes. This is dominantly the family of the {111} planes. Silicon wafers with a surface oriented in either $\langle 001 \rangle$ or $\langle 110 \rangle$ are covered by a thin film of material that etches much slower in the solution used to etch silicon, openings in this film are defined by standard photolithographic techniques, and the sample is immersed into the etchant. A scanning electron microphotograph (SEM) photograph, as seen in Fig. 2.1(a), shows the cross section of the etchpit formed in a certain anisotropic etching solution. This pit contrasts with the pit shown in Fig. 2.1(b), which is obtained when etching silicon through a mask opening in an agent that etches isotropically.

A second striking example that demonstrates the anisotropic character of the etching is seen when a sphere is etched. A sphere exposes all crystallographic orientations to the etchant, and obviously the form of the etched sphere is governed by the fast-etching orientations. The result is shown in Fig. 2.2, which shows a silicon sphere heavily etched in CsOH (for many hours). The flat regions now correspond to fast-etching orientations, and the vertices correspond to slow-etching orientations. One can clearly discern threefold and fourfold axes, corresponding to $\langle 111 \rangle$ and $\langle 001 \rangle$ directions, respectively.

This anisotropic character of etching silicon is of great technological importance. It allows us, starting with a mask opening neatly aligned along crystallographic orientations, to control precisely the shape and the dimensions of microstructures. The technology is now being used to fabricate sensors for pressure, force, flow, and acceleration using the advantages of batch processing. Thus, if the market demands these devices in large numbers, the production costs can be fairly small.

Single crystalline silicon has a few remarkable mechanical properties, making this material attractive for mechanical devices. These are reviewed briefly in Section 2.2. Since the anisotropy of the etching is related to the crystallography of silicon, we will discuss the relevant crystallographic aspects in Section 2.3.

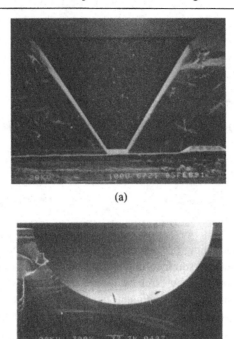

(a)

(b)

Fig. 2.1. Etch pits formed by etching silicon through a mask opening using (a) EDP (ethylenediamine, pyrocathecol, water mixture) and (b) HNO_3:HF:CH_3OOH.

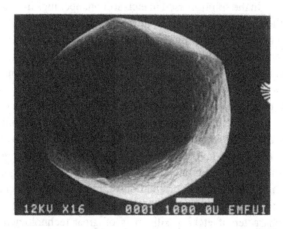

Fig. 2.2. A silicon sphere after heavy etching in CsOH. The facets are now oriented toward the fast-etching orientations, and the vertices point in the slow-etching orientations (i.e., $\langle 111 \rangle$ and $\langle 001 \rangle$) (from Hesketh et al. [1]).

In order to control micromachining using anisotropic wet chemical etching, a number of technologies have to be available. Photolithography is of course indispensable. This is a standard IC technology and is not described here in great detail. Furthermore, wafer cleaning facilities are necessary, and we shall comment on this aspect in Section 2.4 in which we describe a typical process to etch a structure. In this section we also discuss the various experimental techniques to study anisotropic etch rates. In Section 2.6 we review the properties of the etching solutions. We concentrate on KOH, EDP, and TMAH, since these systems are the most widely used and the most thoroughly studied. The following sections are devoted to etchstop mechanisms, mask materials, and corner compensation. In the latter section we also review the etch simulation programs that are currently being developed. The chemical–physical background of anisotropic etching will be treated in Chapter 3.

2.2 Mechanical properties of single crystalline silicon

Single crystalline silicon has a number of remarkable mechanical properties. Being a semiconductor, the chemical binding between the Si atoms is covalent. Covalent binding potentials are strongly anisotropic, and they have a deep minimum. This is in great contrast to metal binding potentials, and consequently, crystal dislocations are much more movable in metals than in semiconductor materials. Metals are ductile and semiconductors are brittle. Shaping of such a material by forging is impossible.

Mechanical engineers often try to avoid the use of brittle material such as glass for very obvious reasons. But ductile materials are easily plastically deformed, meaning that these materials are subject to mechanical hysteresis. Single crystalline silicon can be made perfect – virtually without any defects. Being loaded, there are no dislocation lines that are able to move, and the introduction of a new dislocation line would immediately cause a crack in the material – silicon would break. At room temperature, single crystalline silicon can only be elasticly deformed. There is no mechanical hysteresis.

This would not be so special if the yield strength of silicon were not extremely high. It is comparable in strength to stainless steel. Thus, silicon is a material of strength comparable to steel, but without any plastic deferability and no mechanical hysteresis. That fact makes silicon a material superior to any metal in many applications. Note, however, that stainless steel can stand a much larger strain than silicon. In Fig. 2.3 we show schematically stress-strain curves of steel and silicon.

The elastic deformability is most impressive in microstructures. Figure 2.4 shows an SEM of an etched silicon beam heavily loaded by a stylus [2]. If the load is released, the beam would return to its original shape.

In Table 2.1 (from Petersen [3]) we give a few figures of single crystalline silicon in comparison to other materials.

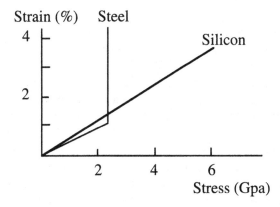

Fig. 2.3. Typical stress strain curves of metals and semiconductors.

Table 2.1. *(Compiled from Petersen [3]).*

	Yield strength (GPa)	Knoop Hardness (Kg/mm²)	Young's Modulus (100 GPa)	Density (1000 Kg/m³)	Thermal expansion (10^{-6}/°K)
Diamond	53	7000	10.35	3.5	1.0
SiC	21	2480	7.0	3.2	3.3
Si_3N_4	14	3486	3.85	3.1	0.8
Si	7	850	1.9	2.3	2.33
Stainless steel	2.1	660	2.0	7.9	17.3
Al	0.17	130	0.7	2.7	25

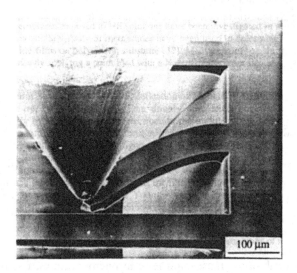

Fig. 2.4. SEM photograph of an etched silicon beam heavily loaded by a stylus force [2].

Gallium arsenide (GaAs) has similar properties [4], however, the yield load is smaller by a factor of two in comparison to silicon. Still, this material is interesting enough for micromechanical applications.

Under certain conditions, in particular at elevated temperature, we have to mention that plastic deformation of single crystalline silicon has been observed [5]. This observation, however, has little practical impact on silicon micromechanics.

If a piece is loaded even well below the yield strength, the material deforms quickly (the time scale is governed by the speed of sound). But after this deformation the piece continues to deform at a much larger time scale, which is measured in minutes. The deformation is reversible: if the load is removed, the material regains nearly all of its original form and dimensions quickly and continues to deform until the original form is exactly reached. This phenomenon is called creep. The effect can be rather large, i.e., the change of strain due to creep can be in the order of several thousandths. For silicon the effect is in the range of a few 10s of ppm. Thus, in this respect, silicon belongs to the superior materials.

No reports seem to exist in which fatigue of silicon has been observed. As an example, from experience in our own labs, it appears that the resonance frequency of a micromachined silicon resonator does not depend on time; if any time dependence was observed, then this was due to temperature instabilities or drifts in the mechanical load of the sample. Since the resonance frequency is of the order of several kHz, the material has undergone several million deformations within one hour.

According to Schweitz [2], polysilicon microstructures have mechanical properties (Young's modulus and yield strength) similar to single crystalline silicon. This is mainly due to the fact that dislocation lines in silicon are quite immobile at room temperature.

Inspection of Table 2.1 shows another material that is frequently used in microtechnology: silicon nitride. As can be seen, this is an exceptionally strong and hard material. Furthermore, it is inert to most acids and bases. As we shall see below, this material has proved itself in many interesting applications.

2.3   Crystallographic properties of silicon

The crystal structure of silicon is of the diamond type with lattice constant $a = 5.43$ Å. The structure is like F.C.C. (face centered cubic) but with two atoms in the unit cell. Take an F.C.C. lattice, cornered at A, and put another atom at B, a point $1/4$ of the way along the main diagonal of the cube (Fig. 2.5).

Crystal planes are characterized by sets of three indices, the so-called Miller indices. They describe vectors normal to the crystal planes in question. For example, in a simple cubic lattice, one finds atoms along the x, y, and z directions in a distance of an integer times the lattice constant a. The vectors $\mathbf{a_x}$ and $\mathbf{a_y}$ span a plane to which the vector $\mathbf{a_z}$ is normal. This plane is referred to as (001). Similarly, the (011) direction is normal to the plane formed by the vectors $\mathbf{a_x}$ and $(\mathbf{a_y} + \mathbf{a_z})$. Finally, (111) is normal to the plane formed by the vectors $(\mathbf{a_x} + \mathbf{a_z})$ and $(\mathbf{a_y} + \mathbf{a_z})$. The Miller indices are the components of the unit cell vectors in the reciprocal lattice. The {100}, {110}, and {111} are sufficient for our purpose. The {...} denotes a set of symmetrically equivalent planes, e.g., (100), (010), and (001) are all members of 100. The (100), (110), and (111) planes are shown in Fig. 2.6. More details can be found in standard textbooks on solid state physics (e.g., [6] and [7]). Crystallographic directions are indicated by $\langle \cdots \rangle$.

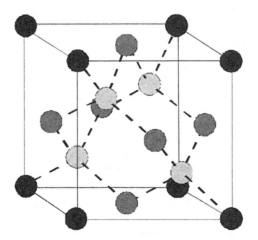

Fig. 2.5. Unit cell of the diamond lattice (e.g. [6]).

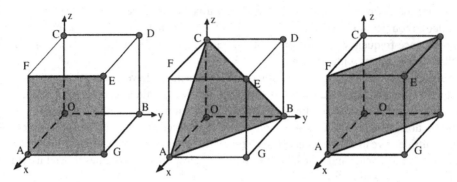

Fig. 2.6. The family of planes AFEG (100), ABC (111), and ABDF (110) in cubic lattices.

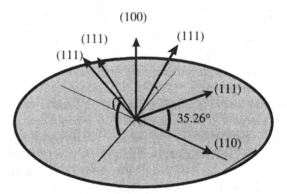

Fig. 2.7. (111) planes in the (100)-oriented wafer.

The orientation of the relevant crystal planes with respect to particular directions is of great importance to anisotropic etching. It is the (111) plane which etches by far the slowest, and therefore one needs to know the angles the (111) planes make with the wafer orientation. For micromechanics, two wafer orientations are of importance: (100) and (110). (111) wafers cannot be micromachined using anisotropic etchants (an exception will be discussed in Section 2.11.1). The orientations can be easily calculated using the inner product of the wafer orientations with the various (111) planes, since

$$\mathbf{A} \cdot \mathbf{B} = |\mathbf{A}||\mathbf{B}| \cos \phi.$$

For example, the angle between (111) and (001) is

$$\arccos\left\{ \frac{(111) \cdot (001)}{|(111)| \times |(001)|} \right\} = \arccos\left(\frac{1}{\sqrt{3} \times 1} \right) = 54.74°.$$

The orientations of the {111} planes of course can be found also by direct inspection of the elementary simple cubic cell. This takes some practice, but the most important results are summarized in Fig. 2.7 ({111} planes in the ⟨100⟩ wafer) and Fig. 2.8 ({111} planes in the ⟨110⟩ wafer).

For a better view, we show in Figs. 2.9 and 2.10 the form of crystals if they are built up of facets of various orientations.

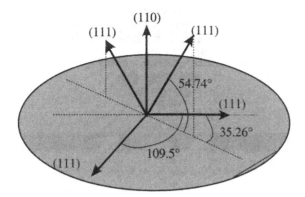

Fig. 2.8. (111) planes in the (110)-oriented wafer.

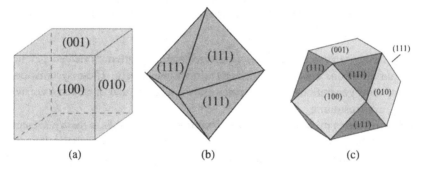

Fig. 2.9. Crystals dominated by: (a) {001} faces, (b) {111} faces, and (c) {111} and {001} faces.

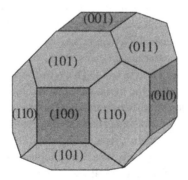

Fig. 2.10. {100} and {110} faces on the cube.

The atomic structure of the surfaces will be shown and discussed in Section 3.2.

2.4   Process of etching

In Fig. 2.11a we show a schematic of a mask for etching membranes of various size. The membranes have to be etched from the backside of the wafer, as shown in Fig. 2.11(b), and are planned to be 30-μm thick. As we shall discuss in Section 2.8, the etching can be stopped just by removing the wafer from the solution in time. Note the difference in the dimensions of the pattern on the mask and in the final result due to the inclined (111)

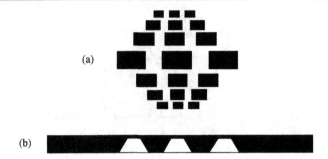

Fig. 2.11. (a) Mask for etching membranes on a ⟨100⟩ oriented wafer. (b) Lateral view of the etch result.

faces. We shall not describe here how such a mask can be generated. There are many mask generating systems on the market. There are only two comments we want to give on mask generation.

(i)   For IC design it is sufficient to generate rectangular patterns oriented parallel to a certain direction. This is insufficient for micromechanics. These systems do not (easily) allow generation of smooth patterns of, e.g., triangular or curved form, which can be very frustrating.

(ii) One has to keep in mind the minimum structure size one wishes to realize. Generally, it is very difficult to fabricate sizes as small as the system in principle allows. The famous electrostatic micromotors have a gap spacing of typically 1–2 mm, and it is by no means trivial to reach this limit.

We assume that we want to etch the black regions as seen on the mask in Fig. 2.11 in a ⟨100⟩ wafer. Let us start with a fresh wafer we got out of a box as delivered by the wafer manufacturer. This wafer will never be as clean as it was before we touched it with our tweezers. We have to deposit a material that can serve as a mask for etching. The simplest way is to grow a thermal oxide in a tube oven. The growth rate of oxide (more precisely, silicon dioxide, but in the lab and in the literature, silicon dioxide is very often referred to as "oxide") is strongly dependent on the temperature. The rate controlling process seems to be the diffusion of oxide atoms or molecules through the growing layer of silicon dioxide. The growth rate of wet and dry oxide is given in Fig. 2.12. In the "wet" process, steam is added to the oxygen that flows through the tube. The simplest way to add steam is to direct the oxygen through (nearly) boiling demineralized water. As seen from Fig. 2.12, water vapor increases the growth rate quite considerably.

The thickness of the oxide film required depends on a number of things. For the following etch step, one has to think of the etch rate of the oxide, which is not insignificant in KOH. This etch rate will be discussed later, but to give a figure, etching through a 300-mm wafer requires an oxide layer of 2 μm.

Wafers covered with a thin film look different. One sees polarization colors, which allow an estimate of the layer thickness of the film. Bare silicon is easily distinguished from wafers covered with oxide (and/or other layers).

On both sides of the oxidized wafer one has to spin negative photoresist. Photoresist manufacturers prefer to use a primer first. The resist does not stick to the wafer very well if the environment is humid or if the wafer has been previously immersed in water. In the

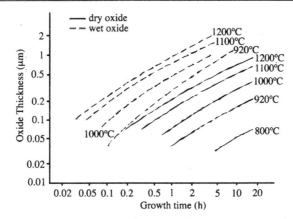

Fig. 2.12. Thickness of a thermally grown oxide layer as a function of the growth time for several temperatures. solid line: dry oxide, broken line: wet oxide (redrawn from Büttgenbach [8]).

latter case the wafer should be annealed well above 100°C for an hour or so. In the former case, using a primer helps, but perfect climate control (40% humidity) is better.

One gets a ca. 1-μm thick resist film when putting ca. 20 droplets resist on the wafer and spinning at 5000 rpm. For exact figures, consult the photoresist manufacturer. The spinning requires some experience.

Typically, one has to anneal the resist at 90°C for 10 minutes before spinning the other side of the wafer or before the illumination step. The annealing can be done on hot plates or in ovens. This step is usually called "prebake."

The next step is illumination of the resist. This often requires alignment of the mask pattern with the flat of the wafer or with patterns already existing on the wafer. Sometimes (in fact quite often) it is necessary to align the mask with patterns on the rear side of the wafer. This is rather convenient if one has a double-sided mask aligner.

Alignment is not trivial. If in a process the alignment has to be better than 10 mm, one will experience some difficulties. Therefore, processes that do not require precise alignment are preferable. Great care must be taken for the design of proper alignment marks on the wafer and the masks. We cannot give general rules here, because this depends very much on the equipment available.

After illumination the wafers are developed, rinsed in water, blown dry or dry spun, and annealed again, typically at 120°C, for 20 minutes. This process step is usually called "postbake." Higher temperature or longer annealing time makes it difficult to remove the resist, and the resist tends to flow. For example, in Twente a process was used, in which the postbake step is at 140°C, with a result that the thickness of the resist layer decreases to zero toward the openings over a distance of ca. 5 μm.

After postbake, the oxide can be stripped in an HF solution. The recipes here are also different; some people use buffered HF ("BHF"), a mixture of 1 part 49% aqueous HF solution and 7 parts NH_4F. This solution does not attack silicon at all. High-concentration aqueous solutions of HF, as used in other labs, seem to etch silicon slightly. The oxide stripping is done at room temperature; the etch rate of oxide in BHF is typically 6 μm/hour. Note that the resist does not like HF too much. Some people repeat the postbake after 15 minutes in BHF to prolong the lifetime of the resist layer.

The process of oxide stripping is quite critical: if the wafer was not dry or clean enough, the resist will be peeled off during the BHF etch.

Note: HF and NH$_4$F are both very toxic. You don't feel HF if you spill it on your skin, but your bones will dissolve. Therefore, great care has to be taken if one works with HF. Wear gloves! Consult, for example, the "Chemie Karten."

Water reacts very differently with an oxidized surface than with a silicon surface. Pure silicon does not like water, water does not wet silicon perfectly, one sees that water contracts to droplets. There are no droplets on oxide layers; water wets oxide perfectly. One says that silicon is hydrophobic and oxide is hydrophilic.

This is used to see if the oxide-stripping process is at an end or not: if the wafer is dipped in demiwater, the water will withdraw from the opened regions, if the oxide layer is perfectly removed. It is advisable to check the etch results under an optical microscope, in order to check if all oxide is removed – if the surface is no longer wetted by water, there may still be tiny islands of oxide. The bare silicon surface looks gray and is easily identified. If the openings in the etch mask are too small, it is quite difficult to see the nonwetting effect.

If the oxide stripping process has satisfactory results, the resist can now be stripped. This can be done in acetone if the postbake was not too long or at too high a temperature. As outlined here, 20 minutes at 120°C is okay. In Twente, where they use a postbake process at 140°C, the resist has to be burned away ("ashed") in an oxygen plasma.

Before etching, cleaning the wafers is recommended by using a standard cleaning procedure. The RCA1 and RCA2 procedures are as follows:

RCA1: 1 part NH$_3$ (25% aqueous solution) in 5 parts water, heat up to the boiling point, add 1 part H$_2$O$_2$, and then immerse the wafer for 10 minutes.

RCA2: 1 part HCL in 6 parts water, heat up to the boiling point, add 1 part H$_2$O$_2$, and then immerse the wafer for 10 minutes.

Note: NH$_3$ etches silicon! The etching is prevented if peroxide is added to the solution. RCA1 removes all organic dirt (resist), and RCA2 removes all metal ions.

The second cleaning process is required to keep the oven tubes for thermal oxidation and indiffusion free of metals. Just for an etch step, RCA2 is not necessary but RCA1 is.

Both cleaning processes (and also other ones such as cooking in NH$_3$) leave the wafers with a thin oxide film. This must be stripped just before etching. This can be done by quickly dipping the wafer in a 1% aqueous HF solution. The solution will not wet bare silicon surfaces. Thus, if the whole wafer is wetted, there is still oxide on it.

The wafer is dipped in demiwater and immersed into the etchant. After etching, the wafer should be inspected using a microscope, a surface profiler, or whatever is available and appropriate, and cleaned (RCA1 + RCA2) before it is processed any further. You will not find it so easy doing it for the first time. The process – and all other processes in the clean room – may be compared with cooking in the kitchen. If you have no experience at all with cooking, you know that you will be unable to prepare a complex meal just following a recipe in a book!

The whole process is summarized in Table 2.2.

A final comment on the RCA cleaning procedure seems to be appropriate. Potassium ions are fatal for integrated circuit processes; KOH etching therefore is not IC compatible. In most labs it is therefore forbidden to process a wafer in an IC lab after the KOH etch, as one fears that potassium ions will spoil, e.g., the oxide oven tubes. However, there are

Table 2.2. *Summary of the process steps required for etching a membrane.*

Process	Duration	Process temperature (°C)
Oxidation	variable (hours)	variable (900–1200)
Spinning at 5000 UPM	20–30 sec	25
Prebake	10 min	90
Illumination	20 sec	
Develop	1 min	25
Postbake	20 min	120
Strip oxide (BHF = 1pHF + 7pNH$_4$F)	variable (ca.10 min)	25
Strip resist (acetone)	10–30 sec	25
RCA1 (1pNH$_3$(25%) + 5pH$_2$O + 1pH$_2$O$_2$)	10 min	boiling
RCA2 (1pHCl + 6pH$_2$O + 1pH$_2$O$_2$)	10 min	boiling
HF dip (2%HF)	variable (10 sec)	25
Etch	variable (minutes up to 1 day)	70–100

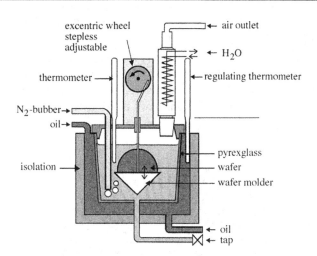

Fig. 2.13. Schematic drawing of an etch apparatus (after Büttgenbach [8]).

labs in which IC processing of KOH-etched wafers is allowed (the etching process itself is performed outside the IC lab) if the wafers are carefully cleaned using RCA1 and RCA2. Obviously, one gets rid of many potassium ions so that the concentration is well below a dangerous level.

For etching, a vessel is required that allows control of the temperature, stirring, escape of hydrogen gas, and – for EDP-based etchants – a hermetic seal to the oxygen in the ambient atmosphere.

All this can be realized with standard glass equipment in a wet chemical laboratory. In Fig. 2.13, however, we show a schematic of a more professional etch apparatus.

2.5   Experimental methods

To observe the plain fact of the anisotropy of the etch rate, it is sufficient to immerse a silicon wafer with mask openings into the etchant. Etch pits occur that are bounded by crystallographic {111} planes. However, it is much more difficult to determine the degree of anisotropy quantitatively. Figures mentioned in the literature range from factors of less than 100 to more than 1000, and it is even stated that a perfectly aligned {111} silicon face does not etch at all in, e.g., KOH solutions, which would mean an inifinitely larger anisotropic ratio. In this section we shall describe the existing main experimental techniques to determine the etch rate as a function of crystallographic orientation. All techniques have their advantages and drawbacks.

2.5.1   Wagon wheel method

The wagon wheel method was developed to determine the rate of etching under a mask that is aligned along a particular crystallographic direction of a wafer of a given crystallographic orientation. Early attempts started with curved mask openings: the idea is that all crystallographic orientations in the zone of the wafer orientation are offered to the etchant, and one hoped to be able to extract the rate at which the mask is underetched for all directions in the zone (see, e.g., Weirauch [9]). However, one encounters the following difficulty: due to the anisotropy of the etch rate, the orientation of the exposed surface under the mask changes direction in the course of time. As a consequence, these experiments cannot be interpreted easily if this is allowed to happen. One has to etch for only a very short time. A short time, however, means that the underetch is small; therefore, the experimental uncertainty is large.

The first difficulty, circumvented straight mask edges as opposed to curved ones, is exposed to underetching. Seidel et al. [10] suggested a mask pattern in a form similar to a

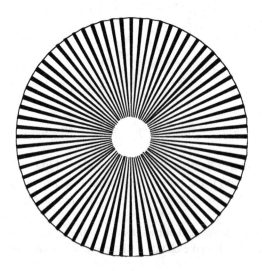

Fig. 2.14. Mask pattern for the wagon wheel method.

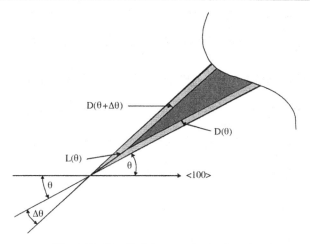

Fig. 2.15. Detail of the wagon wheel mask.

wagon wheel. The mask is shown in Fig. 2.14, which also solves the second problem, as we shall see shortly. The masked regions have the form of long isosceles triangles in place of wagon wheel spokes. In the pattern shown in Fig. 2.14, the spokes have a width of 5°, with a pitch of equal size. For experiments, use is made of width and pitch of 1°, or in some cases even 0.1°.

Let us analyze the geometry of the pattern if it is underetched. In Fig. 2.15 we show a detail of a tip of one of the spokes with an enlarged angle. The mask is underetched from both sides by a distance $D(\Theta)$ and $D(\Theta + \Delta\Theta)$, respectively, and the underetch fronts intersect each other at a distance $L(\Theta)$ to the sharp end of the spoke. If we ignore the underetching in the direction of the long axis of the spokes (along L) and the difference between $D(\Theta)$ and $D(\Theta + \Delta\Theta)$, we have a simple relation between L and D,

$$D = L\sin(\Delta\Theta/2) \approx L\Delta\Theta/2$$

or

$$L \approx 2D/\Delta\Theta.$$

Hence we get an underetch pattern from which the underetching normal to the mask edges (in the direction of D) is hardly visible, if the experimenter takes care not to etch too long, but the undercut is in the direction of L, the long axis of the spokes, is enhanced by factor $2/\Delta\Theta \approx 115$ for $\Delta\Theta = 1°$. The wagon wheel pattern makes anisotropic etching easily visible, and minute underetching of the spokes results in large, easily measurable underetching in a radial direction. Note that the radial pattern is rotated by 90° with respect to the underetching under the spokes, since the direction of L in Fig. 2.15 is normal to the direction of D.

Of course the underetching in the radial direction cannot be completely ignored. The edge is a convex corner which will be bound by fast-etching planes. Therefore, really reliable underetch rates can be only obtained by measuring the distance D.

The disadvantage of this technique is twofold. The first problem is trivial: the underetch rate is determined at isolated orientation, therefore a higher resolution of the underetch rate as a function of the orientation requires more spokes. The second problem is more serious.

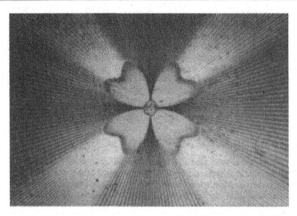

Fig. 2.16. Photograph of the result of underetching a wagon wheel on the ⟨100⟩ oriented silicon wafer ([11]).

The pattern that is seen on the top of the wafer (compare, e.g., Figs. 2.15 and 2.17) is only a projection of underetch rate on the (001) plane. An illustration of the problem is the following: the projection of the ⟨110⟩ vector on the (001) plane has the same direction as the projection of the ⟨111⟩ vector. The underetch rate is determined by the slowest etch rate in these two directions: since $R_{⟨111⟩} \ll R_{⟨110⟩}$ ($R_{⟨hkl⟩}$ for etch rate along the ⟨hkl⟩ orientation), the latter is invisible.

Another illustration is provided by the following example. The wagon wheel pattern after etching in KOH and EDP (see Section 2.6) is very similar. However, the orientation of the etch front when etching in KOH under a mask edge along the ⟨100⟩ direction makes an angle of 90° with the (001) surface; in other words, we encounter here a {001} face. When using EDP, the etch front is inclined by 45°, therefore the etch front coincides with a {110} face. There is a minimum of the etch rate in the {100} orientation using KOH, and we do not know a way to use the wagon wheel method to determine whether there is a minimum in the {110} orientation as well. On the other hand, when using EDP, there is a minimum etch rate in the {110} orientation, and again we do not know a way to find out under these circumstances whether there is a minimum in the {001} orientation. In Fig. 2.16 we show the results of etching under a wagon wheel mask on an Si⟨100⟩ wafer.

In Fig. 2.17 we show the underetch front under spokes in the sector of 90° on a ⟨001⟩ oriented silicon wafer etched in KOH. We see that the etch front is composed mostly of two flat faces, which are the {100} and {111} faces.

2.5.2 Etching of spheres

Etching under a planar mask pattern on a wafer gives information on the projection of the etch rate, but etching a sphere would give complete information. We can think of positive spheres, where the silicon exposes a convex curved surface to the etchant, and of negative spheres, where a concave curved surface is exposed to the etching solution.

The sphere experiment is a well-known tool for crystal growers to study growth rates of single crystals as a function of the orientation. If one immerses a sphere into a supersaturated environment (such as an undercooled melt, a supersaturated solution, or vapor in a CVD equipment), the slowly growing orientations are left behind, and they are very quickly

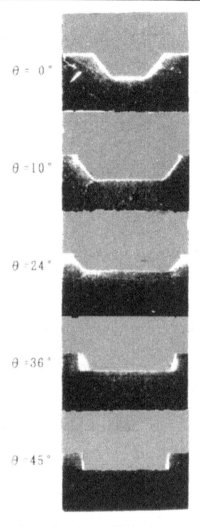

Fig. 2.17. Etch fronts under mask strips on ⟨001⟩-oriented silicon etched in aqueous KOH solution [11].

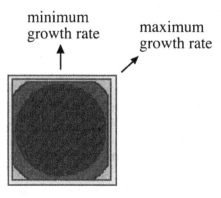

Fig. 2.18. A fast-growing facet disappears from the growth form, and slow-etching facets dominate the growth form.

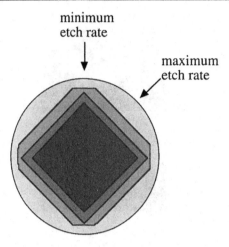

Fig. 2.19. A slowly etching facet disappears from the etch form, the etch form is dominated by the fast-etching facets.

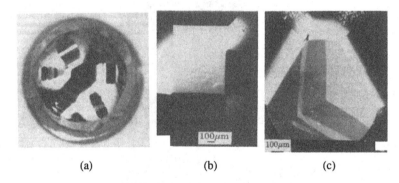

(a)   (b)   (c)

Fig. 2.20. Results of growing a silicon sphere. Optical microscopic photograph. (a) total view of the sphere with a {111} facet on top. (b) magnification of the {111} facet of a. (c) magnified view of a {001} facet (from J. G. E. Gardeniers et al. [12]).

visible. They form facets. If the crystal is allowed to grow further, eventually all fast-growing directions will disappear, as is apparent from the schematic in Fig. 2.18.

In Fig. 2.19 we show a schematic of *etch fronts* starting from a sphere. Here the fast-etching facets emerge on the form while the slow-etching facets disappear and form vortices. When growing a sphere, the vortices on the shape are formed by the intersection of the slow-growing facets.

In Fig. 2.20 we give an example of a sphere experiment for silicon growing in a CVD oven fed by a mixture of $SiH_2Cl_2 + H_2 + HCl$. Figure 2.20(a) shows a photograph of the sphere after growth. The facet on top is a {111} face, as is apparent from the three-fold symmetry. Three relatively narrow bands of facets extend to the (−111), (1−11), and (11−1) faces, and the broader bands extend to the (100), (010), and (001) faces. The bands between the {111} faces consist of facets (331), (551), (110), (55−1), and (33−1), while the broader bands consist of many macrosteps. The latter end has well-developed {001} facets. The experimental results on crystal growth of silicon in this particular CVD system are very complex and far from well understood. The basic reason why there are flat crystal faces at all will be discussed in Chapter 3.

Fig. 2.21. Growth form of a diamond crystal showing {111} and {001} facets (from Giling and van Enckevort [13]).

In Figs. 2.20(b) and 2.20(c) we show magnifications of the (111) and (001) after growth. Both photographs clearly demonstrate the existence of growth hillocks, which are the result of dislocation lines, with a screw component intersecting the crystal surface. Screw disloca- tions ending at surfaces act as continuous sources of steps on the flat surfaces. These steps enable the flat faces to grow, as will be discussed in more detail in Chapter 3. The slope of the hillock in Fig. 2.20(a) is very small (of order of 1°).

When the growth experiment is continued until a stable form is reached, the form is built up from the slowest-growing facets. Unfortunately, we only have an SEM microphotograph of a diamond crystal available, which has the same crystal structure. As can be seen in Fig. 2.21 the form is dominated by {111} and {001} facets. For comparison, see Fig. 2.9, which shows a drawing of a form made up by {111} and {001} facets.

2.5.2.1 Etch experiments starting with positive spheres

If one etches a sphere in, e.g., KOH, one gets a very different structure than the growth form. As explained with the help of Figs. 2.18 and 2.19, in a growth experiment start- ing from a convex form, the slowly growing directions form faces. These directions are slowly attacked by the etchant. When etching, these directions do not form facets but vor- tices. The experimental result is shown in Fig. 2.22. Under the experimental conditions (40 wt% aqueous KOH at 75°C) there are etch rate minima in the following crystallo- graphic orientations: {111}, {001}, and there is a saddle point in {110}, as will become clear below.

In Fig. 2.23 we show a construction of a crystal form where we have vortices in ⟨001⟩ and ⟨111⟩ and their physical equivalent directions, and the facets result simply by connecting the end of the crystallographic vectors. Note the remarkable similarity of the construction and the experimental results. In the drawing we took an arbitrary length for the vectors, but in reality their relative length (⟨111⟩/⟨001⟩) results from the crystallographic orientation of the maxima of the etch rate.

There are important differences between Fig. 2.23 and Fig. 2.22 on one hand, and between Fig. 2.22 and Figs. 2.20 or 2.21 on the other hand. The facets found in the experiment are quite

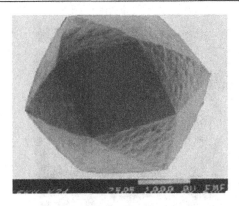

Fig. 2.22. A scanning electron micrograph of a sphere heavily etched in KOH (40 wt%, 75°C) (from [1]).

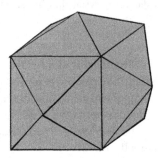

Fig. 2.23. Geometric construction of a crystal anisotropically etched with etch rate minima in ⟨111⟩ and ⟨001⟩ directions.

rough. They are composed of many steps that consist of ⟨111⟩-oriented terraces and rough surfaces without a specific crystallographic orientation. Furthermore, they have a slight convex curvature. The curvature becomes more pronounced if silicon is etched in aqueous CsOH [14]. The difference between the facets obtained in Fig. 2.22 and in Figs. 2.20 or 2.21 is as follows: the facets that form by growth are mirrorlike flat; in fact, they are flat on an atomic scale. Deviation from ideal flatness is realized by individual discernible steps, e.g., in the growth hillock shown in Fig. 2.20(b).

From the experimental result shown in Fig. 2.22, only the crystallographic orientation of the maxima and minima of the etch rate can be determined, but it seems to be difficult to reconstruct a complete etch diagram, i.e., the etch rate as a function of the crystallographic orientation.

2.5.2.2 Etch experiments starting with negative spheres

The results of etch experiments with negative spheres, holes of spherical shape drilled into a silicon crystal, are similar to growth experiments with positive spheres: the slowly etching orientations lag behind, and the resulting shape is governed by these orientations. This is illustrated in Fig. 2.24. The idea of using negative spheres is – if we are informed correctly – due to K. Sato, who was then with Hitachi Ltd. Sato and his co-workers drilled holes of app. 5 cm diameter in thick wafers, etched them, and scanned the surface with a surface profiler.

Fig. 2.24. Schematic of a negative sphere experiment. The dark regions indicate the material that is removed by anisotropic etching.

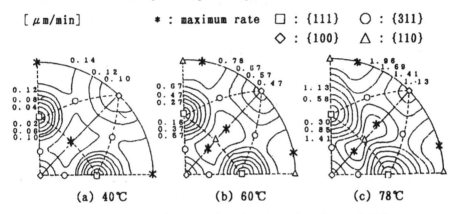

Fig. 2.25. Results of etching a negative sphere (from Sato et al. [15]).

The information obtained from these experiments gave the most complete information so far available; however, the drawback is that the etch rates cannot be determined with the accuracy of the wagon wheel method (because there is no amplification of the etch distance). Results of the scans are shown in Fig. 2.25.

In the scans a new feature of anisotropic etching in KOH becomes apparent: at low temperature the etch rate maximum is parallel to the $\langle 110 \rangle$ direction, and at higher temperature the maximum splits, leading to a saddlepoint at $\langle 110 \rangle$. It seems to be impossible to gain this information from the wagon wheel method. We therefore have the most complete information on anisotropic etching here. Unfortunately, data have only been published at the concentrations used in the experiments shown here, so we have to rely on the much more complete series of experiments using the wagon wheel method.

2.6 Phenomenological properties of anisotropic etching solutions

2.6.1 Introduction

The most recent and complete overview on the know-how of anisotropic etching can be found in the paper of Seidel et al. [10].

There are a number of wet etchants that etch silicon anisotropic. The most important are alkali-hydroxides such as KOH and a mixture called EDP (ethylendiamine, pyrocatechol, and water).

For the other etchants we refer to the literature:

- NaOH [10, 16]
- CsOH [17, 18]

- NH$_4$OH [19, 20]
- Hydrazine [21]
- TMAH (tetramethyl ammonium hydroxide, (CH$_3$)$_4$NOH) [22]

We shall concentrate on KOH, EDP, and TMAH, since these are the most commonly used etchants. Hydrazine is extremely toxic and its vapor is explosive. It is therefore avoided as much as possible. KOH has the disadvantage of being incompatible with IC processes due to the potassium ions. EDP also is toxic, but to a lesser extent than hydrazine.

TMAH is nontoxic and IC compatible, but fewer studies exist on this system than on, e.g., KOH or EDP. TMAH etching results in very flat (001) faces, and the trenching in the etches of membranes seen in Fig. 2.52 is absent. However, there are problems with reproducibility of the etch rates.

Generally, use of a fume box is strongly recommended. Stirring generally increases etch rates and uniformity.

Except at very high concentration of KOH and in TMAH, the etched (100)-plane becomes rougher the longer one etches. This is thought to be due to the development of hydrogen bubbles that hinder the transport of fresh solution to the silicon surface [23].

In the final choice of the etchant, a number of issues have to be considered:

- ease of handling
- toxicity
- etch rate
- topology of the etchground
- IC compatibility
- etch stop
- etch selectivity of other materials
- mask material and thickness of the mask

The last three items will be treated in Sections 2.8 and 2.9.

2.6.2   KOH solutions

KOH is the most commonly used etchant [24]. It is much less dangerous than other etchants, easy to handle, readily available, and etches fast. Note that KOH is fatal to the eyes. The greatest disadvantages are that KOH is IC incompatible and that the selectivity to oxide is rather poor.

As H$_2$ bubbles come up during etching, use of a fume cupboard is recommended.

Compared with EDP, KOH has disadvantages regarding selectivity to oxide – EDP etches oxide slower than KOH by a factor of 100. The etch rate also slows down at higher B$^+$ concentrations as compared with EDP. This will be discussed in more detail in Section 2.8.

In Fig. 2.26 we show lateral underetch rates as a function of orientation on ⟨100⟩ and ⟨110⟩ wafers. For the experimental technique, we refer to Section 2.5. Note the deep minimum at the {111} planes. Typical etch rates far away from the {111} planes are of the order of 1 mm per minute. That means that etching through wafers is a time-consuming process: One needs five hours to etch through a 300-μm thick wafer.

In Fig. 2.27 we show an Arrhenius plot of the etch rate in the ⟨100⟩ and ⟨110⟩ directions using KOH. The temperature dependence of the etch rate is quite large and only slightly dependent on the orientation shown here. In this figure, etch rates are included with isopropyl alcohol added. The activation energy is essentially unchanged. In Table 2.3 the activation energies are also given for the ⟨111⟩ direction.

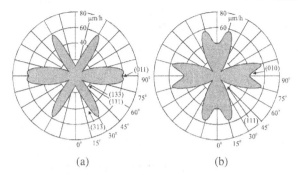

Fig. 2.26. Lateral underetch rates as a function of orientation. 50% KOH solution, 78°.
(a) ⟨110⟩, (b) ⟨100⟩ silicon wafers (from Seidel et al. [10]).

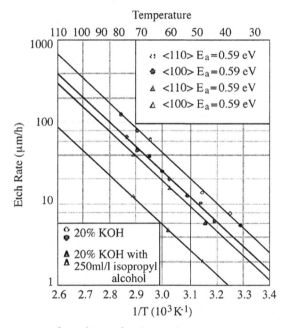

Fig. 2.27. Temperature dependence of etch rates in 20% KOH solutions in ⟨100⟩ and
⟨110⟩ directions, with and without the addition of isopropyl alcohol (from Seidel et al.
[10]).

Note the effect of IPA on the etch rate of the {110} face: this rate is greatly reduced while
the {001} rate is only slightly affected. We shall show in Section 2.7 that this reduction is
associated with a real etch rate minimum in the {110} direction, which can be exploited for
micromachining.

In Fig. 2.28 we give the concentration dependence of the etch rate in aqueous KOH
solutions. The etch rate is proportional to $[H_2O]^4[KOH]^{1/2}$ proposed by Seidel et al. [10],
which gives some support for the chemical reaction scheme we give in Chapter 3. The etch
rate has a maximum at a KOH concentration of 20 wt%.

When etching at low or medium KOH concentrations and in EDP, the (100) surface gets
a rough appearance. This will be treated in more detail in Section 2.8. The roughness of the
etchground generally increases with the etch depth. This is illustrated in Fig. 2.29. If one
wants a flat etchground, one should start the etching from the polished side of the wafer.

Table 2.3. *Activation energies of etch rates in KOH at 20 wt% [10].*

(111):	0.70 eV
(100):	0.57 eV
(110):	0.59 eV

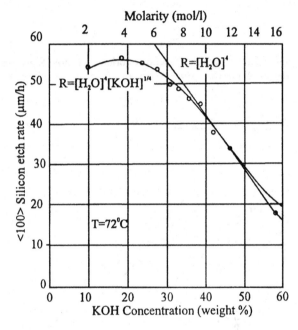

Fig. 2.28. Concentration dependence of the etch rate (from Seidel et al. [10]).

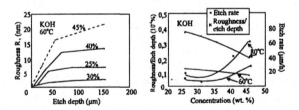

Fig. 2.29. (left) Temporal evolution of the {100} surface roughness for various KOH concentrations [25] (this does not agree with other work). (right) Roughness/etch depth ratio and etch rates in KOH solutions [25].

Usually, this is the backside of the wafer so that the micromechanical structure is located at the frontside of the wafer. Therefore, if etching is stopped after a certain time, usually one needs double-sided polished wafers. The results shown in Fig. 2.29 concerning 40 wt% solutions do not agree with other work.

If one wishes to obtain specula flat (100) surfaces, one must use very high KOH concentrations. At very high KOH concentrations (as high as 10 M), the (100) planes stay perfectly flat. This will be discussed in Section 2.7.

2.6.3   EDP solutions

EDP is not easy to handle. It is toxic, and the solution degrades if it comes in contact with oxygen. The etch pot must be well sealed. If the etchant reacts with oxygen, the liquid turns a red-brown color, and the good properties of EDP disappear.

As the vapor is harmful, use of a well-functioning fume cupboard is strongly recommended. When preparing the solution, the last ingredient added should be water, since by the addition of water the solution starts to react with oxygen.

EDP ages quite quickly. If cooled down after having etched silicon, one gets precipitation of silicates in the solution. Sometimes one gets precipitation on the silicon during etching. This spoils the etch process.

As mentioned, the advantages of EDP are its selectivity to oxide and its lower etch rate of B^+-doped silicon. Therefore the HF dip just before etching is crucial, otherwise the native oxide will prevent any etching. In contrast to KOH, very few bubbles emerge.

For EDP, there are standard concentrations, either for use as a fast etch ("F") or as a slow etch for use at lower temperatures ("S"). If one wishes a smooth (100) surface, an S solution is more appropriate. The concentrations are given in Table 2.4.

Underetch rates are given in Fig. 2.30. Generally, {100} and {110} planes of silicon are etched slower in EDP than in KOH. A further difference is that the minimum at {111} is much steeper in EDP. This has the practical consequence that it is much more important when etching in EDP to align the crystallographic direction (the flat of the wafer) more precisely than when etching in KOH.

The temperature dependence of the etch rate of an S-type EDP solution is given in Fig. 2.31. Generally, the activation energies of the etch rates of EDP are smaller than those of KOH.

Table 2.4. *Concentrations for standard EDP solutions [10].*

	S	F	T	B
Water (ml)	133	320	470	320
ED (l)	1.0	1.0	1.0	1.0
Pyrocatechol(g)	160	320	176	160
Pyrazine (g)	6	6	0	0
Etch rate (μm/min) at 45°C	0.75	1.25		

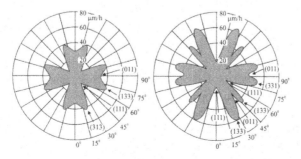

Fig. 2.30. Lateral underetch rates as a function of orientation. EDP solution type T, 95°C. (a) ⟨100⟩, and (b) ⟨110⟩ silicon wafers (from Seidel et al. [10]).

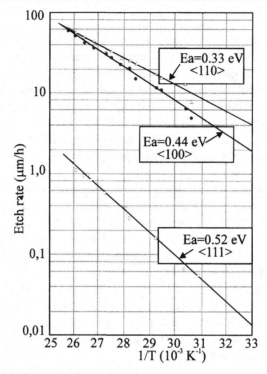

Fig. 2.31. Temperature dependence of etch rates in EDP type solutions in ⟨100⟩, ⟨110⟩, and ⟨111⟩ directions (from Seidel et al. [10]).

In Fig. 2.32 we show the etch rate of silicon in EDP and KOH in the vicinity of the ⟨111⟩ direction, showing the steep dependence of the etch rate as a function of the orientation. Note that the etch rate depends linearly on the misalignment angle. This fact has important consequences for the interpretation of the anisotropy of the etch rate.

Finally, we give a comparison of the activation energies in KOH as a function of concentration and of type S EDP in Fig. 2.33.

Similar to the case of KOH with isopropyl alcohol IPA, the etch rate of {110} faces is much smaller than that of {001}.

2.6.4   TMAH solutions

When TMAH is dissolved in water, it becomes a base (pH > 12, for most of the useful solutions), so one should take care of eye protection. It is not toxic, like EDP, or inflammable, like hydrazine, so it is easy to use, but it reacts to a certain extent with the CO_2 in the air, so etching vessels should not be left open for a long time. One of the most important advantages is that TMAH is an organic material, so it does not contain any metal ions, making TMAH a potentially IC-compatible etching agent. It is known that MF312, the developer of positive photoresist is 5 wt% TMAH, but MF312 contains additives, which influenced the etch characteristics negatively. We recommend the use of the commercially available TMAH of *Riedl-de Haën AG* in Germany.

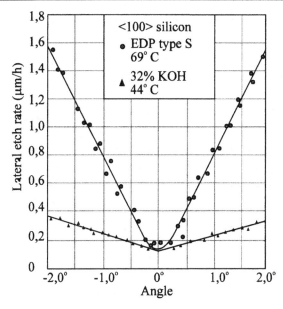

Fig. 2.32. Lateral underetch rates in the vicinity of {111} crystal planes on a ⟨100⟩ wafer (from Seidel et al. [10]).

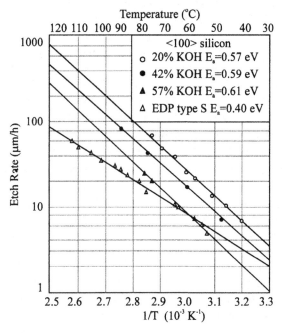

Fig. 2.33. Arrhenius diagrams of the vertical etch rate in ⟨100⟩ wafers for EDP and KOH solutions (from Seidel et al. [10]).

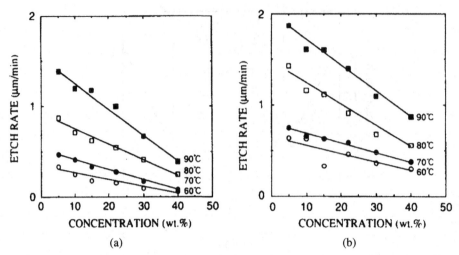

Fig. 2.34. (a) Si ⟨110⟩, and (b) Si ⟨100⟩ etch rates in TMAH as a function of concentration for a range of temperatures (from Tabata et al. [22]).

Tabata et al. [22, 87] and Schnakenberg et al. [63] among others have investigated the etch rates of silicon in TMAH solutions as a function of the temperature and the concentration. They obtained roughly the same rates for equal process parameters. Tabata et al. [22] have presented etch rate measurements at different temperatures, which are reproduced in Figs. 2.34(a) and 2.34(b) for the etch rates of the (100) and (110) planes, respectively.

It is seen that the etch rate decreases for both orientations with increasing concentration and yields values of 1.4 μm/min at 5 wt% and 90°C down to 0.1 μm/min at 40 wt% and 60°C. The activation energy of the etch rate is given by Tabata [22]: It increases from 0.6 eV at 10 wt% to 0.8 eV at 40 wt%. Such a drastic concentration dependency of the activation energy is not reported for KOH etching. There is an anomaly with respect to the activation energy of Si⟨111⟩ etch rates: the activation energy is smaller than the activation energies of etch rates in other crystallographic orientations.

Similar to results found in KOH etching, the roughness of the etch bottom of Si⟨110⟩ increases with decreasing concentration (effectively: decreasing pH). Again, the rough features on the etched ⟨001⟩ faces are constituted by pyramids with faces closely aligned along ⟨111⟩. The change of the topography of the etched ⟨001⟩ surfaces upon a change of the concentration is quite nicely demonstrated in the SEM images in Fig. 2.35. Smooth surfaces start to occur above 22 wt%, at which the measured roughness is 100 nm. It is reported that the roughness does not depend on the temperature.

For micromachining it is quite useful that after having etched a certain minimum amount of silicon, the solution does not attack aluminium [87]. A detailed investigation of this finding has been published by Tabata [88]. It turns out that the pH of the solution is dependent on the amount of Si dissolved in the solution – therefore the roughness depends on the Si concentration.

Finally, we summarize results on the anisotropy of TMAH etching. The degree of anisotropy can be indicated by the (111):(100) etch rate ratio. The smaller this ratio, the higher the degree of anisotropy. Figure 2.36 shows measured values of this ratio for different concentrations of TMAH and for different temperatures. A maximum ratio of 0.05 occurs at approximately 22 wt%.

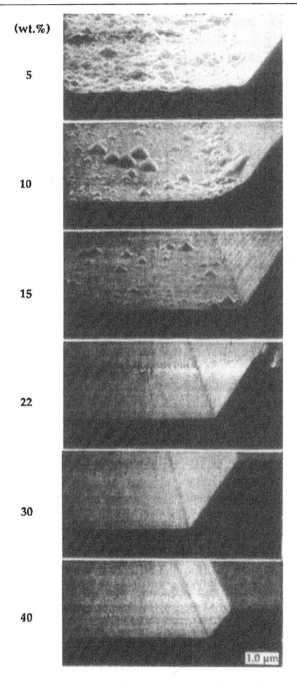

Fig. 2.35. SEM photographs of the surface etched for 90 min at 70°C and concentrations of 5 to 40 wt% (from Tabata et al. [88]).

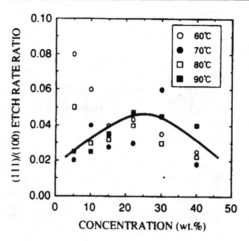

Fig. 2.36. Dependencies of (111):(100) etch rate ratio on concentration and temperature.

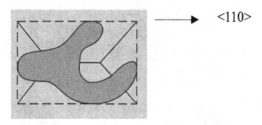

Fig. 2.37. Etch pit under a mask opening of arbitrary shape (redrawn from Petersen [3]).

2.7 Micromachining of $\langle 100 \rangle$- and $\langle 110 \rangle$-oriented silicon wafers

In this chapter we explain how the various slow-etching crystallographic planes emerge when a silicon wafer with a given orientation is anisotropically etched in one of the solutions. We have seen that, when etching silicon in KOH, {111} and {001} faces etch slowly, therefore, structures can be micromachined that are bounded by these crystallographic planes. When using EDP, or when adding isopropanol to KOH, the minimum in the $\langle 001 \rangle$ direction is replaced by the $\langle 110 \rangle$ direction.

2.7.1 Micromachining $\langle 001 \rangle$-oriented wafers

If one wants to exploit the emergence of {111} faces alone, one has to define a mask opening on the wafer with edges along the $\langle 110 \rangle$ direction. Etching then results in an etch pit of pyramidal form. When etching long enough, the form will always be bounded by {111} no matter what the mask opening looks like. This is shown in Fig. 2.37.

If the mask is misaligned with respect to the $\langle 110 \rangle$ direction, the pit will be larger than the mask opening. This effect is illustrated in Fig. 2.38. The width of a rectangular mask opening becomes wider by the factor

$$h \sin \phi,$$

where h is the length of the edge of the mask opening. Typically, the accuracy of the flat is

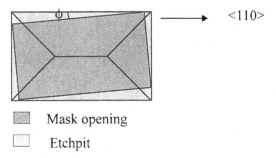

Mask opening

Etchpit

Fig. 2.38. Etch pit formed through a rectangular mask opening with a misalignment.

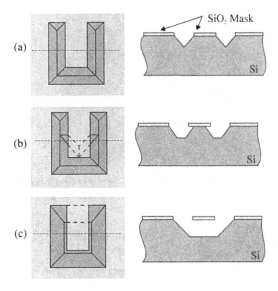

Fig. 2.39. Etch pits resulting from anisotropic etching and mask openings on (100) wafers. The series (a) through (c) shows how the underetch front proceeds in time (redrawn from Petersen [3]).

±2°, and the error in alignment of the mask to the flat is of the same order. An example of an etch pit formed by {111} faces is shown in Fig. 2.1.

Figure 2.39 illustrates the etch pits in (100)-oriented wafers one gets for a few types of masks. The sequence a-b-c in Fig. 2.39 shows how the etch pit under a cantilever beam develops in time. At the tip of the beam there are two convex corners, each of which is formed by the intersection of two (111) planes. These corners are etched quickly; the etchant encounters no (111) planes (c.f., Section 2.10). Thereby the cantilever beam is underetched. If it is desired, the underetching can be prevented by using an appropriate mask design. This will be discussed in some detail in Section 2.10.

The use of the etch rate minimum in the ⟨001⟩ and ⟨110⟩ directions has been demonstrated in a number of papers from the Uppsala Micromechanics Group [83, 84]. A schematic is shown in Fig. 2.40. Features bounded by {001} faces are obtained by etching in aqueous KOH and NaOH solutions [85] orienting the edge of the mask along the ⟨100⟩, i.e., rotated by 45° with respect to the flat (Fig. 2.40(b)). Through the same mask orientation, but using EDP or KOH with IPA, {110} faces appear instead of {001} faces.

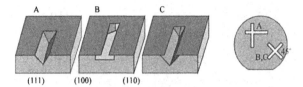

Fig. 2.40. Schematic of etch pits one obtains when etching on a ⟨001⟩-oriented Si wafer through a mask with edges oriented along the (A) ⟨110⟩ direction and the (B and C) ⟨001⟩ direction. At the far right the mask is shown on the wafer (redrawn from Rosengren [84]).

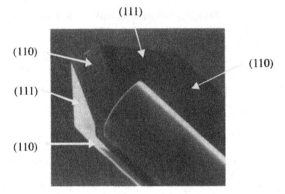

Fig. 2.41. {110} groove on Si⟨001⟩ as a guide for an optical fiber and a mirror for coupling out the light with 90° with respect to the wafer (from Rosengren [84]).

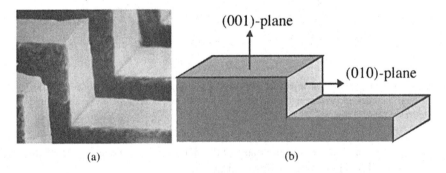

Fig. 2.42. (a) A vertical wall etched in 10M KOH, 80°C, (b) schematic of the crystal planes (from Ternez et al. [82]).

Grooves bounded by {110} faces making an angle of 45° with the ⟨001⟩-oriented wafer surface have interesting application in micro-optics (mirrors). In Fig. 2.41 we show a groove along ⟨100⟩ with a fiber inserted. The groove serves as a guide for the fiber, while the plane close to the opening is a ⟨110⟩ mirror: the light makes an angle of 90° with the wafer surface. When etching at a high concentration of KOH the method B shown in Fig. 2.40 leads to mirror flat sidewalls as seen in Fig. 2.42.

When the mask has the form of a strip oriented along a ⟨100⟩ direction, the underetch front in KOH is again bounded by {100} faces. This way it is possible to micromachine vertical walls standing on a ⟨001⟩ wafer, as shown in Fig. 2.43.

Fig. 2.43. An example of poor etch results. 7M KOH with isopropanol. The etch ratio between the {100} planes and the surrounding directions is too low. Note also the surface roughness of the etch ground (from Ternez et al. [82]).

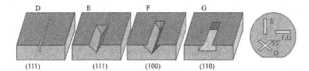

Fig. 2.44. The grooves that can be micromachined on an Si⟨110⟩ wafer (redrawn from Rosengren [84]).

It is also possible to etch mesas standing on the wafer surface. However, mesas have convex corners, and these are bounded by the fast-etching planes. This underetching can be controlled up to a certain degree. This will be discussed in Section 2.10.

2.7.2 Micromachining ⟨110⟩-oriented wafers

As we saw in Section 2.3, there are two different types of {111} faces cutting through the ⟨110⟩ surface: those that make an angle of 34° and those that make an angle of 90°. The possible orientations of mask openings and the result of etching are shown in Fig. 2.44.

The position of the flat on (110)-oriented wafers varies from manufacturer to manufacturer. For micromachining, it should be possible to align the flat of a wafer which is parallel to a crystallographic ⟨110⟩ direction with respect to features on the mask.

On (110) wafers it is possible to etch deep trenches with vertical walls since there are (111) faces inclined by 90°. A good example is shown in Fig. 2.45. This work is due to Uenishi from NTT [86]. This structure is used as an etalon (optical resonator), which consists of two parallel mirror-flat surfaces. The walls are micromachined according to a mask orientation like D in Fig. 2.44.

A further difference between the (100)- and (110)-oriented wafers is that on the (110) wafer it is possible to etch under microbridges crossing a (shallow) V groove formed by (111) planes. The microbridge may cross the V groove with an angle of 90°, provided the bridge is narrow enough so that the (111) planes, which have an angle of 90° with the

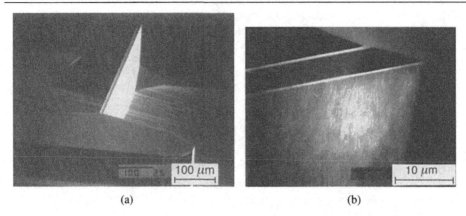

(a)   (b)

Fig. 2.45. {111} walls micromachined on a Si⟨110⟩ wafer. The feature seen in (a) actually consists of two walls as can be clear by seen in (b) (from Uenishi [86]).

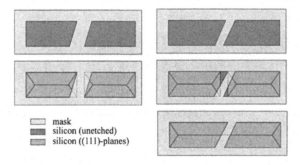

Fig. 2.46. Schematic of how to undercut a microbridge on a (100) wafer. The mask for the bridge is indicated by dashed lines. The bridge on the left cannot be underetched, in contrast to the bridge on the right.

wafer surface but are crossing the V groove under an angle of 30.53°, can undercut the microbridge. This is impossible on Si⟨001⟩ wafers, as illustrated in Fig. 2.46. V grooves in (100) wafers, which cross the groove at 90°, cannot be undercut. In order to undercut a microbridge on a (100) wafer, it has to be inclined with respect to the direction of the groove, the angle being such that the (111) plane undercuts the bridge.

In Fig. 2.47 we demonstrate the undercutting of microbridges on (100)-oriented wafers. It is impossible to underetch the bridge below a certain inclination of the microbridge from the normal over the V groove. This minimum angle obviously depends on the width of the bridge. In Fig. 2.48 we show microbridges that cross V grooves on a (110) wafer. Although they cross the V groove with 90°, they are readily underetched. Note that in this case there is also a maximum width of the bridge that can be underetched (for a given width of the V groove).

2.8   Etch stop mechanisms

2.8.1   Introduction

Use of etching to shape objects is not of much help if one is unable to control the depth and width of the etched third dimension. As we have discussed in the previous chapters, the

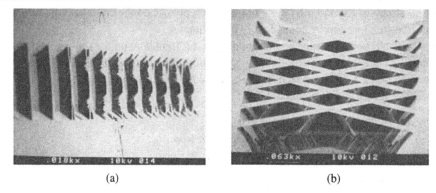

(a)   (b)

Fig. 2.47. Undercutting a microbridge over a V groove in a (100) wafer. If the inclination of the lower two bridges (a) is too small, no underetching occurs; (b) examples of successfully underetching (Legtenberg, unpublished).

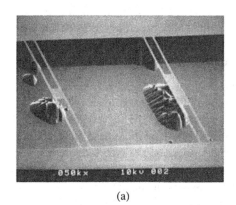

(a)

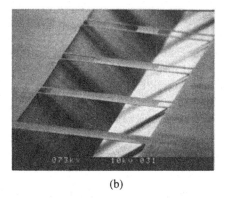

(b)

Fig. 2.48. Microbridges crossing a V groove on a (110) wafer. (a) before, and (b) after completion of the etch process (Legtenberg, unpublished).

strong anisotropy of KOH and EDP etching allows a precise control of lateral dimensions, if the desired structure can be bounded by {111} planes.

The etch depth however must be controlled in a different way unless the whole etch pit is bounded by {111} planes. The following possibilities to control etch depth are used: time etch stop (Section 2.8.2), boron etch stop (Section 2.8.3), thin films (Section 2.8.4), and the electrochemical etch stop (Section 2.8.5).

2.8.2   Time etch stop

The time etch stop is the most trivial way to stop the etching. The wafer is removed from the etch bath after a given time of etching.

The obvious advantage is the simplicity of this procedure. The use of time-controlled etch depth is acceptable, especially if thick structures are required (thicker than, say, 20 Γm). We doubt, however, that time etch stop will ever be an important production method. However, due to its simplicity the method is very useful for the realization of demonstrators.

In order to monitor the result of etching in situ it is useful to observe the light transmitted through a membrane. If the thickness becomes smaller than a few Γm, the membranes appear red-orange.

There are three main problems.

(i)   The etch rate of all etchants is considerably influenced by transport of agents in the solution to the wafer and by transport of etch products away from the wafer. This results in a spread in the etch rates. The etch rate depends slightly on the shape of the etch vessel, on stirring, on how long it has been etched, and on the number of wafers and their spacing. The changes of the etch rate are of the order of a few percent. Therefore, the etch depth cannot be determined by a reproducible etch time with sufficient precision. The spread in etch rates becomes critical if one etches down to the thickness of less than 20 Γm. To etch structures of this thickness, it is necessary to monitor the etch rate and to observe the etching carefully during the last hour. From experience, it is impossible to etch structures down to less than 10 Γm.

(ii)   The thickness of the wafers varies from wafer to wafer, and the thickness of a single wafer is not homogeneous. The variation is typically 5–10 Γm between wafers.

(iii)   The quality of the surfaces that are obtained during etching is poor. Generally, the surface roughness increases during etching, as we have indicated already in Section 2.6 (see Figs. 2.29 and 2.30). An interferogram is presented in Fig. 2.49, which shows the backside of a 10-Γm thick membrane etched in EDP. The interferogram has been taken with laser light of a wavelength of 546 nm; the lines indicate a peak-to-peak roughness of $\approx$2 Γm. The wafer roughness can be considerably reduced if one starts with etching from a polished surface. However, the surface does not become really specula.

In several papers there are comments on the reason for the increase of the surface roughness while etching (see, e.g., [25]). If the (100) surface is rough on an atomic scale (c.f., Chapter 3), any waviness of the surface should be unstable because hillocks feel a greater undersaturation than pits. Therefore, hillocks should etch faster than pits and eventually disappear. The situation is different from growth, since the competition between transport rate to the surface and growth rate can lead to instabilities resulting in extreme cases in whiskers and dendrites. According to a paper by Palic et al. [26], it is possible that the

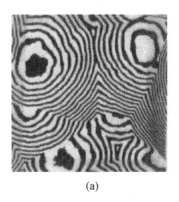 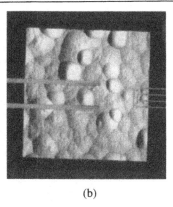

(a)   (b)

Fig. 2.49. View of etch bottom when etching a membrane. The etching has been stopped just by removing the wafer from the etchant. (a) Interferogram of an etched surface (EDP type S) (from Bouwstra et al. [89]). (b) SEM image of a membrane etched in KOH, 25 wt% at 72°C (Legtenberg, unpublished).

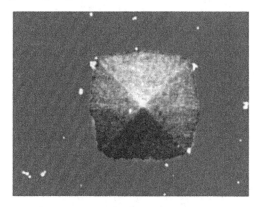

Fig. 2.50. Example of a hillock bounded by {567} faces, as found when etching ⟨001⟩-oriented silicon in low concentration KOH (from Tan et al. [27]).

formation of hydrogen bubbles is responsible for the emergence of rough silicon (100) surfaces. The bubbles tend to stick to the surface impairing the homogeneity by transport of etchants to the surface and of etch products away from it. On the other hand, it has been argued by Tan et al. [27] that hillocks originating from bubbles look quite different from the hillocks found on {001} Si.

For EDP etching of a ⟨100⟩-oriented wafer, a solution of type S gives the smoothest surfaces, and for KOH, concentrated solutions (≥ 10 M) lead to very flat (100) planes.

The appearance of hillocks that form on the {001} face during the etch process is puzzling. An example is shown in Fig. 2.50. Tan et al. [27] performed an extensive study of the hillocks. The hillocks are bounded by faces of an orientation close to {111}; in fact, Tan et al. found faces oriented in {567} orientations. These are forms that are expected from growth rather than from etching processes. Tan et al. even suggested that the hillocks are the result of regrowth [28]. Regrowth would only be concevable if the undersaturation of the etching solution is quite small, so that fluctuations would allow local changes to growth conditions.

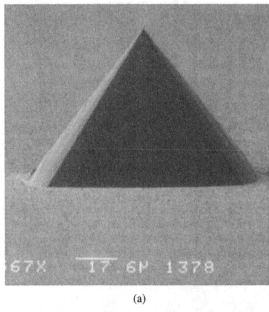

(a)

(b)

Fig. 2.51. (a) Large hillock formed during etching (001) silicon in 30 wt% KOH saturated with IPA at 72°C (Courtesy of Stefan Sanchez, MESA Research Institute). (b) Closeup of the large hillock showing the curvature of the tip (Courtesy of Stefan Sanchez, MESA Research Institute).

Contrary to intuition, these hillocks are stable under certain conditions [32]. We note that formation of flat facets aligned along crystallographic directions during etching or dissolution of monocrystalline material is not unique to silicon – quite similar to the anisotropy of the wet chemical etching process itself – but it is a common phenomenon. It has been observed on quartz [29, 30], it has been mentioned by Tan et al. [28] for a number of materials, and it has been observed even in Monte Carlo simulation of dissolution of simple cubic crystals [31]. Therefore, causes of the formation and the stability of the hillocks cannot be sought in peculiarities of silicon (e.g., being a semiconductor), but one must look for

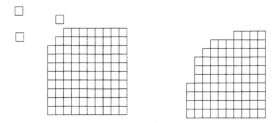

Fig. 2.52. Simple square crystal during etching. Left: The {01} faces are flat and etch slowly, the {11} faces are rough and etch fast. Right: The edge emits steps that run on the {01} faces [32].

solutions to the problem in the dissolution process of crystals. Slightly misaligned {111} facets actually seem to be composed of steps on flat $\langle 111 \rangle$, as clearly can be seen in Fig. 2.51(a). The steps emerge from the edges and from the tips of the pyramid, run to the center of the pyramidal face and form an obtuse edge.

If the steps on $\langle 111 \rangle$ faces run away from the edges faster than they are delivered, and if the tip is etched with a rate similar to the {001} (the base on which the pyramids stand), these features are stable.

To understand this, consider for simplicity a two-dimensional simple square crystal as shown in Fig. 2.52. This crystal etches slowly in the $\langle 01 \rangle$ direction and has a maximum of the etch rate in the $\langle 11 \rangle$ direction. We have drawn the {01} faces perfectly flat, and we ignore any etch rate contribution in $\langle 01 \rangle$. If we deviate from perfect $\langle 01 \rangle$ faces, we have to introduce steps, and the misaligned faces etch by moving step trains. Some consequences of a small etch rate due to spontaneous nucleation of steps have been discussed elsewhere [32].

On the right-hand side in Fig. 2.52 we draw a step that was emitted from the rough edge. If this step moves faster than the {11} face is etched, we obtain a steptrain moving on the {01} facets. The distance λ between the steps is given by the ratio of the speed v of the steps normal to the step, which equals the removal rate of atoms in the step times the length of the atoms, to the rate at which steps are emitted from the edge, times the distance $d_{\langle 01 \rangle}$ between adjacent atoms in the plane. The latter should be close to the $\langle 11 \rangle$ etch rate $R_{\langle 11 \rangle}$ along $\langle 01 \rangle$. The step train leads to a misalignment of the {01} facets, the angle of which[‡] is given by the ratio of the step height $h_{<01>}$ to the step distance. We have for the misalignment angle,

$$\tan \Theta = \left(R_{\langle 11 \rangle} / v_{\langle 01 \rangle} \right) \times \left(h_{\langle 01 \rangle} / d_{\langle 01 \rangle} \right).$$

For steps on silicon $\langle 111 \rangle$ pyramidal faces, emitted from the $\langle 110 \rangle$ edges, we would have

$$\tan \Theta = R_{\langle 110 \rangle} / v_{\langle 111 \rangle} \times h_{\langle 111 \rangle} / d_{\langle 111 \rangle}.$$

We have at standard conditions (25–30 wt% KOH, 80°C) $R_{\langle 110 \rangle} = 2\,\mu$m/min [11, 33] and $v_{\langle 111 \rangle} = 4\,\mu$m/min (as deduced by Elwenspoek [33], see also Chapter 3, from Sato's data [11, 15]). Furthermore, for the height of a double layer on $\langle 111 \rangle$ silicon we have 2.3 Å, and the length of an atom in the steps is 3.1 Å. A similar argument holds for steps emitted from the tip of the pyramid. In this case, $R_{\langle 001 \rangle} = 0.8\,\mu$m/min and we get $\tan \Theta = 0.12$. The total inclination, given by the vector sum of the step trains running from the edges and the tip

---

[‡] In fact, it is the tangents of the angle.

of the pyramid, leads to a face which can be described by $\langle 1\ 1.25\ 1.4\rangle = \langle 5\ 6.3\ 6.9\rangle$, quite close to Tan et al.'s experimental results [27].

This encouraging result means that conditions where no hillocks can be found require either an increased $\langle 110\rangle$ etch rate relative to the step velocity or a change of the mechanism that leads to the formation of the hillocks. Since Si$\langle 001\rangle$ faces immersed in etching solution normally leading to few hillocks are readily attacked from the edges, the first possibility seems to be quite probable.

One may wonder why atoms in steps on $\langle 111\rangle$ are removed faster than atoms in the $\{110\}$ faces. Qualitatively it can be understood in the following way. The steps in the atoms have at least two backbonds (see Chapter 3), so at most two bonds accessible to chemical reaction [34]. A perfectly flat $\{110\}$ face has only one dangling bond per atom! The clue to this conundrum is that $\{110\}$ is not flat on an atomic scale: it is a rough face, as such full of kink positions [35]. A step is a quasi one-dimensional structure and is rough from thermodynamic reasons[‡] [36]. Kink positions along a line are more readily accessible to chemical reaction than kink positions on a surface, so diffusion to and away from the kink position in a step will be faster than that to and from kink positions closely spaced in a surface.

We have not addressed the question of how the pyramids come into life, and we have no good idea currently except that any fluctuation would lead to pyramids because they are stable. Deeper understanding of the mechanism of formation is necessary to understand the dependency of the number of hillocks on the etching conditions and could help us eliminate the problem for micromachining.

The number of hillocks depends strongly on the concentration of the solution. At a temperature of 60°C the hillock density drops to practically zero above a concentration of 30 wt% (see Fig. 2.53). These hillocks are attacked at the edges when they are immersed in a high concentration solution and vanish after some time.

It has been observed that the waviness of surfaces becomes less pronounced if smaller membranes are etched [39]. One reason could be that the probability of nucleation of bubbles per surface area of an etched membrane is such that it becomes very small for small enough membranes (ca. $10 \times 10\ \mu$m). Since the surface tension between H_2 gas and the etchant enters exponentially in the probability of bubble nucleation, a varying surface tension could explain the different surface roughnesses observed in the various etchants.

The vertical $\{100\}$ etch rate close to the (111) planes is somewhat larger than when far away from the side walls. Membranes or double-sided clamped beams ("microbridges") therefore tend to be thinner close to the clamped edges than in the center of the structure. This difference can be as large as 1–$2\ \mu$m, which is quite considerable if one etches 15–20-μm thick structures. This is demonstrated in Fig. 2.54. A sequence of pictures shows how the etching evolves during the last 10 minutes before etching "stops" at a highly boron-doped layer, leading to a 50-nm thick membrane [40]. It is clearly seen in Fig. 2.54a that the etch front reaches the B^+ region first at the edges of the membrane and that there is a hillock in the center. Also, the overall surface roughness is clearly visible. In Figs. 2.54(b) and (c) the membrane has a thickness of less than $1\ \mu$m. Most of the contrast seen in the figures is due to polarization colors, which indicate a nonuniform thickness of the membrane. In the last picture, all undoped silicon has been etched away, and a 50-nm thick membrane remains.

[‡] There is a thermodynamic phase transition between rough and smooth faces, the so-called roughening transition (see, e.g., [37, 38]). Accordingly at some temperature T_R there is a phase equilibrium between a rough and a smooth face. In one-dimensional systems, no phase equilibrium can exist, and it follows that steps are always rough. In other words: the roughening transition is at $T = 0$.

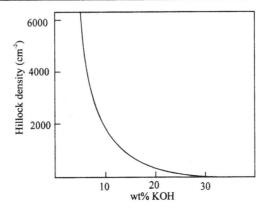

Fig. 2.53. Hillock density as a function of the KOH concentration (redrawn from Tan et al. [27]).

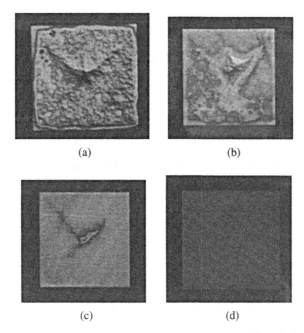

(a)　　　　　　　　　　　　(b)

(c)　　　　　　　　　　　　(d)

Fig. 2.54. Sequence of pictures showing how the etch front evolves during the last few minutes of the etch process, before the etching stops at a B^+ region (from de Boer [39]).

2.8.3 B^+ etch stop

A widely used etch stop mechanism is the B^+ etch stop. The etch stop is based on the fact that heavily boron-doped silicon etches much slower than low doped silicon.

The etch stop has been studied thoroughly by Seidel et al. [41]. The etch stop works in all alkaline-based etchants so far investigated. At a B concentration around 2–3×10^{19} cm^{-3}, the etch rate starts to slow down. According to observations, the etch rate is asymptotically proportional to the inverse fourth power of the boron concentration. The decrease of the etch rate is nearly independent of the crystallographic orientation. The critical concentration at which the etch rate starts to decrease is slightly temperature dependent: the activation energy at this concentration is about 25 meV. The temperature dependence of the etch rate

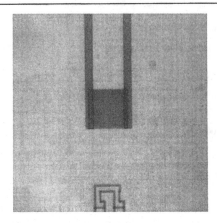

Fig. 2.55. View on a membrane obtained by the boron etch stop showing texture along ⟨110⟩ (Legtenberg, unpublished).

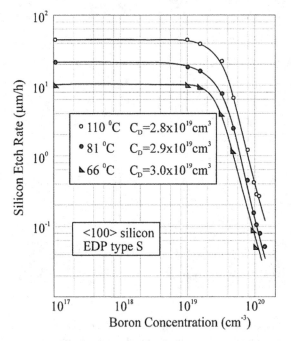

Fig. 2.56. Etch rate of ⟨100⟩ silicon as a function of boron concentration for various temperatures. EDP type S (from Seidel et al. [41]).

is larger for heavily doped silicon than for low doped silicon – the activation energy seems to increase by ≈0.1 eV at high doping levels.

In Fig. 2.55 we show an effect that one usually finds when stopping at a high boron concentration layer: striations along the ⟨110⟩ direction. This comes from the high tensile stress in the layer. During the high-temperature step (during the indiffusion), dislocations may enter the silicon wafer, which give rise to the phenomenon.

In Fig. 2.56 we show the etch rate of silicon in an EDP solution type S as a function of the doping level for three temperatures. The etch rate drops by a factor of 100 at a boron

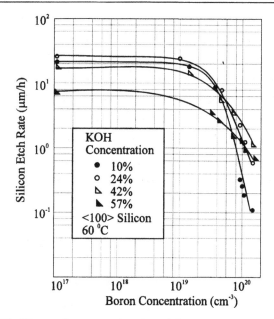

Fig. 2.57. (100) silicon etch rate as a function of the boron concentration for various aqueous KOH solutions at 60°C (from Seidel et al. [41]).

concentration of $7-8 \times 10^{19}$ cm^{-3}. Typically the thickness of an indiffused boron layer is a few μm; in a solution of type S it would take 200 minutes to etch 1 μm. Here one has quite a safe etch stop; it is not very critical when the operator takes the wafer out of the etchant.

The boron etch stop is strongly degraded in EDP solutions that are allowed to react with atmospheric oxygen.

In Fig. 2.57 we show the dependence of the etch rate on the boron concentration using KOH as an etchant. The decrease of the etch rate is steepest for low KOH concentrations; however, one needs a concentration of 10^{20} cm^{-3} in order to reduce the etch rate by a factor of 100. The etch stop is much less effective for higher KOH concentrations. In KOH solutions saturated with IPA, the B^{+} etch stop improves considerably. A disadvantage, however, is the emergence of large hillocks. Generally, therefore, many workers prefer to use EDP as an etchant if the boron etch stop is used.

The B^{+} etch stop has also been studied when using TMAH as an etching solution [42, 43]. The results seem to be even worse than when using KOH. There is a reduction of the etch rate of a factor of up to 50 at high TMAH concentration, and the dopant concentration needs to be considerably larger (several 10^{20}/cm^{3}) than in EDP or KOH.

Other impurities have been tried as an etch stop in anisotropic etchants. The results of these investigations are quite disappointing. Doping with germanium had no notable influence on the etch rate either in KOH solutions or in EDP. At a doping level as high as 5×10^{21} cm^{-3} the etch rate is only reduced by a factor of 2.

Doping with phosphorous has a somewhat greater effect on the etch rate (reduction of a factor of 5 at a doping level of 3×10^{20} cm^{-3}), however, this is still too small for practical use.

Heavy doping with boron has two important drawbacks. First, the conductivity of silicon is greatly enhanced by dissolving boron in silicon. The material is no longer suitable for electrical purposes (integration of ICs), and indiffusion of resistors is no longer possible.

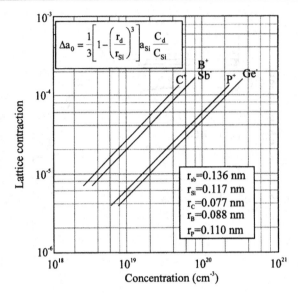

Fig. 2.58. Relative change of the silicon lattice constant for various doping elements (from Heuberger [44]).

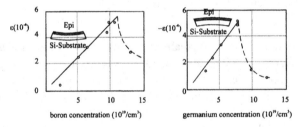

Fig. 2.59. Elastic strain ε of silicon wafers doped with (left) boron and (right) germanium (from Heuberger [44]).

The second drawback is sometimes more severe. Since the boron atom is smaller than silicon atoms, the lattice constant decreases if boron is dissolved in silicon. Doping with germanium increases the lattice constant. In Fig. 2.58 the change of the lattice constant of doped silicon is given as a function of the dopant concentration for a few important impurities. Accordingly, the doped layer is strained. Membranes and microbridges fabricated using the boron etch stop are under tensile stress.

The resulting strain in a doped layer for doping with boron and germanium is shown in Fig. 2.59 [45, 46].

At a critical impurity concentration the stress leads to plastic deformation in the doped layer, resulting in a reduction of the strain at very high doping levels.

An interesting point regarding the etch stop is that silicon doped with both boron and germanium still etches much slower than undoped silicon. Thus, while the etch stop is still working, a reduction of stress in the layer can be achieved by doping with boron and germanium. A stress-free and dislocation-free, slowly etching (ca. 10 nm/min) layer is obtained at combined doping levels of 10^{20} cm^{-3} boron and 10^{21} cm^{-3} germanium [41, 44].

Seidel et al. propose the following mechanism for the etch stop [41]: at a doping level exceeding $2.2\ 10^{19}\ \mathrm{cm}^{-3}$ the Fermi level drops, touching the valance band. Heavily doped silicon behaves more like a metal rather than a semiconductor. As a consequence, the electrons injected into the solid silicon during the chemical reaction of silicon with OH^- groups (see the discussion in Chapter 3) are no longer localized at the silicon–solution interface. Therefore the electrons are no longer available for the reduction step which is required for the generation of new hydroxide ions at the surface. If one assumes that the number of electrons available for the reaction at the silicon surface is proportional to the reciprocal number of holes, which in turn is proportional to the number of boron atoms, the dependence of the etch rate on the boron concentration is easily understood.

Since, according to Seidel [10], each reaction needs four electrons (c.f., Chapter 3), the reaction rate should be proportional to the fourth power of the electron density, thus proportional to the inverse fourth power of the boron concentration, as is observed asymptotically.

For the technology of indiffusion we refer to standard IC-technology textbooks (e.g., Sze [47]).

The concentration profile of the impurities can be calculated using SUPREME. Our experience with this program is that the calculation of impurity levels leads to quite reliable results. Our experience shows that a simulation of the etch profile is indispensable for micromachining. Leaving out the simulation step in the whole planning process will inevitably lead to frustration.

After indiffusion of boron, a boron silicate glass (BSG) layer has grown on the silicon wafer. This material must be removed before the process is continued. It can be stripped in buffered HF (c.f., Table 2.2), but this may take hours. If the wafers are put into a wet oxidation oven at 1000–1100°C for half an hour, the BSG etches much faster in BHF. During the oxidation process, microcracks form due to mechanical stress in the thin oxide. The cracks apparently facilitate the attack of the HF. The stripping still may take half an hour.

We conclude this section with an example of an ingenious device [48]. In Fig. 2.60 we show "nonreturn valves." The design was inspired by the valve in the heart. Smith first etched deep V grooves from one side of a double-sided polished (100) wafer. If the etch opening has the correct size, the V groove is a little bit less deep than the thickness of the wafer. The second step is a deep indiffusion of boron (8 μm), and a V groove is etched from the other side of the wafer. The etching stops at the B^+ layer, leaving V-shaped membranes. The V is opened finally (by etching or simply – less elegantly – by cracking mechanically), and we have a valve with a small fluidic resistance in one direction and a large one in the other.

2.8.4 Thin films

Anisotropic etching solutions do not attack a great number of materials. Hence, a thin film of such a material can be used as an etch stop. The consequence is, of course, that the material of the micromechanical component is no longer silicon but a thin film material.

Thermal oxide is useless for most applications, since thermal oxide is under a great compressive stress. Membranes and microbridges buckle; this material is only useful when buckling is the objective.

A thin film material that has been used as a mechanical construction material, exploiting the fact that it is not affected by anisotropic etchants, is silicon nitride ($\mathrm{Si_3N_4}$). Silicon nitride is a very strong, hard, chemically inert material. In our laboratories we were able to fabricate 1-μm thick silicon nitride membranes of 2 by 2 cm.

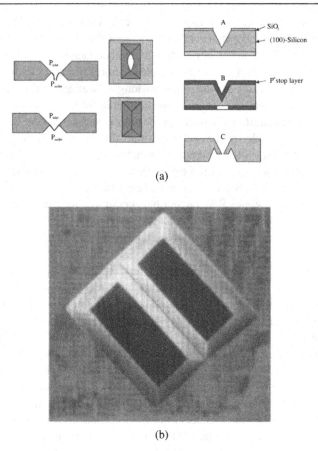

(a)

(b)

Fig. 2.60. Demonstration of the use of the B^+ etch stop. (a) Shows the design of nonreturn valve. Flexible flaps deflect under a pressure difference, so that in one way the valve
opens and in the other way it closes. The sequence A-B-C describes the principal process
flow, (b) shows an SEM image of the valve. The thin membranes of the nonreturn valve
are membranes of highly boron doped silicon (from Smith [48]).

It is possible to control the lateral stress in the film not growing at the stochiometric
concentration of silicon and nitrogen to obtain silicon-rich thin films. The silicon coagulates
to crystallites in the thin film, and at a given concentration the stress turns from tensile
(stochiometric silicon nitride films) to compressive stress in silicon-rich films. For details,
c.f., e.g., Bouwstra [49]. Low tensile stress, silicon-rich silicon nitride films have been
used by Bouwstra [49] for the fabrication of a resonant microbridge mass flow sensor (see
Chapter 5).

Several researchers have tried to implant oxygen and nitrogen in silicon. At energies
high enough, the implanted ions are buried $1/2$–1 μm deep. If the dose is high enough, the
etching stops in that region. It turned out that it is not necessary to implant so much oxygen
or nitrogen that one gets the stochiometric concentration, but a dose lower by a factor of 2
or 3 is sufficient.

After implantation, it is necessary to anneal the wafer because the implantation destroys
the crystal structure at the surface of the wafer. Thereafter an epitaxial layer can be grown

to achieve structures of the desired thickness. This is necessary because the implantation depth is limited to $1/2$–1 μm. This technology is not very well established yet.

2.8.5 Electrochemical etch stop

The last method we describe here is referred to as electrochemical etch stop. It involves doping and biasing of silicon during anisotropic etching. The advantage is that low doping levels are sufficient for the etch stop; therefore microstructures machined with this method still are suitable for integrated circuits, and there is no problem with stresses. The disadvantage is that the method is much more complicated than the boron etch stop.

An introduction into electrochemistry of semiconductors is out of the scope of this book. We describe here only some phenomenological aspects of etching silicon under a bias potential. In order to control the voltage drop across the interface between the etching electrolyte, a three-electrode setup is necessary (see Fig. 2.61). By controlling the potential of a reference electrode so that the current through this electrode equals zero, the potential of the electrolyte is defined. Omitting this electrode leaves the electrolyte potential floating, and one gets unreproducible results with regard to both etching and current voltage characteristics. The material of the reference electrode is usually platinum. The counterelectrode is a standard calomel electrode. Voltages mentioned in this chapter refer to this electrode. (Note, however, this electrode is sometimes reported to be unstable at T < 70°C; an alternative is Ag/AgCl as a counterelectrode [50, 51].)

The silicon substrate must be contacted. This requires either a high-doping step with B (this layer can be contacted by Al) resulting in an ohmic contact, or a metal which makes ohmic contact to silicon (a Ga/In eutectic is suitable; see, e.g., [51, 52]). In most cases the contact corrodes in the etching solution; therefore one needs a holder which protects the backside and part of the frontside of the wafer. A holder is described in some detail in Section 2.11.

The typical current/voltage characteristics of ⟨001⟩-oriented silicon wafers is given in Figs. 2.62 and 2.63. The i/U curves have the following features: There is a maximum current

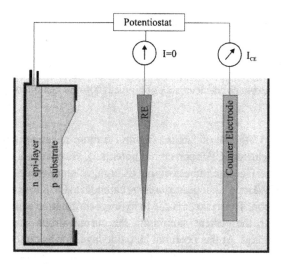

Fig. 2.61. Three-electrode configuration for a p/n junction etch stop.

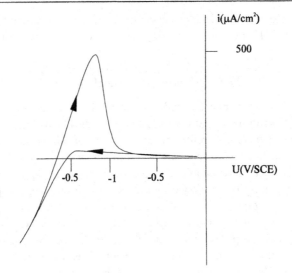

Fig. 2.62. Current/voltage characteristics of n-type ⟨100⟩-oriented silicon in 40% KOH at 60°C. Arrows indicate the direction of the voltage sweep [53, 54].

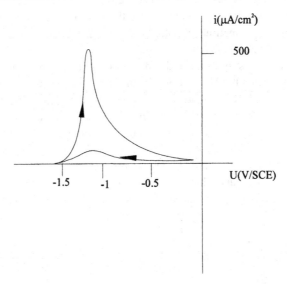

Fig. 2.63. Current/voltage characteristics of p-type ⟨100⟩-oriented silicon in 40% KOH at 60°C [54].

close to -1.2 V/SCE. Cathodic of this maximum (to more negative voltage), the current drops to zero at the so-called OCP (open circuit potential). We have this voltage drop across the silicon–solution interface one etches without external voltage source. Note the difference between p-doped and n-doped silicon: At a voltage cathodic to OCP we get a negative current when using n-type silicon. This current is due to hydrogen formation at the silicon electrode. Anodic with respect to the current maximum, the current drops sharply at a potential called passivation potential. At this potential an oxide layer forms on the silicon electrode, which gives rise for a high electrical resistance. This oxide layer is also responsible for the asymmetry of the i/U curves with respect to the sweep: coming from the anodic side we start

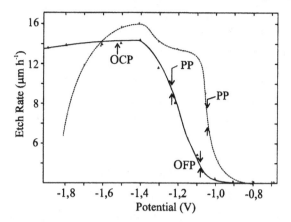

Fig. 2.64. Potential dependence of the etch rate of ⟨100⟩ silicon. Full line: p-type, broken line: n-type silicon in 40% KOH at 60°C.

with an oxide layer. This must be etched away before the resistance falls. Coming from the cathodic side there is no oxide layer. The oxide layer can be exploited for micromachining.

In Fig. 2.64 we show the Si⟨001⟩ etch rates as a function of the bias. It is seen that the current maximum roughly corresponds to a maximum in the etch rate, and that anodic to the maximum, close to the passivation potential, the etch rate drops down together with the current. Again there is a difference between n-type and p-type silicon: In n-type silicon the etch rate becomes very small at more negative bias, while the etch rate of p-type silicon remains more or less constant.

Waggener proposed in 1970 [55] to use the anodic oxidation for etch stop purposes. Since then it has frequently been used in micromachining [56–62]. The method involves growing an n-type epitaxial layer on a p-type substrate or to dope silicon accordingly by indiffusion or implantation to form a p/n junction. The aim is to etch away the substrate selectively by terminating the etching at the junction.

This is done in the following way. A positive voltage is applied to the epitaxial n-type layer (positive with respect to a reference electrode in the etch solution). This voltage maintains the epitaxial layer at a passivating potential to prevent its etching. Simultaneously, the p/n junction is reversely biased, and the substrate floats to the OCP and is etched. When the etch front reaches the p/n junction, the etching stops. The thickness of the remaining silicon membrane is thus defined exactly by the thickness of the epitaxial layer. Initial irregularities at the backsurface have no influence on the final result. Thus the method is very precise, reproducible, and reliable.

Best results are obtained when a four-electrode setup is used, to force the potential of the substrate to the OCP [59, 62].

2.9 Mask materials

2.9.1 Introduction

In principle, all materials that etch slowly enough in the relevant etchants are suitable as mask material. Here we concentrate on the mask materials that are most commonly used. These are silicon dioxide and silicon nitride.

The following metals are resistant against KOH and EDP: gold, chromium, platinum, silver, copper, and tantalum. Aluminum is etched readily.

The choice of a particular mask material depends on a number of considerations. The main issues are availability in the laboratory, ease of the process (such as processing time, complexity of the process, and reliability), selectivity of the etch process with respect to silicon, and mechanical properties.

In this section we summarize the most important properties of these materials relevant for their use as mask materials.

2.9.2   Silicon dioxide

Silicon dioxide can be thermally grown (c.f., Section 2.4) and deposited in a chemical vapor deposition (CVD) process. Thermal oxide is easily obtained if the thickness required is not larger than 2 μm. Thermal oxide is under strong compressive stress due to the fact that in the oxide layer one silicon atom takes nearly twice as much space as in single-crystalline silicon. This has to be kept in mind for further processing. If the oxide is stripped on one side of the wafer, it will bend. Wafer curvature can be fatal for processes such as direct bonding.

APCVD oxide ("AP" refers to "atmospheric pressure"), being a process as simple as the growth of thermal oxide, tends to have pinholes and etches much faster in KOH than thermal grown oxide. Annealing of the CVD oxide removes the pinholes, but the etch rate in KOH remains greater (by a factor of 2–3) than that of thermal oxide.

LPCVD ("LP" refers to "low pressure") oxide is a mask material of comparable quality to thermal. The disadvantage of oxide is that the etch rate in KOH is rather large. In Fig. 2.65 we give the etch rate dependence of oxide as a function of concentration and temperature. There is a distinct maximum at 35 wt% KOH of nearly 100 nm/min. A 2-μm thick oxide film is etched away after 3–4 hours, a time shorter than that required to etch through a wafer. Fortunately, the etch rate of oxide drops to ca. half the value at 20 wt%, making the use of oxide as a mask material practical. The activation energy of the etch rate depends slightly on the concentration and lies between 0.8 and 0.9 eV [10].

In Fig. 2.66 we give the etch rate ratio of ⟨100⟩ silicon to oxide both for KOH and EDP. Here the second great advantage of EDP over KOH becomes apparent. The etch rate of oxide is smaller in EDP by two orders of magnitude.

The oxide is relatively easily removed in HF solutions. The etch rate in BHF is ca. 1 μm/10 min.

With respect to oxide as a mask material, TMAH is somewhat superior to KOH: the etch rate is nearly one order of magnitude smaller than in KOH solutions. In Fig. 2.67 we show results of etching oxide in TMAH [22].

2.9.3   Silicon nitride

Silicon nitride is grown in an LPCVD process. This process is much more involved than thermal growth of oxide or an APCVD/annealing process. However, silicon nitride is much less attacked by KOH than oxide; the etch rate is so small that it has not been determined yet.

The disadvantage of nitride is that it is difficult to remove the layer in the lithographic process sequence. Büttgenbach reports an etch rate of Si_3N_4 of 10 nm/min if H_3PO_4 at 180°C [8]. Nitride layers are under a tensile stress of 10^9 Pa.

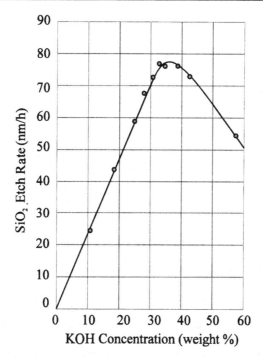

Fig. 2.65. SiO$_2$ etch rate in aqueous KOH solutions at 60°C as a function of concentration.

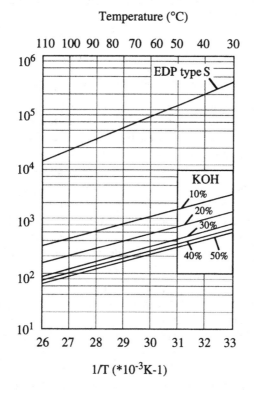

Fig. 2.66. Etch rate ratio of ⟨100⟩ silicon for EDP and KOH solutions (from Seidel et al. [10]).

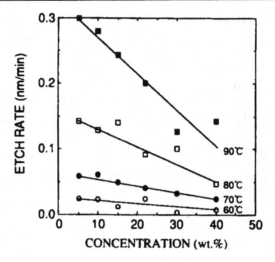

Fig. 2.67. SiO$_2$ etch rate versus TMAH concentration for different temperatures [22].

Tabata et al. [22] did not measure any etch rate within their accuracy of measurement, so this works even better as an etch stop layer. Schnakenberg et al. [63], however, did publish a measured selectivity (ratio of (100) silicon etch rate over the etch rate of silicon nitride) of 24.4×10^3 for 4 wt% TMAH at 80°C. Merlos et al. [42] published a rather extensive study of the etch rate of Si$_3$N$_4$, as well as etching of low-temperature deposited silicon oxide (LTO) and thermally grown oxide.

2.10   Corner compensation

2.10.1   Introduction

So far we have concentrated on structures obtained by anisotropic etching that are bounded by concave forms. As we discussed in Section 2.5, the forms resulting from etching in concave structures are dominated by the slowly etching planes. These forms can be easily controlled by appropriately positioning the mask opening with respect to the ⟨110⟩ direction (for forms bounded by ⟨111⟩ planes). However, when mesas and bent V grooves are needed, convex corners cannot be avoided. The resulting forms are then dominated by the fast-etching planes. Since, in general, the etch rate depends on the etchant, the temperature, the concentration of the etchant, the doping of the substrate, and stirring (see, e.g., [64]), the crystallographic orientation of the etch rate maxima also depends on the conditions. In the literature, many fast-etching directions are identified, and the disagreement is probably mostly due to the strong dependency on the etch conditions.

It is important to stress the following point: the planes that emerge under convex corners are not compatible with the planes that we find in the etch rate minima. The latter are smooth on an atomic scale, and any misorientation is realized by step trains of one or many atomic layers in height. This smoothness enables us to describe these facets with a low number set of indices.

The planes normal to the etch rate maxima are not flat on an atomic scale. They mostly break up into small terraces made by low-index planes such as {111} planes, and the overall direction is not so well defined. Therefore, care must be taken with the description of the

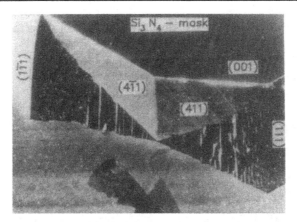

Fig. 2.68. Corner undercutting of mesa (from Offereins et al. [67]).

etch rate maximum: the direction is not necessarily a low-index orientation. Rather, from the reports in the literature, the orientation seems to vary, e.g., when the temperature of the etchant changes (an impressive example is provided by Sato's work, see Fig. 2.25 [11]). Therefore, a physically sound description of the crystallographic orientation of etch rate maxima by small indices is impossible. Nevertheless, in agreement with the ideas currently in use we use low indices to indicate the orientation of the maxima.

It is possible to compensate for the underetching of convex corners by a suitable mask design. This chapter presents several such mask designs based on literature that exists on this subject. Since most of the compensation structures are derived from the fast-etching planes, these will be determined first, followed by the derived compensation structures. In the last paragraph, some results achieved with those compensation structures will be shown.

In most cases the mask will be aligned in the $\langle 110 \rangle$ direction unless mentioned otherwise. The photograph in Fig. 2.68 shows the rounding off of convex corners due to underetching of the uncompensated mask of quadratic form.

2.10.2 Fast-etching planes

In order to determine the fast-etching planes, many researchers have etched a mesa structure without corner compensation. After etching for some time they tried to measure the appearing planes at the corners of the mesa structure. Since the anisotropy of the etch rate depends on the etchant, the temperature, the concentration of the etchant, the doping of the substrate, and stirring [64], different researchers find different fast-etching planes. The following results are reported in the literature (with $\langle 100 \rangle$ silicon as the substrate).

Potassium hydroxide (KOH)

Wu and Ko report that the fast-etching planes are {212} planes [65]. They use n-type silicon wafers with 100 g KOH + 320 g water + 80 g normal propanol. They show that the fast-etching planes are {212} planes and not {331} or {211} planes, because the intersecting lines of these planes at the surface make angles of 127° (see Fig. 2.69) with each other and of 48.1° with the surface.

Fig. 2.69. Etching a pyramid formed by {411} planes (from Offereins et al. [67]).

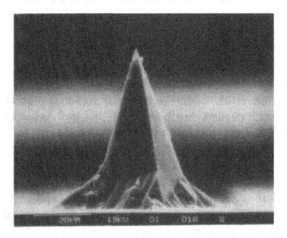

Fig. 2.70. Pyramid formed by {411} planes (from Offereins et al. [67]).

However, Puers and Sansen have found that the {130} direction is dominant for etching [66] (they remark that "it was virtually impossible to determine the fast etching planes because beveling planes intersect"). They report no concentrations but use a temperature of 80°C.

Offereins et al. found the fast-etching planes to be {411} planes [67, 68, 72]. They used aqueous KOH without any additions in a range of 15–50 wt% at temperatures of 60–100°C. In the case of small beams, these planes only occur directly under the mask because they are overlapped by rugged surfaces (therefore they are not very useful to derive corner compensation structures). This is shown by the pyramids formed by {411} planes (see Fig. 2.70). Their work is confirmed by Dizon et al. [69] and Trimmer et al. [70]. In the latter paper the pyramids are used as microprobes to implant DNA into plant and animal cells.

Bean has found {331} planes using 250 g KOH + 800 g deionized water + 200 g normal propanol at 80°C [71].

Table 2.5. *Fast-etching planes for various etchants.*

{110}	KOH, 40°C	Sato
{212}	30 wt% KOH + 25 wt% propanol	Wu
	80% Hydrazine + 20% H_2O	Wu
	KOH + IPA	Puers
	EDP S, B	Wu
{130}	KOH 80°C	Puers
{331}	EDP	Beau
	EDPB	Bäcklund
	30 wt% KOH + 25 wt% propanol	Beau
{411}	KOH (15–50 wt% KOH, 60–100°C)	Sand

KOH with isopropyl alcohol (IPA)

Puers and Sansen report {212} planes for n-type silicon also by measuring the angles of the corner intersecting lines at the surface [66]. They report no concentrations but use a temperature of 80°C.

Ethylenediamine-pyrocatechol-water (EDP)

In the same way as with KOH, Wu and Ko report that the fast-etching planes are {212} planes [65]. They use a B etch and an F etch (c.f., Section 2.6). In agreement with this, Abu-Zeids also reports of {212} planes but with small ledges on {323} planes, which he further neglects [73]. He used 55 vol% ethylenediamine in deionized water at 85°C.

However, Bean reports to have found {331} planes using 255 cm^3 ethylenediamine + 45 g pyrocatechol + 120 cm^3 water [71]. In agreement with Bean, Bäcklund and L. Rosengren [64] also determined {331} planes, but only at the top of the etched mesa. At the bottom they could not identify the planes.

Hydrazine

In the same way as with KOH, Wu and Ko report that the fast-etching planes are {212} planes [65]. They use 80% hydrazine + 20% water by volume.

Table 2.5 gives a summary of these results. The discrepancy of the experimental results is quite surprising.

2.10.3 Mask designs for corner compensation

Based on the fast-etching planes, different compensation structures can be derived. These compensation structures will be described below.

Triangles

Because Wu and Ko found that the {212} planes were the fast-etching planes and intersect the (001) surface at {210} orientations, they added triangles bounded by {210} directions to the corners of a square mask pattern (see Fig. 2.71). The distances indicated in Fig. 2.70 are

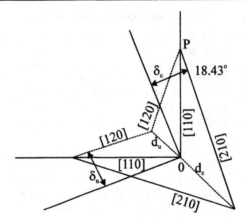

Fig. 2.71. Triangle for mask compensation.(from Wu and Ko [65]).

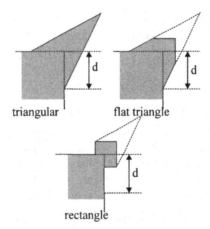

Fig. 2.72. Triangle, flat triangle, and rectangle (d = relative dimension) (from Puers and Sansen [66]).

related to the cystallographic properties of the cubic lattice. The following equation gives the design rule for the compensation structure:

$$d_c = \sqrt{5} \cdot \delta_c, \quad d_u = 1/2\sqrt{5} \cdot \delta_u, \quad d_c = 2 \cdot d_u, \quad op = \sqrt{2} \cdot d_c.$$

d_u is the distance between points of the corner before and after etching, without compensation. d_c is that distance with compensation. p marks where the undercutting line and the compensation line intersect on the (110) side of the mask pattern. Puers and Sansen [66] also introduce a flat triangle (see Fig. 2.72).

Offereins and coworkers have made triangles bounded by {410} planes [72]. But this resulted in a convex corner only directly below the mask. A drawback is that this compensation structure needs a lot of space.

Bäcklund and Rosengren introduce triangles bounded by {310} planes [64], but they use mask patterns which are oriented along the ⟨100⟩ orientation.

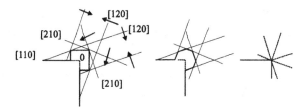

Fig. 2.73. Compensation principle for a rectangle (from Puers and Sansen [66]).

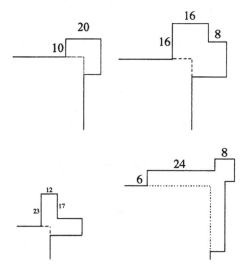

Fig. 2.74. Compensation patterns attaining the same results. (from Abu-Zeid [73]).

Rectangle

Puers and Sansen introduced a rectangle [66]. Figure 2.73 shows the compensation principle for this rectangle based on {210} directions.

In order to reduce the spatial requirement of the rectangle, Offereins and coworkers have added {110} oriented beams to the convex corners of the rectangle. This is done in order to delay the undercutting of those convex corners. The etch front can be controlled by making the beam of a certain corner larger than the beams of the others.

Abu-Zeid reports four compensating patterns that give the same results for a mesa height of 25.25 μm [73] (see Fig. 2.74). These patterns are based on calculations derived from the etching shapes.

Strips in (110) direction

Bao and coworkers used ⟨110⟩ strips for corner compensation [74]. To solve the difficulties others have found, they used folded strips to control the etching front. The folded strips are asymmetrical so that the etching front of the shortest branch reaches the turning point first, which results in an asymmetrical etching front under the strip part attached to the corner. This reduces the corner undercutting. It is also possible to use more than one turning (see Fig. 2.75). The undercutting is reduced best by making only one turn, as shown in the right side of Fig. 2.75.

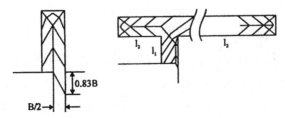

Fig. 2.75. (left) Nonfolded strip with symmetrical etching front, (right) asymmetrical folded strip reducing corner undercutting (from [74]).

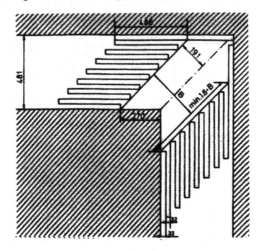

Fig. 2.76. Fanned beams in (100) direction (from Mayer et al. [68]).

The protection time corresponds to an etch depth of $L_{\text{eff}}/2.7$ because the etch rate in the $\langle 110 \rangle$ direction is 2.7 times faster than in the vertical $\langle 100 \rangle$ direction at the conditions used by Bao et al.

Offereins introduced double beams in order to reduce the beveling of the etched convex corner due to the rugged surfaces which appear at the bottom [72]. The beveling is only decreased, it does not disappear.

Strips in (100) direction

In [67, 68, 72], Offereins and coworkers introduced strips in the (100) direction. They used beams that are fanned out at the long sides, as shown in Fig. 2.76. Between the narrow beams there are only concave corners, therefore the fast-etching (411) planes are stopped by the slower-etching (100) planes and align with these planes. For 33% KOH the width of the beam must be 1.6 times smaller to prevent the beam being underetched by the (411) planes before the etching depth is reached. The narrow beams prevent the edges and the corners of the $\langle 010 \rangle$ beam being underetched. The top convex corner is protected by connecting it to the external mask.

The length of the narrow beams is given by

$$L = (H - B/2) \cdot V \cdot 1.03 \cdot 1.94,$$

where V is the ratio of etch rates between the {411} and {100} planes.

The factor 1.03 takes into account the inclination of the {411} planes with respect to the vertical line. The factor 1.94 results from the angle between the ⟨411⟩ and ⟨010⟩ direction.

The width of the beams should vary between 15 μm and 30 μm to guarantee that the {411} planes align with the {100} planes and to avoid premature undercutting due to lateral effects.

The advantage of the (100) strips is that the compensation structure is mainly undercut by (100) planes and not by (411) planes, and no undefined rugged surfaces pose problems concerning reproducibility. Therefore an exactly defined edge formed by {111} planes is possible [72].

2.10.4   Results

Triangle

With the mask corner compensation, Wu and Ko have made perfect convex corners without visible distortion at the bottom [65]. They show results for EDP only.

With the triangular compensation mask and the pattern oriented along the ⟨100⟩ direction, Bäcklund and Rosengren obtained convex corners which are still slightly rounded off by unidentified planes in EDP but bounded with {100} side-planes at 90° to the surface. With pure KOH (30%) they obtained a perfect cubical stud (walls and top made of {100} planes), as shown in Fig. 2.77. The maximum height is 27% of the width. Note that the etch front under the mask edges are {100} planes, which etch approximately as fast as the bottom, so a serious undercut of the mask cannot be avoided.

Rectangles

With the compensation structures introduced by Abu-Zeid [73], mesas were made. At the corners, cones appear with the shape depending on the compensation structure. See Fig. 2.78 as an example for square compensation patterns.

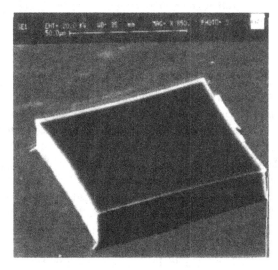

Fig. 2.77. Perfect cubical stud with enlargement of one corner (from Bäcklund and Rosengren [64]).

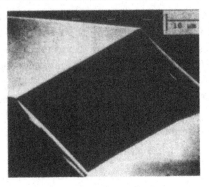

Fig. 2.78. Square compensating pattern resulting in cones (from [73]).

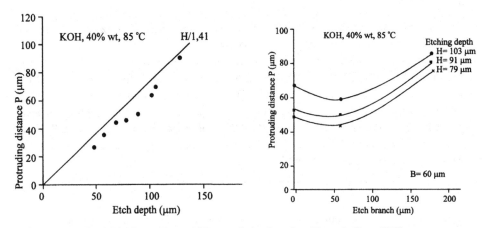

Fig. 2.79. Dependence of P on etch depth and end branch (from [74]).

The cones result from the difference in the inclinations of the {111} and {212} planes. The cones disappear when the etching height is increased by 0.185 times its initial value.

Strips in (110) direction

Bao and coworkers [74] concentrate on the development of the etch surface at the bottom, because for application of mesas in accelerometers as seismic mass, the clearance between mass and frame on the bottom is more important than the appearance of the surface corner.

The horizontal distance of {111} sidewalls, S, is $0.707H$ (H is the etch depth). P is the protruding distance of the etch front in the direction of the strip. Thus, $P = S$ means a well-shaped corner, $P < S$ means overetching of the bottom, and $P > S$ means underetching of the bottom when the surface corner is critically compensated. In Fig. 2.79 the dependence of P on the etch depth and the end branch is shown. Bao and coworkers recommend the length of the end branch to be about the width of the strip in order to control surface etching effectively and for the requirement of a small P.

The convex corner attained with the double beams is not a perfect convex corner. The corner is formed by a somewhat flat surface.

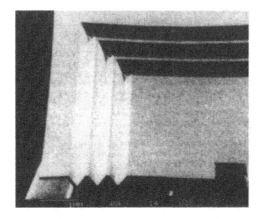

Fig. 2.80. Bent V grooves (from [68]).

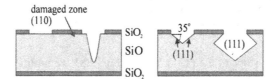

Fig. 2.81. (left) Schematic diagram of a ⟨110⟩ silicon wafer with a laser-induced damaged zone. The hatched region is an etch mask. (right) The resulting etch pit (from Alavi et al. [78]).

Strips in (100) direction

A convex corner made with a (100) oriented strip [72] is shown in Fig. 2.80. With the fanned strips, sequentially bent V grooves are made. The grooves are not etched entirely to the bottom, because otherwise mechanical stress could cause problems concerning damage to the structure. The size of these remnants is determined by the beam width.

2.11 Miscellaneous

2.11.1 *Laser assisted etching*

There are several ways in which laser light can assist micromachining. An overview can be found in Büttgenbach's book [8]. The heating power of the light can be used to enhance locally the nucleation rate of some material. This way it is possible to control the growth of structures in the three-dimensional space. This technique has been demonstrated by Westberg et al. [75] to fabricate spirals from polycrystalline boron and silicon in an LPCVD furnace. Alternatively, the light can be used for local polymerization. The nonpolymerized region can be etched away, leaving a truly three-dimensional structure. For examples we refer to the literature [76]. In some gaseous environments, the etch rate of silicon or germanium is greatly enhanced if the material is illuminated, allowing local dry etching [77].

   A further alternative works as follows. Using intensive IR radiation, the crystalline structure of silicon can be destroyed; the silicon crystal melts and recrystallizes in polycrystalline modification. This material, being isotropic, is etched isotropically in anisotropic etchants.

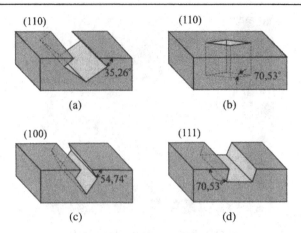

Fig. 2.82. Schematic view of possible microchannels in silicon formed by anisotropic etching after laser damage of the silicon. (a) and (b) ⟨110⟩ wafers, (c) ⟨100⟩ wafer, (d) ⟨111⟩ wafer (from Alavi et al. [78]).

Eventually, the etching stops at the (111) face. In Fig. 2.81 we outline the principle idea, showing the damaged region and the resulting etch pit.

The structures that can be fabricated by this method are schematically shown in Fig. 2.82. The pattern defined by the area irradiated with the laser beam can be written directly by a computer controlled movement of the beam.

Note that by using this method it is possible to use ⟨111⟩ wafers for micromachining. Details for the process can be found in [78, 79].

The depth of the destroyed area depends on the intensity of the light and the velocity. If a Nd:YAG laser is used with a power of 50 W, the velocity of the writing is in the range of a few mm/sec. The light of the Nd:YAG laser is very suitable for this micromachining technique due to the photon energy (1.17 eV, just exceeding the band gab energy of silicon, which is 1.12 eV). Experimental results are given in Fig. 2.83 [78].

In our own labs we used RIE (reactive ion etching) to etch trenches in silicon. When etching trenches in KOH solutions, similar structures can be fabricated as shown in Fig. 2.83.

2.11.2   Backside protection

In many cases it is necessary to protect the backside of the wafer from the anisotropic etchant, either because the protective layer is too thin or because there are parts that cannot be protected at all. Although the solution to this problem is rather difficult and quite important, there is no literature on this issue. Therefore we can only speak from our own experience.

In principle, there are two ways for backside protection: (i) a protective layer and (ii) mechanical protection.

In the Uppsala laboratories, people tried to use different waxes deposited on the wafer backside. Wax can be spun on the wafer and easily removed with the help of organic solvents and it is insoluble in KOH and EDP. The results are poor: the yield is quite low because the wax does not stick perfectly to the wafer in KOH or EDP.

In the Twente labs, research is currently being done on alternative organic materials that can be spun on the wafers, but it is too early to report on the results. A general problem

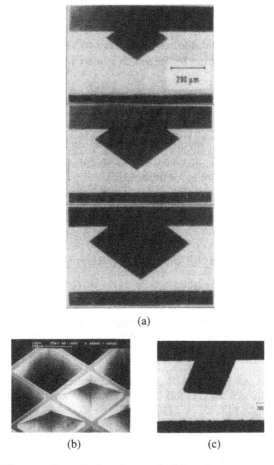

(a)

(b)   (c)

Fig. 2.83. (a) Cross sections of microchannels in ⟨110⟩ silicon formed by laser melting at different process parameters. (b) Cross section of a microchannel in ⟨111⟩ silicon. (c) Vertical-walled shafts fabricated in ⟨110⟩ silicon using laser micromachining and anisotropic etching. The shallow pits were produced without laser processing (all photographs from [78]).

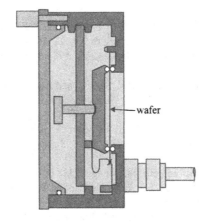

Fig. 2.84. Schematic of a teflon holder for backside protection (from Aukema [80]).

with spinning is that if there are structures on the backside, the spinning process does not cover all features, which is absolutely necessary.

Mechanical protection of the backside can be achieved with a holder. An example is given in Fig. 2.84. The holder is made from teflon, which is inert with respect to the anisotropic etchants.

The holder shown here allows an electrical connection of the wafer, e.g., for the electrochemical etch stop. The wafer is fixed between teflon-coated O-rings, which are carefully aligned in order to avoid mechanical stress in the wafer by the mount.

Kwa et al. [81] have shown that stainless steel contaminates the etching solution, which results in poor reproducibility of the electrochemical etch stop and rough surfaces. We therefore recommend the use of teflon for the holder.

References

[1] P. J. Hesketh, C. Ju, S. Gowda, E. Sanoria, and S. Danyluk, J. Electrochem Soc. **140**, 1080 (1993).

[2] J.-Å. Schweitz, MRS Bulletin **17**, 34 (1992).

[3] K. E. Petersen, Proc. IEEE **70**, 420 (1982).

[4] F. Ericson, S. Johansson, and J.-Å. Schweitz, J. Mater. Sci. Eng. **105/106**, 131 (1988).

[5] F. Mazeeh and S. D. Senturia, Sensors and Act. **A21–A23**, 861 (1990).

[6] J. M. Ziman, *Principles of the Theory of Solids* (Cambridge University Press, Cambridge 1972), p. 124.

[7] C. Kittel, *Introduction into Solid State Physics* (J. Wiley & Sons, New York, 1986), p. x.

[8] S. Büttgenbach, *Mikromechanik* (Teubner Studienbücher, Stuttgart, 1991), p. x.

[9] D. F. Weirauch, J. Appl. Phys. **46**, 1478 (1975).

[10] H. Seidel, L. Csepregi, A. Heuberger, and H. Baumgärtel, J. Electrochem. Soc. **137**, 3612 (1990).

[11] A. Koide, K. Sato, and S. Tanaka, Proc. MEMS, Nara, Japan, 1991, p. 216.

[12] J. G. E. Gardeniers, W. E. J. R Maas, R. Z. C. van Meerten, and L. G. Giling, J. Crystal Growth **96**, 832 (1989).

[13] L. J. Giling and W. J. P. van Enckevort, Surf. Sci. **161**, 567 (1985).

[14] C. Ju and P. J. Heketh, Sensor and Act. **A33**, 191 (1992).

[15] K. Sato, A. Koide, and S. Tanaka, JIEE Technical Meeting Micromachining and Micromechatronics, IIC-89-30, p. 9 (1989).

[16] J. Pugacz-Muraszkiewicz, IBM J. Res. Develp. **16**, 523 (1972).

[17] L. D. Clark Jr., J. L. Lund, and D. J. Edell, Tech. Digest IEEE Solid State Sensor and Actuator Workshop, Hilton Head Island, S. Carolina, June 6–9, USA, 1988, pp. 5–8.

[18] U. Lindberg, Micromechanics-Fabrication Processes and Fluid Components, PhD Thesis, Uppsala University (1993).

[19] U. Schnakenberg, W. Benecke, and B. Lochel, Sensors and Act. **A21–A23**, 1031 (1989).

[20] U. Schnakenberg, W. Benecke, B. Lochel, S. Ullerich, and P. Lange, Sensors and Act. **A24**, 1 (1991).

[21] M. J. Declercq, L. Gerzberg, and J. D. Meindl, J. Electrochem. Soc. **122**, 545 (1975).

[22] O. Tabata, R. Asahi, H. Funabashi, K. Shimaoka, and S. Sugiyama, Sensors and Actuators **A34**, 51 (1992).

[23] T. Baum and D. Schiffrin, J. Micromece. Microeng. **7**, 338 (1997).

[24] J. B. Price, in *Semiconductor Silicon*, eds. H. R. Huff and R. R. Burgess, The Electrochemical Society Softbound Proceedings Series (Princeton, 1973), p. 339.

[25] G. Findler, J. Muchow, M. Koch, and H. Münzel, Proc. MEMS '92, 62 (1992).

[26] L. Ternez, PhD Thesis, University of Uppsala (1988).

[27] S. Tan, H. Han, R. Boudreau, and M. L. Reed, Proc. MEMS '94, Oiso, Japan, January 25–28, 1994, p. 229.

[28] S. Tan, M. L. Reed, H. Han, and R. Boudroau, J. Microelectromechanical Systems 5, 66 (1996).

[29] Y. Bäcklund, Uppsala University, private communication (1986).

[30] K. Hjort, Uppsala University, private communication.

[31] D. G. Vlachos, L. D. Schmidt, and R. Aris, Phys. Rev. B 47, 4896 (1993).

[32] M. Elwenspoek, J. Micromech. Microeng. 6, 405 (1996).

[33] M. Elwenspoek, Proc. ASME Dynamic Systems and Control Division, Vol 2, DSC-Vol. 57–2 (1995), p. 901.

[34] P. Allongue and J. Kasparian, Microsc. Microanal. Microstruct. 5, 257 (1994).

[35] P. Hartman (ed.), *Crystal Growth, An Introduction* (North Holland, Amsterdam, 1973), p. 367.

[36] L. D. Landau and E. M. Lifshitz, *Statistical Physics* (Pergamon Press, Oxford, 1980), p. 537.

[37] J. M. Kosterlitz and P. J. Thouless, J. Phys. B 13, 4942 (1976).

[38] P. Bennema, J. Crystal Growth 69, 182 (1984).

[39] M. de Boer, unpublished.

[40] W. M. van Huffelen, M. J. de Boer, and T. M. Klapwijk, Appl. Phys. Lett. 58, 2438 (1991).

[41] H. Seidel, L. Csepregi, A. Heuberger, and H. Baumgärtel, J. Electrochem. Soc. 137, 3625 (1990).

[42] A. Merlos, M. C., Acero, M. H. Bao, J. Bausells, and J. Esteve, Sensors and Actuators A 37–38, 737 (1993).

[43] E. Steinsland, M. Nese, A. Hanneborg, R. W. Bernstein, H. Sandmo, and G. Kittisland, Proc. Transducers '95, Stockholm, June 25–29, 1995, Vol. I, pp. 190–193.

[44] A. Heuberger, *Mikromechanik* (in German) (Springer Verlag, Heidelberg, 1989), p. 218.

[45] Y. T. Lee, N. Miamoto, and I. Nishizawa, J. Electrochem. Soc. 122, 530 (1975).

[46] H. J. Herzog, C. Csepregi, and H. Seidel, J. Elecrochem. Soc. 131, 2696 (1984).

[47] S. M. Sze (ed.), *VLSI-Technology* (McGraw Hill, London, 1985).

[48] L. Smith and B. Hök, Proc. Transducers '91, 1991, p. 1049.

[49] S. Bouwstra, Resonating Microbridge Mass Flow Sensor, PhD Thesis, University of Twente (Enschede, The Netherlands, 1990).

[50] A. Götz, J. Esteve, J. Bausells, S. Marco, J. Samiter, and J. R. Mortante, Sensors and Act. A37–A38, 477 (1993).

[51] J. Rappich, H. L. Lewerenz, and H. Gerischer, J. Electrochem. Soc. 140, L187 (1993).

[52] P. M. M. C. Bressers, S. A. S. P. Pagano, and J. J. Kelly, J. Electroanl. Chem 391, 159 (1995).

[53] P. M. M. C. Bressers, Silicon Etching, Au Electrochemical Study, PhD Thesis, Utrecht (1995).

[54] O. J. Glembocki, R. E. Stahlbusch, M. Tomkiewicz, J. Electrochem Soc. 132, 145 (1985).

[55] H. A. Waggener, Bell Systems Technical J. 49, 473 (1970).

[56] R. L. Smith, B. Kloek, N. F. de Rooij, and S. D. Collins, J. Electroanal. Chem. 238, 103 (1987).

[57] T. N. Jackson, M. A. Tischler, and K. D. Wise, IEEE Electron Device Lett. EDL-2, 44 (1981).

[58] P. M. Sarro and A. W. van Herwaarden, J. Electrochem Soc. 133, 1724 (1986).

[59] B. Kloeck, and N. F. de Rooij, Proc. Transducers '87, 1987, p. 116.

[60] M. Hirata, S. Suwazono, and H. Tanigawa, Sensors and Act. A13, 63 (1988).

[61] Y. Linden, L. Ternez, J. Tiren, and B. Hok, Sensors and Act. A16, 67 (1989).

[62] B. Kloeck, S. D. Collins, and N. F. de Rooij, R. L. Smith, IEEE Trans. Electron. Devices 36, 663 (1989).

[63] U. Schnakenberg, W. Benecke, and P. Lange, Proc. Transducers '91, San Francisco (1991) pp. 815–819.

[64] Y. Bäcklund and L. Rosengren, J. Micromech. Microeng. 2, 75 (1992).

[65] X. P. Wu and W. H. Ko, Sensors an Act. A18, 207 (1989).

[66] B. Puers and W. Sansen, Sensors and Actuators A21–A23, 1036 (1990).

[67] H. L. Offereins, K. Kühl, and H. Sandmaier, Sensors and Act. A25–A27, 9 (1991).

[68] G. K. Mayer, H. L. Offereins, H. Sandmaier, and K. Kühl, J. Electrochem. Soc. 137, 3947 (1990).

[69] R. Dizon, H. Han, and M. L. Reed, Proc. MEMS '93, Ft. Lauderdale, Florida, USA, Feb. 7–10, 1993, pp. 48–52.

[70] W. Trimmer, P. Ling, C. Chin, P. Orton, T. Gaugler, S. Hashmi, G. Hashmi, B. Brunett, and M. Reed, MEMS '95, Amsterdam, The Netherlands, Jan. 29–Feb. 2, 1995, pp. 111–115.

[71] K. E. Bean, IEEE Trans. Electron. Devices **ED-25**, 1185 (1978).

[72] H. L. Offereins, H. Sandmaier, K. Marusczyk, K. Kühl, and A. Plettner, Sensors and Materials **3**, 127 (1992).

[73] M. M. Abu-Zeid, J. Electrochem. Soc. **131**, 2138 (1984).

[74] M. Bao, Burrer, J. Esteve, J. Bausells, and S. Marco, Sensors and Act. **A37–A38**, 727 (1993).

[75] H. Westberg, M. Boman, S. Johansson, and J.-Å. Schweitz, Proc. Transducers '91 (1991), p. 516.

[76] K. Ikuta and H. Hirowatari, Proc. MEMS, Ft. Landerdale, Florida, USA, Feb. 7–10 1993, p. 42.

[77] T. M. Bloomstein and D. J. Ehrlich, Proc. MEMS '91, p. 202.

[78] M. Alavi, S. Büttgenbach, A. Schumacher, and H.-J. Wagner, Sensors and Act. **A32**, 299 (1992).

[79] M. Alavi, S. Büttgenbach, A. Schumacher, and H.-J. Wagner, Proc. Transducers '91, San Francisco, USA, June 23–27, 1991, p. 512.

[80] J. Aukema, PECVD Silicon Nitride and Electrochemical Etch Stop for Integration of Force Sensor and hT- B:CMOS Process, Master Thesis, University of Twente (1992).

[81] T. A. Kwa and R. F. Wolffenbuttel, MME '94, Pisa, Italy, Sept. 5–6, 1994, pp. 36–39.

[82] L. Ternez, Y. Bäcklund, J. Tiren, and J. O'Connell, Sensors and Materials **1–6**, 313 (1989).

[83] F. Bäcklund, Silicon Micromachining-Methological Aspects and Applications in Biomedical Pressure Sensors, PhD Thesis, Uppsala University (1992).

[84] L. Rosengren, Silicon Microstructures for Biomedical Sensor Systems, Thesis, Uppsala University (1994).

[85] W. Choi and J. G. Smits, J. Microclectromechanical System **2**, 82 (1993).

[86] Y. Uenishi, M. Tsugai, and M. Mehregani, IEEE Workshop on Microelectromechanical Systems, Oiso, Japan, Feb. 1994 (MEMS '94), p. 319.

[87] O. Tabata, R. Asahi, H. Funabashi, and S. Sugiyama, Proc. Transducers '91, San Francisco, June 23–27, 1991, pp. 811–814.

[88] O. Tabata, Proc. Transducers '95, Stockholm, June 25–29, 1995, Vol. I, pp. 83–86.

[89] S. Bouwstra and T. S. J. Lammerink, in J. C. Lodder, (ed.) *Sensors and Actuators*, Kluwer Technical Books (Deventer, The Netherlands, 1986), p. 199.

3

Chemical physics of wet chemical etching

3.1 Introduction

This section reviews wet chemical etching of single crystals from the viewpoint of the science of crystal growth. In spite of a large body of literature on this subject, there is currently no generally accepted theory for a mechanism that explains the great anisotropy of the etch rate of silicon and other monocrystalline materials such as quartz and GaAs. Research on silicon etching is mainly concentrated on the mechanisms of the chemical reaction that transfers a silicon atom into a molecule that is soluble in the etching solution. There is also no consensus concerning the chemical reaction, and until now no mechanism could be found that would account for an anisotropy.

As we have seen in Section 2.5, the *growth rate* of crystals is anisotropic as well, otherwise crystals would have a spherical form. The degree of anisotropy is also very large. Usually, crystals are bounded by a small number of flat, low-index faces; these faces are flat on an atomic scale. If a crystal is broken, sawn, or polished so that high-index faces are exposed to a supersaturated environment, one finds that the growth rate of these faces is orders larger than the growth rate of the flat faces that occurs naturally on crystals. The theory of crystal growth is well established, and there is a satisfactory explanation of the anisotropy of the growth rate, the crystallographic orientation of the growth rate minima, its temperature dependence, the dependency of the type of supersaturated environment (vapor, solution, or melt), the role of impurities, stability of crystal faces, etc. The major phenomena of crystal growth and crystal etching are so similar that it is quite natural to use these theories to explain etching.

The starting point is that there are smooth and rough crystal surfaces. In a first approximation, the flat faces that belong to a particular crystal can be found from its structure. The analysis of the diamond crystal structure reveals that the {111} face is the only smooth face in this lattice – other faces might be smooth only because of surface reconstruction or adsorption.

The kinetics of smooth faces is controlled by a nucleation barrier that is absent on rough faces. The idea that a nucleation barrier is responsible for the slow etching of the {111} face of silicon has been introduced by Elwenspoek [1]. The rough faces therefore possibly etch faster by orders of magnitude. Steps can be created by two distinct mechanisms: two-dimensional nucleation or screw dislocations that outcrop a crystal face, thereby giving rise to a permanent step source. This step deforms while etching into a spiral. In both mechanisms the key parameters governing the etch rate are the free energy required to create steps and the change of the Gibbs free energy by transferring a silicon atom from the crystalline state to a state in which it is bonded in a molecule that is dissolved in the etching solution. The ratio of these two energies is proportional to the size of a critical nucleus and it dominates the average steady state distance between the arms in a spiral. We show that this picture explains major features of isotropic and anisotropic wet chemical etching of silicon. There is one important aspect that seems to differentiate growth and etching: the chemical potential difference during (controlled) crystal growth is believed to be usually much smaller than during etching. Etching is much further away from equilibrium than controlled crystal growth. This is not, of course a natural law. Crystals also grow at very large supersaturation. However, the growth front becomes unstable and one finds dendrites and cellular crystal forms. This instability does not occur when etching: the instability results from concentration gradients close to the crystal surface through which an incident waviness grows. The top feels a larger supersaturation and grows faster than the valleys. While destabilizing the growth front, this mechanism has a stabilizing effect on the etchfront. As we shall see, it is not the chemical potential difference itself that matters, but the ratio of the energy required to create steps on flat surfaces square over the chemical potential difference to kT.

We stress that the relation between growth and anisotropic wet chemical etching of crystals was made earlier. Frank and Ives [2], Jaccodine [3], and Weirauch [4] referred to etching as a negative crystal growth and pointed out the importance of the free energy of surfaces. However, Frank and Ives considered the case of small chemical potential difference, and did not apply the theory to the materials of interest here. In the work of Weirauch and Jaccodine, there is neither a detailed discussion of the etch mechanisms nor an attempt to explain the etch phenomena. This situation is quite different when etching from the gas phase, where a large body of literature exists. The relation to crystal growth has been made in great detail. In these papers, issues such as topography of dissolution spirals [5, 6], macrostep formation [7], and step train orientation [8] have been addressed. Apparently, there is a large gap in the information flow from the crystal growth community to the micromachining community.

With reference to theory of crystal growth, a number of experimental results can be understood quite easily. We emphasise that these theories, and the resulting predictions and explanations of the phenomena, are independent of the material: the theory is valid for etching of all single crystals. The following points follow directly from standard theory of crystal growth:

- The isotropy of the etch rate in certain solutions and the anisotropy in others can be understood by comparing reaction enthalpies and the undersaturation of the etch products in the solutions; criteria can be given that determine whether the etch rate is isotropic or anisotropic, and the criteria can be compared to experimental results.
- The crystallographic orientations of the minimum etch rate (and growth rate) can be deduced from the crystal symmetry in first approximation.
- The ratio of the etch rate of different flat faces and its temperature dependence can be estimated.
- The shape of the etch rate diagram, i.e., the variation of the etch rate as a function of the crystallographic orientation is at least qualitatively predictable.

A large part of this chapter is devoted to summarizing the crystal growth view on wet chemical silicon etching in order to try to overcome the gap mentioned above. A few predictions of the theory are tested experimentally, and the results support the crystal growth view. In particular, we looked with greater detail at the dependency of the etch rate of silicon etched in KOH on the crystallographic orientation close to the $\langle 100 \rangle$ direction, and we studied a transition to anisotropic silicon etching in $HF:HNO_3:CH_3COOH$.

We emphasize that the approach to anisotropic etching is rather new, and therefore much work has to be done to answer a number of questions, which will be asked in this chapter. We hope that future research will clarify some of them. Note also that the view outlined here is subject to discussion in the micromachining community as well as in the electrochemistry community. A large body of experimental data has been obtained using the wagon wheel method (see Section 2.5) which consists of radially diverging mask segments separated by a given angle. These masks have been applied on {100}- and {110}-oriented wafers. An alternative method has been used by Koide et al. ([9], see also Section 2.5), who performed etch experiments in hollow hemispheres drilled in thick {100}-oriented single crystalline silicon. Etching of positive silicon spheres has also been studied (e.g., [10]). The latter is suitable to determine the maximum etch rates (amount and crystallographic direction), while the former is more suitable to determine etch rates in the minima.

The second method gives the most complete information. In all anisotropic etching agents, the slowest-etching planes are {111}. Additionally, there are minima in ⟨001⟩ when etching is done in aqueous KOH, NaOH, LiOH, CsOH, TMAH, and NH_4OH, which are sometimes steep enough to employ them for micromachining. When etching in KOH at low temperature (<40°C), there is a maximum of etch rate in the ⟨110⟩ direction. By increasing the temperature, a third (shallow) minimum in the ⟨110⟩ direction appears [9]. When adding 10–30% IPA (isopropyl alcohol), a deeper minimum of the etch rate in the ⟨110⟩ directions occurs (e.g., [11]). A minimum in the ⟨100⟩ direction has not been reported when IPA is added. When etching in EDP, the minimum in the ⟨110⟩ direction is also observed [11,12]. The temperature dependence of the slowest-etching planes is largest. The minimum close to ⟨111⟩ is very steep; the etch rate varies linearly with the misalignment angle and sometimes extrapolates to a zero etch rate at {111} (see Figs. 3.18, 3.19, 3.21, and 3.23 and [9, 12, 13, 14]).

3.2 Crystallography revisited: Atomic structure of surfaces

The flat character of the {111} faces in the diamond lattice follows directly from the analysis of the crystal structure. In Fig. 3.1 we repeat the unit cell of the diamond. The crystal structure of silicon is of the diamond type with lattice constant $a = 5.43$ Å.

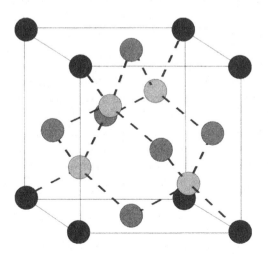

Fig. 3.1. Unit cell of the diamond lattice.

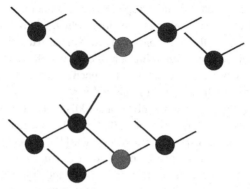

Fig. 3.2. Atomic configuration of the {001} face of silicon. Only the dangling bonds are shown. (top) Flat configuration, (bottom) configuration with one excitation.

The following analysis is due to Hartman and Perdock [15], and has been extended for complex crystal structures by, e.g., Bennema and van der Eerden [16]. We shall now draw the atomic configuration of a {001} face (Fig. 3.2). In the upper part we show a flat, undisturbed {001} face. Only the dangling bonds are shown. We see that each atom has two dangling bonds and two backbonds. Atoms that have an equal number of bonds to the solid phase as to the fluid phase are said to be in a kink position. They are in a position where they can change the phase most easily. In the theory of crystal growth (e.g., Burton, Cabrera, and Frank [17]), kink positions are assumed to be sinks of atoms in the case of growth, and sources of atoms in the case of dissolution. Obviously, the {001} face of silicon has only kink positions. Furthermore, it is easily seen that this surface should be rough (in reality, sometimes it is not rough, but this has to do with surface effects, such as adsorption or reconstruction [18]; we treat this phenomenon below). This is demonstrated in the lower part of Fig. 3.2, where we have taken one atom from the face, a process that requires the energy two times the energy of a covalent bond Φ_{SiSi}, and replaced it somewhere else. The replacement gives us the energy of twice the covalent bond energy, and the total procedure can be performed without doing work. Evidently, the {001} face must be rough in equilibrium, since the entropy of a face full of these excitation pairs is much larger than that of a perfectly flat face (the entropy of which is exactly equal to zero). The {001} face of the diamond structure is an example for a "K-face." K stands for kink, in the nomenclature of Hartman and Perdock.

Figure 3.3 shows the atomic structure of the {110} face. Here we see that the {110} face is built up by parallel running bond chains. The bond chains are bonded with a third backbond per atom to the body of the crystals and there is one dangling bond per atom. Note that the chains do not cross: there is no direct link between the chains. This fact allows us to cut a whole chain from the crystal, which costs us one bond per atom, and replace the chain somewhere on the crystal surface, thereby regaining the energy of one bond per atom. Again, we have an operation that does not require work. The surface will therefore be stepped along the ⟨110⟩ direction. Such faces are called "S-faces," with S being derived from steps.

Finally, let us analyze the {111} face (Fig. 3.4). Here the three bond chains form a triangular connected net. Each atom is connected with three bonds to his next neighbor in the face, and every second atom has a dangling bond. The other atoms have one bond that connects the two-dimensional net to the body of the crystal. The excitation, i.e., taking one

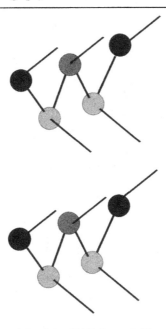

Fig. 3.3. Atomic configuration of the {110} face of silicon. The dangling bonds are shown and those that connect the atoms in the {110} face directly to each other, leading to a chainlike structure. The third backbond is not shown, which connects the chains to the bulk of the crystal.

atom out of the face and replacing it somewhere else, now definitely costs energy, twice the bond energy of one covalent bond, which is −2.34 eV per covalent bond. Therefore, this is a flat face. In fact, the ⟨111⟩ face is the only flat face in the diamond lattice (starting from crystallography alone, thus neglecting additional surface effects such as adsorption and surface reconstruction). The (111) face of silicon is called an "F-face," with F being derived from flat.

The energy to create the first excitation state of the surface ΔE divided by kT is known as the α factor of Jackson [19]:

$$\alpha = \Delta E / kT. \tag{3.1}$$

The {111} face exposed to vacuum at room temperature, ignoring the 7 × 7 surface reconstruction, would have $\alpha_{(111)} \approx 200$!

It is useful to mention the following: both the (110) and the (111) have one and one half, respectively, dangling bond per atom. The number of dangling bonds alone cannot account for the differences in the etch rate. For the same reason, it has proved very difficult to find a satisfactory purely (electro)chemical model to explain the anisotropy of the etch rate. It has been argued that the bonds running parallel in the (110) face must not be counted as backbonds, but we do not see any reason why this should be. Why could the bond energy of these bonds be so much smaller than the bonds to the body of the crystal? Why should this be so much different for the (111) face, where three bonds run nearly parallel to the surface?

The analysis we described here for the diamond lattice is universal. Since the papers of Hartman and Perdock appeared some 30 years ago, a vast number of crystal structures have been investigated, correctly predicting the flat faces of crystals. Deviations could mostly be

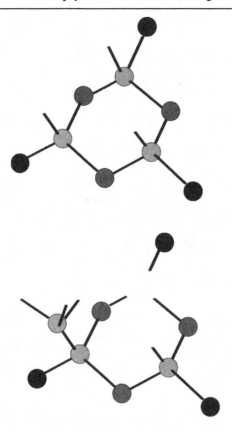

Fig. 3.4. Atomic structure of the {111} face. (top) The completely flat face, (bottom) with one excitation. Each second atom has three backbonds and one dangling bond. The other atoms have four backbonds. Only the zig-zagging backbonds that connect the surface atoms directly are shown, as well as the dangling bonds.

understood in terms of surface effects. In particular, quartz has been analyzed by Hartman, leading to prediction of all flat faces except one. These are precisely the faces where etch rate minima occur.

We are left to estimate the α factor for the silicon etching situation. With our current knowledge this is a formidable task. According to Ref. [20], the {111} face is covered by a layer of hydrogen; the bond energy per hydrogen atom is -3 eV. The first excitation offers in total four more bonds to react with hydrogen than the flat face. This would give

$$\Delta E = 4\Phi_{SiH} - 2\Phi_{SiSi} \approx -8\,\text{eV}, \qquad (3.2)$$

with Φ_{xy} the bond energy between x and y. A negative ΔE would mean that the surface breaks up spontaneously. Two effects make ΔE considerably larger: (i) There are certainly no free hydrogen atoms in the solution, so they have to be created, which costs energy to be added to the right-hand side of Eq. (3.2). If this came from the dissociation of water, the energy would be very large. The reaction enthalpy of the formation of one water molecule from $1/2O_2 + H_2$ equals 2.4 eV, and the dissociation of H_2 in 2H requires another 4.4 eV. Hydrogen could also result from dissociation of OH ions. The energy required to form an

H atoms to adsorb at the silicon surface must be somewhere between 1 and 3 eV. (ii) There is certainly not enough space in the hole formed in the lower part of Fig. 3.4 for three hydrogen atoms to bond to the silicon. At most there can be two, maybe even only one hydrogen atom. Points (i) and (ii) would lead to

$$\Delta E = \xi \Phi_{SiH} - \xi \Phi_{HX} - 2\Phi_{SiSi}, \qquad (3.3)$$

where ξ is between 2 and 3, and X is the partner of the hydrogen before it adsorbs at the silicon. Φ_{HX} for HX = 1/2H$_2$O would equal 4.6 eV, resulting in $\Delta E = 1.6\xi + 4.7$ eV.

Note that this is a crude approximation. The chemical reaction does not lead to SiO$_2$, but to Si(OH)$_4$. We are not able to find the appropriate reaction enthalpy in the literature. Furthermore, when water is decomposed, we have to include the activity of water and the vapor pressure of the oxygen.

From growth experiments described in Section 2.5 we know that the {111} faces of silicon are not the only ones that are flat. Apparently, here we have to deal only with the {001} and {110} faces – apart from the {111}. In the discussion in the literature the appearance of flat {001} and {110} faces has been ascribed to surface reconstruction and adsorption. There is some evidence that the {001} face is reconstructed [18]. The dangling bonds feel each other, and with a rather small change of the bond angle they can combine. This would give rise to an additional bond parallel to the {001} surface (see Fig. 3.5).

Gardeniers et al. suggested a picture in which each atom can only have one of his two dangling bonds attached to that of his neighbor, because the configuration in Fig. 3.5 would cost too much energy [21, 22]. The difficulty of the surface reconstruction in Fig. 3.5 is that we only get bonds running in one direction, we do not get a connected net. A face like that in Fig. 3.5 would be stepped, an S-face, not an F-face. This difficulty has been resolved by van der Eerden et al. [64], who showed that the stress resulting in the dimer formation also gives rise to periodic bond chains (PBC) normal to those shown in Fig. 3.5.

The distance between the dangling bonds on the {110} face is too large for the formation of dimers. However, adsorption of molecules that are long enough to bridge the distance between the dangling bonds would change the face into an F-face. The distance is just equal to the lattice constant, 5.43 Å. Interestingly, slowly etching {110} faces are observed in solutions that contain relatively large molecules: isopropyl alcohol (CH$_3$—CH(OH)—CH$_3$), which has a length of app. 5 Å, and ethylenediamine (NH$_2$—CH$_2$—CH$_2$—NH$_2$), with a length of app. 6.5 Å.

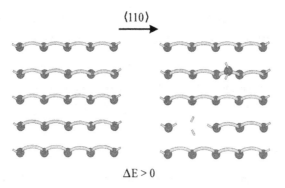

$\Delta E > 0$

Fig. 3.5. Dimer forming on the {001} face of silicon [18].

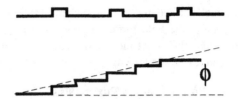

Fig. 3.6. Schematic of (bottom) a misoriented crystal face in contrast to (top) a perfectly oriented one.

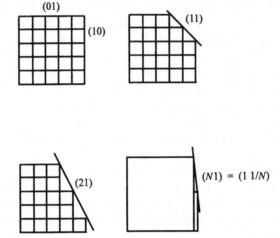

Fig. 3.7. (01), (11), (21), and (N1) surfaces of a simple square crystal.

3.3 Surface free energy and step free energy

The equilibrium form of crystals is controlled by the surface free energy $T^{\ddagger}$. It can be found by minimizing the surface free energy at constant temperature, pressure, and mole number. The surface free energy of crystals has a very special property: there are crystallographic orientations with a minimum of the surface free energy, and the minimum has a form of a cusp. This can be easily seen in Fig. 3.6. When the surface is misaligned, a step strain has to be added on the surface. This costs a finite amount of energy, independent of the direction of the step train.

The surface energy can be estimated from the number of bonds that must be broken to cleave a crystal. If N bonds are broken, the energy of Nf (with ϕ the broken bond energy) must be divided over two surfaces. Therefore a surface with N dangling bonds has a surface energy of $N\phi/2$.

From Fig. 3.7 we can see that the surface energy ε of a (01) face of a simple square lattice equals ϕ/a, with a being the lattice constant. The surface energy of an (11) face is obviously $\sqrt{2}\phi/a$. In Fig. 3.8 we show a plot of the surface energy as a function of the orientation. Note that the surface energy is larger than the surface free energy if there is a finite entropy.

$\ddagger$ A free energy F is defined as the reversible work required to bring a system from one state to another at constant temperature and pressure. It is related to the energy U by $\mathrm{d}F = \mathrm{d}U - T\mathrm{d}S$, with T the absolute temperature and S the entropy.

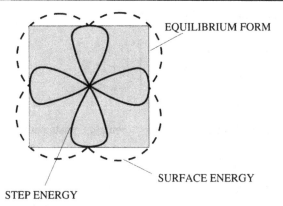

Fig. 3.8. Equilibrium form, surface energy, and step energy of the simple square crystal.

The (11) face is rough: each atom has two dangling bonds and two backbonds, and an operation similar to that in Fig. 3.2 costs no energy.

The step free energy γ is the reversible work required to create a step at constant volume. It is related to the derivative of Γ at the orientation of a flat face,

$$\gamma = \left.\frac{\partial \Gamma}{\partial \Theta}\right|_{\Theta=\Theta_0}. \tag{3.4}$$

Note that it is not the minimum of the surface tension that directly causes the etch rate minimum. Orientation variations of the surface tension are rather small, with a factor of two, as has been noted by several authors (see, e.g., [10, 12, 21, 22]). Orientation variation of the step free energy, however, is very large.

3.4   Thermodynamics

3.4.1   Chemical reaction

Several schemes of chemical reactions have been proposed (see, e.g., [12–14, 23, 24]), and experimental work of Palick and coworkers has revealed that OH^- groups play a key role in the chemical reaction [25, 26]. Based on Palic et al.'s suggestions [27], Seidel et al.[‡] proposed the following reactions [12]:

The first step is a reaction of a silicon atom, attached with two bonds to the solid and thus having two dangling bonds, with OH^- from the solution,

$$\begin{array}{c} Si \\ \diagdown \diagup \\ Si \\ \diagup \diagdown \\ Si \end{array} + 2OH^- \rightarrow \begin{array}{c} Si \quad OH \\ \diagdown \diagup \\ Si \\ \diagup \diagdown \\ Si \quad OH \end{array} + 2e^-. \tag{C1}$$

This reaction delivers two electrons to the solid which are thought to be confined to the surface of the silicon. The second step would be an ionization of the $Si[OH]_2$ complex,

---

[‡] Seidel's model seems to be widely accepted in the micromachining community, but not in the electrochemistry community.

whereby two further electrons are delivered to the solid:

$$\mathrm{Si\bullet} \begin{bmatrix} \mathrm{Si} \begin{matrix} \diagup \mathrm{OH} \\ \diagdown \mathrm{OH} \end{matrix} \end{bmatrix}^{++} + 2\mathrm{e}^-. \qquad \mathrm{Si\bullet}$$

(C2)

In the third step, the positively charged $\mathrm{Si[OH]}^{++}$ complex reacts with two OH^- ions from the solution:

$$\mathrm{Si\bullet} \begin{bmatrix} \mathrm{Si} \begin{matrix} \diagup \mathrm{OH} \\ \diagdown \mathrm{OH} \end{matrix} \end{bmatrix}^{++} + 2\mathrm{OH}^- \rightarrow \mathrm{Si(OH)_4} + \underset{\mathrm{solid}}{\mathrm{Si}}.$$

(C3)

$\mathrm{Si(OH)_4}$ is soluble. The molecule can leave the surface. However, at the extremely low pH of the solution it is not stable, but reacts according to

$$\mathrm{Si(OH)_4} \rightarrow \mathrm{SiO_2(OH)_2^{--}} + 2\mathrm{H}^+$$
$$2\mathrm{H}^+ + 2\mathrm{OH}^- \rightarrow \mathrm{H_2O}.$$

(C3)

Finally, the four electrons at the silicon surface react with water:

$$4\mathrm{H_2O} + 4\mathrm{e}^- \rightarrow 4\mathrm{OH}^- + 2\mathrm{H_2}.$$

(C4)

The second step is considered to be the rate-determining step. The difference between the (111) and the other faces is explained by Seidel et al. [12] by the fact that on the (111), the silicon atoms are bonded with three bonds to the solid, instead of only two bonds.

On the (100) face we have

$$\begin{matrix} \mathrm{Si} \\ \diagdown \\ \mathrm{Si} \diagup \end{matrix} \mathrm{Si} \begin{matrix} \mathrm{OH} \\ \diagdown \\ \diagup \mathrm{OH} \end{matrix} \rightarrow \mathrm{Si\bullet} \begin{bmatrix} \mathrm{Si} \begin{matrix} \diagup \mathrm{OH} \\ \diagdown \mathrm{OH} \end{matrix} \end{bmatrix}^{++} + 2\mathrm{e}^- \qquad \mathrm{Si\bullet}$$

(C5)

while on the (111) face we have

$$\begin{matrix} \mathrm{Si} \\ \diagdown \\ \mathrm{Si-Si-OH} \\ \diagup \\ \mathrm{Si} \end{matrix} \rightarrow \mathrm{Si\bullet} \begin{bmatrix} \mathrm{Si\blacksquare} \\ \mathrm{Si\bullet} \end{bmatrix} \mathrm{Si-OH} \end{bmatrix}^+ + \mathrm{e}^-.$$

(C6)

The dots indicate that the attachment of an OH radical changes the energy of the covalent bonds.

The anisotropy of the etch rate is now assumed to result from this energy change: it could depend on the number of OH radicals attached to the silicon atom. If there is only one OH radical attached, Seidel et al. assume that the bond energy will be larger than in the case when there are two OH radicals. This explanation for the anisotropy of the etch rate is not satisfactory. Since flat faces are assumed in this reaction scheme, the main point should be the number of backbonds that gives the anisotropy. A flat {110} silicon face also has three backbonds per atom and should etch with an etch rate and an activation energy comparable to the {111} faces. Any attempt to explain the anisotropy by the number of backbonds alone will fail.

Seidel's scheme seems to fit the observations of Palick et al. quite well. It provides in principle an explanation of the B^{++} etch stop. But it is an electrochemical reaction, the rate of which should depend strongly on the density of electrons at the surface and therefore of the doping. Up to a concentration of 10^{18} atoms/cm^3 (donor or acceptor doping), the etch rate is independent on the doping (c.f., Fig. 2.55). Glembocki et al. [23] and Allongue et al. [61] conclude that there must be two parallel reactions: one chemical, without participation of charges, and one electrochemical.

Glembocki et al. even suggest that at the OCP the reaction is entirely chemical. This assumption is supported by analysis of current voltage data close to the OCP. As shown by Allongue et al. [61], the exchange current at the OCP (an anodic current for silicon oxidation and dissolution and a cathodic current for hydrogen evolution, which cancel each other) is so small that it cannot account for the etch rate: the etching at open circuit potential is practically unaccompanied by a charge transfer. So at the OCP the OH$^-$ ions do not contribute directly to the chemical reaction, but serve as catalysts. The chemical reaction is due to attack of the Si—Si bonds by water.

Furthermore, any proposal for a chemical reaction sequence has to take into account that at least the {111} faces of silicon are hydrogenated during etching [20]. This can be the case only if the rate-determining step consists of a replacement of the H by an OH or by direct rupture of a backbond [61].

In view of these observations, the chemical reaction consists of several steps again. The first could be one of the following three alternatives [61, 63]:

$$\begin{array}{ccc}
\text{Si}\diagdown \quad \diagup\text{H} & & \text{Si}\diagdown \quad \diagup\text{OH} \\
\quad\text{Si} & + \text{H}_2\text{O} \xrightarrow{\text{OH}^-} & \quad\text{Si} \quad + \text{H}_2 \\
\text{Si}\diagup \quad \diagdown\text{H} & & \text{Si}\diagup \quad \diagdown\text{H}
\end{array} \qquad (C7)$$

$$\begin{array}{ccc}
\text{Si}\diagdown \quad \diagup\text{H} & & \text{Si}\diagup^{\text{OH}}\diagdown\text{H} \\
\quad\text{Si} & + \text{H}_2\text{O} \xrightarrow{\text{OH}^-} & \quad\text{Si} \\
\text{Si}\diagup \quad \diagdown\text{H} & & \text{Si}\diagup \quad \diagdown\text{H} .
\end{array} \qquad (C8)$$

The third initial step is a repetition of the first one, resulting in the surface state

$$\begin{array}{c}
\text{Si}\diagdown \quad \diagup\text{OH} \\
\quad\text{Si} \\
\text{Si}\diagup \quad \diagdown\text{OH} .
\end{array} \qquad (C9)$$

Allongue et al. [61] and Bressers et al. [63] regard the first of these possibilities as the most probable (dominating) one. The OH group will polarize the backbonds, thereby facilitating the rupture of one of the backbonds by the attack of water:

$$\begin{array}{c} Si \\ \backslash \\ Si \\ / \backslash \\ Si H \end{array} \begin{array}{c} OH \\ / \\ \end{array} + H_2O \xrightarrow{OH^-} \begin{array}{c} H \\ Si OH OH \\ \backslash | / \\ Si \\ / \backslash \\ Si H \end{array} \quad . \tag{C10}$$

A last reaction step with water leads to the primary etch product $HSi(OH)_3$. This molecule is soluble and may leave the surface. In solution it will react further to the final $[SiO_4H_2]^{2-}$. The initial step (C8) would suggest $H_2Si(OH)_2$ as the primary soluble etch product.

This chemical reaction scheme does not provide any explanation of the anisotropy, similar to Seidel's proposal. In the following we assume the overall chemical reaction to be [28]

$$\begin{array}{c} Si H \\ \backslash / \\ Si \\ / \backslash \\ Si H \end{array} + 2H_2O \xrightarrow{OH^-} \begin{array}{c} Si-H H OH \\ \backslash / \\ Si \\ / \backslash \\ Si-H H OH \end{array} \quad . \tag{C11}$$

The most crucial assumption regarding the anisotropy is that all attempts to explain the anisotropy of the etch rate start with the picture that all faces are flat on an atomic scale. It is a difficulty of all purely (electro)chemical-based models. In Fig. 3.9 we show a scanning tunnel microscopic (STM) picture of a high step in a $\langle 111 \rangle$ silicon face. Note the reconstruction of the surface; a simple cut would suggest a honeycomb structure. This is the famous seven by seven reconstruction. The $\langle 111 \rangle$ face is perfectly smooth on an atomic scale. Shallow and high steps can be clearly discerned in this picture. This is in contrast to the physical state of the high step: at least part of it is obviously rough on an atomic scale. This is, to our knowledge, the first STM picture of an atomically rough surface. The image in Fig. 3.9 was obtained in a high vacuum; the presence of a liquid will change the surface state, as outlined below.

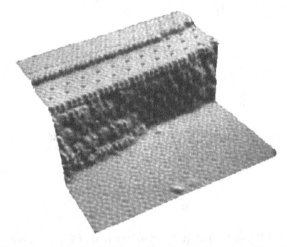

Fig. 3.9. STM picture of the $\langle 111 \rangle$ face of silicon with a macrostep (from Wiesendanger et al. [29]).

3.4.2   Chemical potential difference

A crystal dissolves or grows only in a nonequilibrium situation. There must be a difference in chemical potential, $\Delta\mu$, of the atoms in the liquid phase and in the crystalline phase. Etch and growth rates depend on $\Delta\mu$, and the type of dependency differs greatly for smooth and rough faces (see, e.g., [30, 31]). In anisotropic wet chemical etching of silicon, the chemical potential difference consists of two parts: one is for the chemical reaction itself, and the other one has to do with the dissolution of the etch product.

The free energy associated with the reaction in (C11) is not known. The oxidation enthalpy of silicon is very large, close to 3 eV. Generally, it is assumed that the formation of the $Si(OH)_4$ gives an energy gain of the same order; however, it must be less because of the dissociation of water. The formation of a water molecule from $1/2O_2$ and H_2 costs 2.4 eV, therefore $\Delta\mu_{react}$ should be much smaller than -3 eV.

From elementary thermodynamics the contribution of dissolving the chemical etch product to the chemical potential is

$$\Delta\mu_{diss} = \mu_l - \mu_s = kT \ln a/a_{eq}, \tag{3.5}$$

in which l and s refer to liquid and solid, respectively, and a is the activity of the etch product in the solution close to the crystal surface; a_{eq} is the activity that corresponds to thermal equilibrium, i.e., the situation when the solution is saturated with etch products. In Eq. (3.5) it is assumed that the etching is divided into two steps: the chemical reaction which leads to a solid compound(s), which is then dissolved.

The total change of the chemical potential is therefore

$$\Delta\mu = \Delta\mu_{diss} + \Delta\mu_{react}. \tag{3.6}$$

Note that in this definition, $\Delta\mu$ is negative for etching. In ideal solutions, $a(X) = X$, with X the mole fraction of the dissolved species (in our case the etch products). In an ideal solution this results in

$$\Delta\mu_{diss} = kT \ln X/X_{eq} \sim kT(X - X_{eq})/X_{eq}. \tag{3.7}$$

The latter equality applies if one is close to the equilibrium situation. We see that $|\Delta\mu|$ is large if X is small; since X increases during etching, the undersaturation will decrease in the course of time. $|\Delta\mu|$ will decrease more slowly if the solubility of the etch products in the solution is large. Therefore the undersaturation will generally be larger in solutions that do not saturate so easily. The undersaturation will be largest in $HF:HNO_3$ solutions, and smaller, e.g., in EDP as compared to KOH, since EDP ages quicker than KOH. The solubility also depends on the temperature. It will be larger at higher temperatures, implying that the undersaturation will then be larger. $\Delta\mu$ is also called the driving force for growth and etching. In the Monte Carlo simulations of Camon et al. [32, 33], this parameter is omitted.

3.4.3   The roughening transition

There is a phase transition at a temperature T_R between a smooth and a rough crystal face, above which γ vanishes (for a review, see, e.g., [30]). The roughening transition was anticipated by Burton, Cabrera, and Frank in 1951 [31]. For simple systems this is

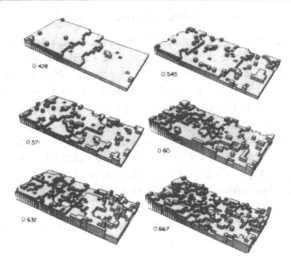

Fig. 3.10. Interface structure of the {001} face of a simple cubic crystal at different temperatures $\Theta = T/\phi_{ss}$ (from van der Eerden et al. [64]).

a continuous phase transition (see [34–37] for theoretical results and, e.g., [38–40] for experimental evidence).

An impression of the roughening transition is given in Fig. 3.10 [64]. Here a {001} face of a simple cubic crystal has been simulated using a Monte Carlo simulation method [41]. The state of the surface in equilibrium with its vapor is shown as the temperature increases. The temperature is normalized to the interaction energy of nearest neighbours in the solid phase, $\Theta = T/\phi_{ss}$. Two steps have been created on the surface to show how their appearance changes when increasing the temperature. At a low temperature, the steps are rather straight and more or less aligned parallel to the $\langle 010 \rangle$ direction. On the terraces between the steps there are only a few ad atoms and vacancies. By increasing the temperature, the number of ad atoms and vacancies increase, and the alignment of the steps along the $\langle 010 \rangle$ becomes less distinct. Finally the steps become invisible in the noise of ad atoms. Somewhere between $\Theta = 0{,}632$ and $\Theta = 0{,}667$ the surface becomes rough.

The analysis of the diamond lattice resulted in faces that we called flat: on these faces one needs a finite energy to create a step. The analysis was based on a model in which only nearest-neighbor interaction has to be taken into account. The next-nearest-neighbor interaction is certainly much smaller that the former, but taking this into account one would find more flat faces, however, with a much smaller α factor.

We can define an α factor that corresponds to roughening of the face,

$$\alpha_R = \Delta E / k T_R. \tag{3.8}$$

Based on Monte Carlo simulation of crystal surfaces, van de Eerden has shown that α_R only depends on the symmetry of the face in question [42]. For the diamond lattice, he found

$$\alpha_{R\langle 001 \rangle} = \infty; \qquad \alpha_{R\langle 110 \rangle} = \infty; \qquad \alpha_{R\langle 111 \rangle} = 4.2.$$

This means that at a sufficiently high temperature, all faces become rough.

Van der Eerden's results are in agreement with our analysis of the (110) and the (001) faces. An infinitely large α_R means $T_R = 0$, the face is rough at all temperatures. If $\alpha > \alpha_R$, the face is flat.

The state of the surface, i.e., whether being flat or rough, plays a key role in the kinetics of crystal growth and dissolution.

3.5 Kinetics

In kinetics of crystal growth, active sites for growth and dissolution play a key role. These active sites are atoms with as many bonds to the crystal as to the liquid (or gaseous) environment. Such a site is called a kink site. Pioneers in creating these notions were Stranski, Volmer, Kossel, and Kaishev; see [34] for a historical account. An atom in a kink site in a simple cubic lattice is shown in Fig. 3.11a. The shaded atom in the kink position has three bonds to the crystal and three bonds to the liquid. In a dissolution situation it is commonly believed that this atom will diffuse over the surface (Fig. 3.11b), until it either finds a kink position again or it desorps and diffuses away from the crystal in the liquid phase (indicated in Fig. 3.11c). In a growth situation an atom diffuses from the liquid to the crystal (Fig. 3.11c), and it diffuses over the crystal surface (Fig. 3.11b) until it is either desorped or it finds a kink site (Fig. 3.11a).

Kinetic rates (for growth and dissolution) thus depend critically on the number of kink sites on a crystal surface. This aspect has been neglected so far in the discussion of etch rates of single crystals such as silicon, quartz, and GaAs. Only parts of the total process have been considered: the chemical reaction rate [12, 43–45] (which is important for the adsorption and desorption process and the kink (des-)integration process, erroneously considered anisotropic), diffusion in the liquid solution (isotropic), and the thickness of the boundary layer. In our view, the most prominent anisotropy effect is due to the number of kink sites.

3.5.1   Nucleation of etch pits

In order to remove atoms from a smooth face, one has to create a step. Consider a cavity that contains N vacancies (N atoms have been removed, Fig. 3.12). This implies a gain of

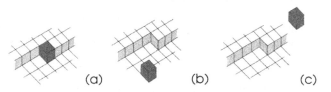

Fig. 3.11. Illustration of kink sites and the elementary kinetic processes for etching (from a to c) and for growing (from c to a) crystals.

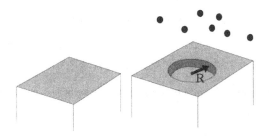

Fig. 3.12. Illustration of etching a cavity into a flat face.

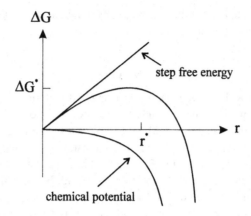

Fig. 3.13. Free energy as a function of the size of a cavity in a flat crystal face.

the free energy of the system which is equal to $N\Delta\mu$. Obviously, N is proportional to the surface of the cavity, and for a cavity of circular shape therefore proportional to the square of the radius.

The free energy change ΔG_k, by adding an island or digging a cavity (of circular shape in an isotropic face) of radius r, for a face with orientation $\mathbf{k} = (hkl)$ is given by

$$\Delta G_k = N_k \Delta\mu + 2\pi r \gamma_k, \tag{3.9}$$

where N is the number of atoms in the island respectively removed from the cavity and the solution, and γ is the step free energy. We have

$$N_k = \pi r^2 h_k \rho, \tag{3.10}$$

where h_k is the height of a step and ρ is the density (atoms per cm^3) of the solid material. The result is

$$\Delta G_k = -\pi r^2 \rho \Delta\mu + 2\pi r \gamma_k. \tag{3.11}$$

Note that $\Delta\mu$ is counted positive and γ is positive in any case. $\Delta G(r)$ is shown in Fig. 3.13. Equation (3.11) has the interesting feature that it has a maximum at

$$r^* = \gamma/\rho h_k \Delta\mu. \tag{3.12}$$

At r^*, the free energy is

$$\Delta G_k^* = \Delta G_k(r^*) = \pi \gamma_k^2 / h_k \rho \Delta\mu. \tag{3.13}$$

Classical nucleation theory predicts the following formula for the nucleation rate J_k for a crystal face of square symmetry [46]

$$J_k \sim \kappa(\Delta\mu/kT)^{1/2} \exp -\{4\gamma_k^2/(3kT\Delta\mu)\}, \tag{3.14}$$

where κ is the kinetic constant, containing the activation energy $\Delta\varepsilon$. For the elementary chemical reaction,

$$\kappa \sim \exp -\{\Delta\varepsilon/kT\}. \tag{3.15}$$

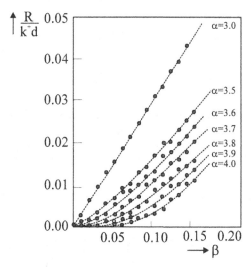

Fig. 3.14. Growth rate of a {100} face of a simple cubic crystal as a function of the supersaturation for different temperatures [48].

The total etch rate is given by Gilmer [47]

$$E_{\mathbf{k}} = h_{\mathbf{k}}\left(\pi J_{\mathbf{k}} v_{\mathbf{k}}^2/3\right)^{1/3}. \tag{3.16}$$

Here, v is the velocity of steps. Note also that $v_{\mathbf{k}}$ contains the activation energy in Eq. (3.15). Equation (3.16) is a good approximation if $\Delta G_{\mathbf{k}}^*$ is not too small, i.e., in a situation where the mutual interaction of nuclei and steps can be neglected.

If $\Delta G_{\mathbf{k}}^*/kT$ is large, the etch rate will be very small. This is the case for small undersaturation, small temperature, and large step free energy. Consequently, flat faces of perfect crystals etch at a very small rate compared to rough faces having zero step free energy.

The growth curves of the surfaces similar to those in Fig. 3.10 are given in Fig. 3.14. These curves show clearly the effect of roughening on the kinetics. Growth of facets of an ideal crystal is practically blocked by the nucleation barrier at small supersaturation, while at large supersaturation the growth rate becomes linear in $\Delta\mu$. If the face becomes rough, the crystal can grow at an infinitely small supersaturation.

Over wide ranges of supersaturation or undersaturation, the growth and etch rate varies as [48]

$$R \propto (1 - \exp -\beta)\exp -\{\psi^2/\beta\}, \tag{3.17}$$

where we have normalized the relevant energies to the temperature,

$$\beta = \Delta\mu/kT$$

and

$$\psi = \zeta\gamma/kT,$$

with ζ being a numerical factor of order 1. The first term on the right-hand side of (3.17) stems from the phenomenological equations that state that a current density must be proportional to a chemical potential difference. The second term describes the nucleation barrier.

Therefore $kT\psi^2/\beta$ represents the activation energy necessary to overcome the nucleation barrier.

The kinetics of rough faces are not controlled by a nucleation barrier. Every atom or molecule at the solid–fluid interface that feels a smaller chemical potential can change its thermodynamic phase without the need to find a step. If $\psi = 0$ on a particular face, the kinetics are obviously much faster than if $\psi > 0$. This observation constitutes the large anisotropy in the kinetics – whether growth or etching – of all single crystals. The anisotropy of the growth rate and etch rate is a universal phenomenon. In most cases the minima in the etch rate and in the growth rate occur in identical crystallographic directions.

In the second term we have to deal with the ratio of ψ^2 to β. Nucleation becomes unimportant if this ratio is of order 1 or smaller. In this case, one speaks of kinetic roughening [49, 50], and the etch rate (and growth rate) starts to approach the maximum possible rate (i.e., controlled by diffusion of material and heat). This is the case for isotropic etching.

The classical two-dimensional nucleation theory requires that it is possible for an atom or molecule to choose the state in which it prefers to be: If the nucleus is below the critical size there must be a way back from a dissolved molecule to the solid. In other words, the chemical reaction must be reversible, or the total chemical potential difference must not be too large – thermal fluctuations must enable the molecules to return to the solid state. Since we do not know how large $\Delta\mu$ is, the classical nucleation picture must be taken with some care.

3.5.2 Etch pits formed by screw dislocations

There is a second mechanism for creating steps, thereby creating kink positions, which makes the etching of smooth faces possible. In Fig. 3.15(a) we depict a screw dislocation. Any screw dislocation that crosses a surface gives rise to a step. This step moves if the crystal grows Fig. 3.15(b) or if it is etched Fig. 3.15(c). This mechanism for growth and etching of crystals was first proposed by Burton, Cabrera, and Frank [31], and is now established as the most important growth mechanism of crystals [30, 34]. The spiral mechanism is of much more practical importance for crystal growth and etching than the nucleation mechanism, because practically all single crystals contain screw dislocations. It is very important to understand that the kinetic rate of growing or etching does not depend on the number of dislocations that intersect a smooth surface. In a steady state situation the whole surface of the crystal is covered by a spiral that is caused by single dislocation, if this is the only one that exists. Steps interact in such a way (see below) that the number of steps is not increased if there are more spiral centers.

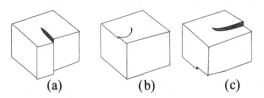

Fig. 3.15. (a) A screw dislocation gives rise to a step on the crystal surface. The step moves when the crystal (b) grows and if it is (c) etched.

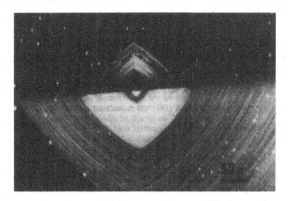

Fig. 3.16. Differential interference contrast microphotograph of a shallow etch pit on ⟨001⟩ silicon after gas phase etching in the H-I-Si CVD system (from Ref. [8]).

When looking at a crystal surface with a phase contrast or interference contrast microscope, one usually sees the resulting growth spirals. An example is given in Fig. 3.16. They can also be seen on crystals during etching (see, e.g., [18, 51, 52]).

In modeling the kinetic rate, it is assumed in the classical paper of Burton, Cabrera, and Frank [17] that on the crystal surface the steps act as sources or sinks (for etching and growth, respectively) of crystal units (silicon atoms in our case). These ad atoms diffuse over the surface according to their concentration gradient. Far away from the steps it is assumed that the density of ad atoms is given by the transport through the liquid. The resulting differential equation for the concentration of ad atoms finally leads to an equation for the velocity of steps, and to the form of the spiral, which is almost Archimedal. The resulting growth/etch rate of imperfect crystals then given by [17] is

$$E_\mathbf{k} = C(\Delta\mu/kT)^2/\sigma_\mathbf{k} \tanh(\sigma_\mathbf{k}/\Delta\mu). \tag{3.18}$$

The important parameter $\sigma_\mathbf{k}$ is given by

$$\sigma_\mathbf{k} = \xi_\mathbf{k}^2 \gamma_\mathbf{k}/\chi. \tag{3.19}$$

$\xi_\mathbf{k}$ is the surface diffusion length (in principle also $\mathbf{k}$-dependent). Again the step free energy plays a key role for the kinetics. We see that for small $\Delta\mu$ the rate is proportional to $(\Delta\mu)^2$, while for very large $\Delta\mu$, a linear relationship results. However, if $\Delta\mu$ becomes considerably larger than $\sigma_\mathbf{k}$, the nucleation mechanism takes over, mainly because in this case the nucleation rate becomes appreciably large.

The radius of the critical nucleus is also important for spiral growth. The average distance between the spiral arms is proportional to $r_\mathbf{k}^*$ (see, e.g., [34]):

$$r_{\mathrm{sk}} = 19 r_\mathbf{k}^*. \tag{3.20}$$

A rough crystal face etches or grows with a rate

$$\mathrm{E} \sim 1 - \exp -(\Delta\mu/kT), \tag{3.21}$$

which is proportional directly to $\Delta\mu$ if it is not too large (e.g., [53]), and independent of $\Delta\mu$ for $\Delta\mu/kT \gg 1$.

3.5.3 Kinetic roughening

The energy of the critical nucleus depends on the supersaturation (Eq. (3.7)), which in turn depends on the concentration of dissolved etch products in the solution (c.f., Eqs. (3.6) and (3.8)). If the undersaturation becomes so large that $\Delta G^* \ll kT$, the nucleation barrier breaks down. All single-atom cavities act as nuclei. The face in question etches with a rate comparable to the etch rate of a rough face. This situation is known as kinetic roughening [49, 50]. If all faces are kinetically rough, the etch rate becomes isotropic. We speculate that this is the case in $HF:HNO_3$-based etchants, since apparently in this system the free energy of the chemical reaction may be more negative and we know that the solubility of the etch products is large. A change of the degree of anisotropy by a change of $\Delta\mu$ can be expected in two cases: (i) in the very beginning of, e.g., KOH etching the undersaturation is very large, the etching should be isotropic. The time however might be too short in order to observe this effect. (ii) Etching in $HF:HNO_3$-based etchants could show some anisotropy if one etches for a sufficient length of time, so that the undersaturation decreases below the level of kinetic roughening. There are some observations that can be interpreted in these terms.

In Fig. 3.17 we show one of these phenomena. Circular oxide + CrAu masks on a 2″ (111)-oriented silicon wafers were underetched in $HF:HNO_3:CH_3COOH$ (2:15:5) at 25°C in a stirred solution. The volume of the solution was 400 ml. The wafers were placed horizontally in the beaker. While in the beginning the underetched columns are perfectly cylindrical, after 20 minutes, having etched 100 μm deep into the (111)-oriented wafer, vertices start to develop showing a sixfold symmetry.

Similarly, it has been observed by Hashimoto et al. [54] that isotropic etching etchants may etch anisotropically under certain circumstances. They etched silicon through small holes in the mask, trying to etch semispherical cavities. If the hole in the mask was large enough, they succeeded but they observed that the cavities revealed some anisotropy if the holes in the mask were very small. This is a case where aging of the solution may be visible, since the small hole in the mask hinders the transport of fresh solution to the silicon surface. These results give indirect evidence that when decreasing undersaturation the etch rate becomes anisotropic.

At least under conditions that have been studied by Allongue et al. (20 wt% NaOH at 25°C) the (111) faces are flat, as evidenced from their in situ STM imaging. So, the chemical potential difference cannot be very large, otherwise one would not observe flat faces.

Fig. 3.17. SEM photograph of a column etched under a circular mask in a (111)-oriented wafer after etching 100-μm deep in the silicon. In earlier stages the column stays cylindrical as one expects for isotropic etching. Etching time was 40 minutes.

3.6 Etch rate diagrams

3.6.1 Close to the etch rate minimum

A misalignment close to smooth faces implies steps; there is no need of nucleation in order to etch. Since the density of steps is proportional to the angle of misalignment (if the angle is small enough), the etch rate should be proportional to the angle, provided the distance between steps is not too large, so that nucleation of new cavities is very probable. This has been observed for the etch rate close to {111} [12–14], and we reproduce Seidel et al.'s results [12] in Fig. 3.18, and Uenishi et al.'s results [13] in Fig. 3.19.

The variation of the etch rate with the misalignment angle can be easily derived as follows (compare Fig. 3.20):

On a flat face of some crystallographic orientation, there are steps present, giving rise to a misorientation Θ. The etching proceeds by movement of the steps (height h), and the average distance between the steps is λ. Steps move from the left to the right in Fig. 3.20.

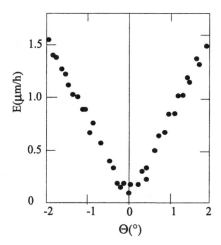

Fig. 3.18. Underetch rate variation in $\langle 001 \rangle$-oriented silicon wafers close to {111}, EDP type S, 69°C (redrawn from Fig. 11 of Seidel et al. [12]).

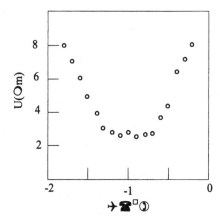

Fig. 3.19. Underetching close to the $\langle 111 \rangle$ orientation of $\langle 110 \rangle$-oriented silicon wafers in 40 wt% KOH at 70°C (redrawn from Uenishi et al. [13]). The etch depth was 350 μm. $\Theta = 0°$ refers to the orientation of the wafer flat.

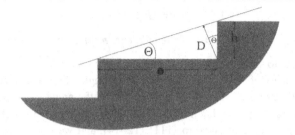

Fig. 3.20. Illustration for the derivation of the relations between step height h, distance between steps λ, angle of misorientation λ, and the proceeding of misoriented crystal front D, when the step has moved the distance λ [65].

The speed of the steps v is given by

$$v = \lambda/\tau, \qquad (3.22)$$

where τ is the time a step needs to move the distance λ.

After time τ the crystal surface has moved a distance D, and the etch rate E is given by

$$E = D/\tau. \qquad (3.23)$$

Elementary geometry now gives the following expression of the etch rate:

$$E = v \sin \Theta \approx v\Theta. \qquad (3.24)$$

The last equality applies for a small misalignment.

Koide et al.'s data [9] are suitable to test Eq. (3.24) for the variation of the etch rate over a wider range of angles. In Fig. 3.21 we redraw the etch rate as a function of $\sin \Theta$, where Θ is the angle of misalignment with respect to the ⟨111⟩ toward the ⟨110⟩ direction.

As we see, the data points lie on a straight line within the experimental accuracy, as expected for moving steps. From the data, the temperature dependence of the velocity of the steps can be extracted, and we find $\Delta E_{\langle 111\rangle \text{steps}} = 0.6$ eV. This agrees perfectly with the idea that the etching of the {111} face is blocked by a nucleation barrier: the activation energy of the ⟨111⟩ etch rate is 0.7 eV in KOH, and of the fast(er) etching faces is 0.6 eV, close to the activation energy of the velocity of the steps. We could conclude that the energy of a critical nucleus equals the difference between the activation energy of the ⟨111⟩ etch rate and the activation energy of the steps, i.e., $\Delta G_k^* = 0.1$ eV. If we assume $\Delta\mu = 3$ eV, the step free energy would be of the order of $\gamma = 0.6$ eV.

3.6.2   Form of the minimum [65]

The distance between the steps becomes very large very close to the etch rate minimum (e.g., the (111)). If it is so large that nucleation between the steps is possible, or that the distance between the misalignment step is larger than the distance between the steps of spiral arms, then the two step sources compete. Steps interact in a very special way (see Fig. 3.22). Steps annihilate each other or add in height as a consequence of a "collision." As a consequence, the etch rates resulting from the different step sources (misorientation and nucleation/dislocations) do not add up. The etch rate is governed by the step source

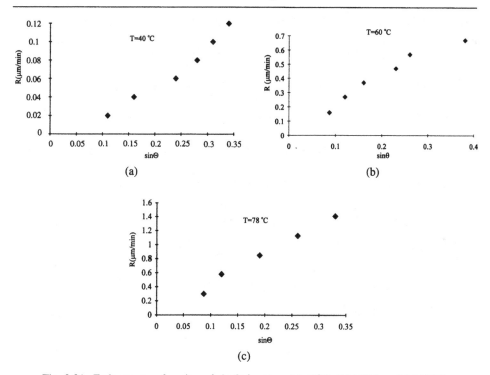

Fig. 3.21. Etch rate as a function of sin Θ for T = (a) 40°C, (b) 60°C, and (c) 78°C (recalculated from Koide et al. [9]).

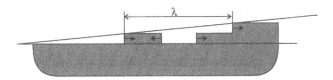

Fig. 3.22. Illustration for the interaction of steps: annihilation and addition [65].

that leads to the largest density of steps. Close to the minimum of the etch rate, there is a flat portion of a finite etch rate that belongs to the nucleation/dislocation step source. At sufficiently large angles the steps due to misorientation take over and the etch rate starts to vary linearly with the angle. A striking example of this effect can be seen in Uenishi et al.'s paper [13], in which the studied the underetch rate of silicon in KOH (40 wt%, 70°C) close to the (111) orientation on (110)-oriented wafers. Their results are redrawn in Fig. 3.19.

The width of the flat portion Θ_{ck} in the plot of E versus Θ close to the orientation of the smooth face can be estimated in an intuitive way as follows. The nucleation rate J gives the number of nuclei of critical size per cm$^2 \times$ sec. In order to have a good chance that the etch rate is governed by steps originating from a misalignment, we must have

$$J\lambda^2\tau < 1. \tag{3.25}$$

In the opposite case, $J\lambda^2\tau > 1$, we have the situation that steps originating from misorientation play a minor role for the etch rate.

We have $\lambda/\tau = \nu \approx d/\tau^*$, with d the width of an atom in a step and $\tau^* = 1/\kappa$ the chemical reaction rate to remove a single atom from the step. λ is related to the misorientation by

$$\lambda = h/\tan\Theta \approx h/\Theta. \tag{3.26}$$

The nucleation rate at very large undersaturation is given by

$$J \propto \tau^{*-1} \exp{-\Delta G^*/kT}. \tag{3.27}$$

The transition from R constant to $R \propto \Theta$ will occur if $J\lambda^2\tau = 1 = J\lambda^3\tau^*/d$ or

$$\Theta_{ck} \propto \exp{-\Delta G^*/3kT} \tag{3.28}$$

for the temperature dependence of the flat portion in the E versus Θ plot close to the orientation of a flat face. Experimental determination of $d\Theta_c/dT$ will give us directly the free energy of the critical nucleus.

Alternatively, the etch rate of the smooth face is dominated by steps originating from dislocations, and Θ_{ck} should be related to the average distance r_{sk} between the spiral arms and the misorientation steps. r_{sk} is given by Eq. (3.20).

In this case we must have

$$\lambda_{ck} = 19 r_k^*, \tag{3.29}$$

from which we find

$$\Theta_{ck} = 2h_k/19 r_k^*, \tag{3.30}$$

or in terms of the step free energy

$$\Theta_{ck} \approx 2\Delta\mu h_k/19\xi_k\gamma_k. \tag{3.31}$$

The interesting point of this equation is that the measurement of the flat portion of E versus Θ curve gives information about γ.

3.6.2.1   Results for Si {111}

With the results of Koide's experiments, assuming a nucleation mechanism for etching, we found $\psi^2/\beta \approx 4$. With the help of Eq. (3.28) and the activation energy of the temperature dependence of $\Theta_{c\{111\}} \approx 0.03$ eV. Unfortunately, there is no data to test this result for Si $\langle 111\rangle$.

3.6.2.2   Results for Si {001}

The situation close to the $\langle 100 \rangle$ directions has been studied by Elwenspoek et al. [55]. In the $\langle 100 \rangle$ direction there is also a minimum if one etches silicon in KOH [11, 56], CsOH [57], and NaOH [58]. Mirror-like flat {100} faces have been reported [56] at particular etching conditions (35 wt% KOH, 80°C). Possibly the {100} are smooth under these conditions, in which case Eq. (3.24) would apply.

To investigate this question, a mask was designed that consisted of rectangular openings ($250 \times 50\,\mu$m), the long side of which made an angle with the $\langle 100 \rangle$ direction on an $\langle 001 \rangle$-oriented wafer. The angular resolution was $0.5°$. Etching in KOH results in grooves bounded

Table 3.1. *Temperature, the respective etch times, and some results.*

T (°C)	t (min)	E_{min} (μm/min)	$dE/d\Theta$ (μm/rad $\times$ min)	Θ_c (°)
30	300	0.048	0.15	3
70	15	0.56	2.0	3.5
108	3	3.8	10	4

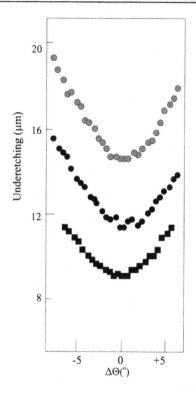

Fig. 3.23. Underetching close to $\langle 100 \rangle$ on a $\langle 001 \rangle$-oriented silicon wafer in aqueous KOH (40 gr/100 ml). Open points: 30°C, 300 min; full squares: 70°C, 15 min; full points: 108°C, 3 min [55].

by {001 + $\Delta\Theta$} planes. The experiments were performed in aqueous KOH (40gr/100ml). Temperature, etch times, and some results are given in Table 3.1.

The wafers were placed vertically in the etch vessel. The results of the etch rates are shown in Fig. 3.23. The underetch rates vary linearly with the angle for all temperatures except at a small 3°–4° region around the $\langle 001 \rangle$. The underetch rates are slightly asymmetrical with respect to the minimum, a trend that is more pronounced at higher temperatures. This is possibly due to convection in the etchant. Being placed vertically, the flow of hydrogen bubbles may cause convection that is different on either side of the grooves. The temperature dependence of the etch rate and of the etch rate slope is shown in Fig. 3.24. The activation energy of the minimum etch rate is E_{min}: 0.55 eV, and the slope of the etch rate ($dE/d\Theta$) is

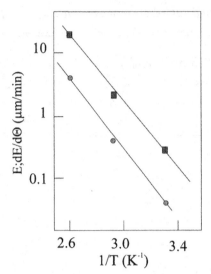

Fig. 3.24. Temperature dependence of the minimum etch rates (cycles) and the slope of the etch rate $dE/d\Theta$ (squares). The lines are fits to the data.

equivalent within the experimental accuracy. No formation of macrosteps ("ledges") could be observed. The minimum of the etch rate at the $\langle 100 \rangle$ direction implies, in our view, that the {100} faces are smooth when they are in contact with KOH solutions, as investigated in our experiments. This is only possible if the surfaces are either reconstructed or if there is an adsorption layer on the {100} faces that stabilises the surface in some way [7]. Certainly, the step free energy on {100} is considerably smaller than on the {111} faces, as evidenced by the large etch rate of the {100}. This is consistent with the absence of high steps on slightly misoriented {100} faces. These high steps are created by "collisions" of steps, a process that is strongly enhanced by adsorption of impurities at the surface.

The temperature dependence of the slope of the etch rate, $dE/d\Theta$, is dominated by the velocity of steps; no nucleation barrier should be visible in the temperature dependence. On the other hand, the nucleation barrier should be visible at the small flat portion of the plots in Fig. 3.19, if nucleation plays the decisive role for etching. However, the activation energies 0.53 eV for the step velocity and 0.55 eV for the minima of the etch rate (Seidel et al. report 0.58 eV [12] for the underetch rate in $\langle 100 \rangle$ direction in KOH) are very similar, hence the nucleation barrier is very small[‡].

The temperature dependence of the width $\Theta_{c(001)}$ of the flat parts in Fig. 3.23 is rather small; it changes only by 25% in the temperature range investigated. The small temperature dependence is in agreement with the nucleation picture, since the activation energy of Θ_{ck} is three times smaller than the energy of a critical nucleus, which is too small to be detected by the experiments described in [55].

An etch rate dominated by spirals would have the following consequences: with $h_{(001)} = a/4 = 1.36$ Å and $\xi^2_{(001)} = a^2/2 = 14.75$ Å^2, Eq. (3.30) gives $r^*_{(001)} = 2.4$ Å, which is quite small, the critical radius contains just one or two atoms. Consequently, the {001} face would

[‡] Neglecting the very unprobable possibility that dissolution spirals dominate the etch mechanism in the $\langle 100 \rangle$ direction.

be very close to kinetic roughening, a fact that accounts for the fast etching of this face. With the interpretation of Uenishi's experiment and the results for {001} faces, we can determine the ratio of the step free energies, which turns out to be close to $\gamma_{(111)}/\gamma_{(001)} = 50$[‡].

In order to check if this is enough to explain the large etch rate ratio of {111} and {001} silicon faces under these conditions, we inspect Eq. (3.18). The problem is that $\Delta\mu$ and ξ are not known. In the case that $\sigma_{(001)}/\Delta\mu \gg 1$, $E_{(111)}/E_{(001)} \approx \xi^2_{(001)}\gamma_{(001)}/\xi^2_{(111)}\gamma_{(111)} = 1/73$. The order of magnitude seems to be okay. But bearing in mind that the radius of the critical nucleus on the {001} face is very small, we may assume that $\sigma_{(001)}/\Delta\mu \ll 1$. In this case, $\tanh \sigma_{(001)}/\Delta\mu \approx \sigma_{(001)}/\Delta\mu$ and $E_{(001)} \sim C\,\Delta\mu/kT$. Etching of {111} is quadratic in $\Delta\mu$, while etching of {001} is linear. Consequently, we have $E_{(111)}/E_{(001)} = \Delta\mu/kT/\sigma(001)$, which can be considerably smaller than 1/73.

In [55] the etch depth was not more than $10\,\mu$m. A spiral mechanism would imply that the density of dislocations with a spiral component must not be much less than $1/(10\,\mu\text{m}^2)$. This seems to be too large to be realistic. Tsukamoto et al. put forward arguments that the density of dislocations after high-temperature processing is of the order of 1 per $100\,\mu\text{m}^2$, but this figure cannot account for the results in [55]. Koide et al. [9] used the whole wafer for their experiments, with extensions of flat portions of {001} and {111} of the minimum order of $100\,\mu\text{m}^2$, but no pretreatment is reported. Uenishi etched right through $\langle 110 \rangle$-oriented wafers, so we cannot strictly exclude that the etch rate was accurately dominated by a spiral mechanism in this case. The possibility of two competing etch mechanisms could account for the large spread of the etch rate ratio of Si{111} and Si{001}.

3.7 Direct evidence for steps

So far we have presented indirect evidence for a step mechanism for anisotropic etching. Using in situ STM imaging, Allongue et al. observed the steps directly. The experiments were performed on n-Si$\langle 111 + 0.7° \rangle$, etched under a cathodic bias in NaOH at room temperature. The STM technique does not allow one to monitor the etching at the OCP, but the scans must be performed negative to the OCP.

The results of the experiments were as follows: at a misalignment of $0.7°$ to the (111) the distance between the steps corresponds to an average step distance of 360 Å. In Fig. 3.25 we show a sequence of STM images of the misaligned $\langle 111 \rangle$ silicon face while being etched at a voltage negative to the OCP. Steps can be clearly discerned. They have a distance in the expected range and they move, indicating the etching.

So there is little doubt that there are steps, the movement of which is responsible for etching at misalignments. Since STM imaging is impossible at a voltage close to the OCP or even positive to the OCP, close to the passivation potential, the experiments were conducted in a different way under these conditions. The STM tip was removed from the sample, the silicon was allowed to etch under the desired conditions, and the voltage was subsequently set to a level convenient for scans. An example of the resulting scans is shown in Fig. 3.26.

The etching becomes aggressive. The main feature is that etch pits between the steps are formed. All etch pits have a depth of one atomic (double) layer. In Fig. 3.27 we reproduce a plot that shows etch rates and the number of etch pits as a function of the bias. Apparently

[‡] If the etch rates were governed by a two-dimensional nucleation mechanism, a gigantic etch rate ratio between {001} and {111} could result (if, e.g., $kT\Delta\mu \approx \gamma^2_{(001)}$, the ratio would be $\exp 15 \approx 10^6$). The actually observed etch rate ratio of ~ 100 would then point to $kT\Delta\mu \approx 3\gamma^2_{(001)}$.

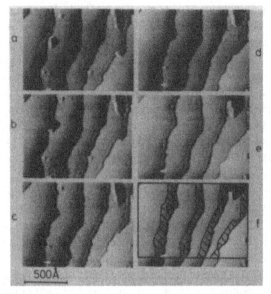

Fig. 3.25. Sequence of in situ STM scans of Si$\langle 111 + 0.7° \rangle$. The shaded region in f indicates the material that has been removed by the aqueous NaOH solution. T = 20°C [24].

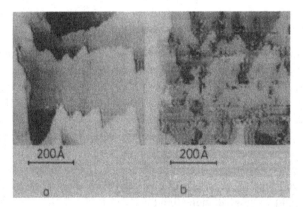

Fig. 3.26. STM scan similar to Fig. 3.25, but at a voltage anodic to the OCP [24].

we have

$$N \propto \exp V/V_o. \tag{3.32}$$

It also turns out that $E \propto N$, in agreement with Eq. (3.16).

Although at the smallest voltage that is experimentally possible with STM imaging, the surfaces look rather rough (c.f., Fig. 3.26) they are quite flat: the density of pits was only 2×10^{11} pits/cm^2. This corresponds to 1 pit per 10,000 sites!

When we plot the same data as in Fig. 3.28, log (pit density) versus $1/V$, the data points again lie on a line (see Fig. 3.29). This must be expected from Eq. (3.14), if the electrical

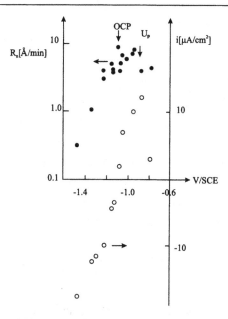

Fig. 3.27. Etch rate of Si⟨111 + 0.7°⟩ and current as a function of the bias. Data have been taken from Allongue et al. [24].

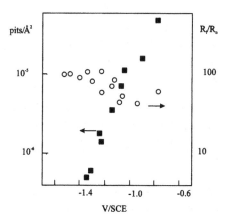

Fig. 3.28. Etch pit density corresponding to Fig. 3.27 [24].

potential has something to do with the chemical potential. This would be the case if (i) charge transfer is associated with the chemical reaction, i.e., if the reaction is electrochemical, and (ii) if there were a large voltage drop across the interface. (i) is not established, but an electrochemical route parallel to a chemical route of the reaction sequence must exist [23, 24]. With regards to assumption (ii), the difficulty is that generally one believes that the voltage drops across a space charge layer at the surface, so the voltage drop directly at the surface is thought to be very small. The latter might not be true. If we identify the voltage with $\Delta\mu$, we would arrive at γ of the order of 0.3 eV, in the range required for layer-controlled etching.

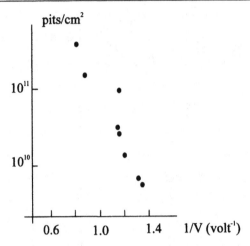

Fig. 3.29. Etch pit density on $\langle 111 + 0.7° \rangle$-oriented silicon as a function of reciprocal voltage with respect to an SCE electrode. Data taken from Ref. [24].

Alternatively, the voltage could increase the bond strength of the hydrogen atoms adsorbed at the surface. This would lead to two effects. First

$$\gamma = \gamma_0 - \delta eV, \tag{3.33}$$

with δ a (small) numerical factor, e the elementary charge, and V the voltage/SCE (standard calomel electrode). Equation (3.33) would lead to the observed dependency of the etch pit density together with Eq. (3.16) and would equally account for the drop of the etch rate at increasing anodic voltage. Second, the activation energy of the elementary chemical reaction $\Delta\varepsilon$ (Eq. (3.15)) would increase by an amount of 8 eV.

Both interpretations are in a somewhat premature state, and more research is required to support one of them (or to discard both). An interesting experiment would be to study the etch rate diagram using the wagon wheel method with n-silicon as a function of the voltage. The anisotropy should increase with the anodic voltage, and the maximum etch rate in the $\langle 113 \rangle$ orientation should drop slower with the anodic voltage.

The third alternative would be that the activation energy is voltage dependent. In this case, the anisotropy would not change very much with the bias.

3.8 Summary

The theory outlined here is rather complex, and many questions remain unanswered. The existence of steps on etching surfaces is not in doubt, and the {111} face, and other faces under certain conditions, are flat and are etched below the critical undersaturation for kinetic roughening. A crystal face can only be flat if it has a finite step free energy. This in turn automatically includes a nucleation barrier.

The theory has the following consequences for wet chemical etching of silicon:

(i)   Rough crystal faces etch much faster than smooth crystal faces. Elementary analysis predicts that the only smooth face of the diamond lattice is {111} [59]. There may be more flat faces due to surface reconstruction and/or absorption, prominent candidates being the {001} and {110}. The {111} remains the face with the largest step free energy and should etch with the slowest rate. Single silicon crystals grown in a CVD reactor develop more faces (see, e.g., [21, 22]).

(ii)   The activation energy of the etch rate of smooth faces of perfect crystals contains the free energy of a critical nucleus. The energy barrier of the chemical reaction and transport in the liquid contribute to the activation energy. The latter contributions are isotropic. The former is anisotropic and absent on rough faces. The larger the step free energy, the larger the activation energy and the smaller the etch rate. The {111} faces should have the largest activation energy. This agrees with experimental results [12].

(iii) The key parameters $\Delta\mu$ and γ both depend on the type of etchant. These parameters might provide the clue in understanding the variation of the etch rate, degree of anisotropy, and temperature dependence with the etch solution. Further research is required.

(iv)   Analysis of etch data can provide clues on the physical state of real crystal surfaces. As we pointed out above, the {111} faces of silicon may not look like the honeycomb structure shown in Fig. 3.2, but there is a famous 7×7 surface reconstruction [28]. There is also evidence that the {001} faces are reconstructed. As demonstrated by Allongue et al., there is no surface reconstruction of the Si⟨111⟩ face under etching conditions [24].

(v)   The equilibrium form of a crystal is dominated by the smooth faces, which are the faces having the smallest surface free energy ("surface tension") [60]. These faces have large step free energies, they grow and etch slowly. Therefore, once the surface tension or the equilibrium form is known, one has a good idea of the growth form and of the faces that etch slowly.

(vi)   The theory gives one the chance to understand the influence of the various solvents and additives used in etching. For example, we know that the results of etching in KOH, NaOH, and CsOH are similar but not identical. The cations obviously play a role. This role is not necessarily only chemical – thermodynamics could also play a role. Both $\Delta\mu$ and γ will depend on the nature of the etchant, i.e., the cation. We addressed above the point of adsorption, which can result in stabilization of a particular face or, in other words, which could lower the roughening temperature of a face. This might be the case if one adds IPA to the solution. In this case we know that the maximum etch rate is no longer in the ⟨110⟩ direction, even though there appears to be a minimum etch rate in this direction. Perhaps the IPA molecules bridge the parallel running ⟨110⟩ PBCs, so that a bond exists between these PBCs. The same effect might play a role when one etches in EDP, which also contains large molecules capable of bridging the distance between the PBCs on the {110} face. Some of Koide's results (their appearance of (110) faces in KOH etching, if the temperature is increased), however, cannot be simply understood in this way.

References

[1] M. Elwenspoek, J. Electrochem Soc. **140**, 2075 (1993).
[2] F. C. Frank and M. B. Ives, Appl. Phys. **31**, 1996 (1960).
[3] R. J. Jaccodine, J. Appl. Phys. **33**, 2643 (1962).
[4] D. F. Weirauch, J. Appl. Phys. **46**, 1478 (1975).
[5] K. Tsukamoto, L. J. Giling, and T. Yasuda, J. Crystal Growth **57**, 412 (1982).

[6] K. Tsukamoto and L. J. Giling, J. Crystal Growth **60**, 338 (1982).

[7] W. P. J. van Enckefort and L. J. Giling, J. Crystal Growth **45**, 90 (1978).

[8] L. J. Giling and W. J. P. van Enckevort, Surf. Sci. **161**, 567 (1985).

[9] A. Koide, K. Sato, and S. Tanaka, IEEE Workshop on Microelectromechanical Systems (MEMS '91), Nara, Japan, Feb. 1991, p. 216.

[10] P. J. Hesketh, C. Ju, S. Gowda, E. Sanoria, and S. Danyluk, J. Electrochem Soc. **140**, 1080 (1993).

[11] Y. Bäcklund and L. Rosengren, J. Micromech. Mircroeng. **2**, 75 (1992).

[12] H. Seidel, L. Csepregi, A. Heuberger, and H. Baumgärtel, J. Electrochem Soc. **137**, 3612 (1990).

[13] Y. Uenishi, M. Tsugai, and M. Mehregani, IEEE Workshop on Microelectromechanical Systems (MEMS '94), Oiso, Japan, Feb. 1994, p. 319.

[14] D. L. Kendall, J. Appl. Phys. **26**, 195 (1975).

[15] P. Hartman and Perdak, in *Crystal Growth, An Introduction* (North Holland, Amsterdam, 1973), p. 367.

[16] P. Bennema and J. P. van der Eerden, in *Morphology of Crystals*, ed. I. Sunagawa (Terrapub, Tokyo, 1987), p. 1.

[17] W. K. Burton, N. Cabrera, and F. C. Frank, Phil. Trans. Roy. Soc. London **A243**, 299 (1951).

[18] L. J. Giling and W. J. P. van Enckevort, Surf. Sci. **161**, 567 (1985).

[19] K. A. Jackson, in *Crystal Growth*, ed. H. S. Peiser (Pergamon Press, Oxford, 1967), p. 17.

[20] J. Rappich, H. J. Lewerenz, and H. Gerischer, J. Electrochem Soc. **140**, L187 (1993).

[21] J. G. E. Gardeniers, W. E. J. R. Maas, R. Z. C. van Meerten, and L. G. Giling, J. Crystal Growth **96**, 821 (1989).

[22] J. G. E. Gardeniers, W. E. J. R. Maas, R. Z. C. van Meerten, and L. G. Giling, J. Crystal Growth **96**, 832 (1989).

[23] O. J. Glembocki, R. E. Stahlbush, and M. Tomkiewicz, J. Electrochem Soc. **132**, 145 (1985).

[24] P. Allongue, V. Kosta-Kieling, and H. Gerischer, J. Electrochem Soc. **140**, 1009 (1993).

[25] E. D. Palik, H. F. Gray, and P. B. Klein, J. Electrochem. Soc. **130**, 956 (1983).

[26] E. D. Palik, V. M. Bermudez, and O. J. Glembocki, J. Electrochem. Soc. **132**, 871 (1985).

[27] E. D. Palik, V. M. Bermudez, and O. J. Glembocki, J. Electrochem. Soc. **132**, 135 (1985).

[28] J. Kelly, University of Utrecht, personal communications (1995).

[29] R. Wiesendanger, G. Tarrach, D. Bürgler, and H.-J. Güntherodt, Europhys. Lett. **12**, 57 (1990).

[30] P. Bennema, J. Crystal Growth **69**, 182 (1984).

[31] W. K. Burton, N. Cabrera, and F. C. Frank, Phil. Trans. Roy. Soc. London **A243**, 299 (1951).

[32] H. Camon, D. Esteve, M. Djafari-Rouhani, and A. M. Guè, Proc. Micromechanics Europe (MME) '90, Berlin, Germany, Nov. 1990.

[33] H. Camon, A. M. Guè, J. S. Danel, and M. Djafari-Rouhani, Sensors and Act. **A33**, 103 (1992).

[34] P. Bennema, J. Crystal Growth **69**, 182 (1984).

[35] H. van Beijeren, Phys. Rev. Lett. **38**, 993 (1977).

[36] J. P. van der Eerden and H. J. J. Knobs, Phys. Lett. **66A**, 334 (1978).

[37] J. M. Kosterlitz and P. J. Thouless, J. Phys. **B13**, 4942 (1976).

[38] A. Passerone and N. Eustatopolos, Acta Metall. **30**, 1349 (1982).

[39] P. E. Wolf, F. Gallet, S. Balibar, E. Rolley, and P. Nozières, J. Pysique **46**, 1987 (1985).

[40] J. E. Avron, L. S. Balfour, C. G. Kuiper, J. Landau, S. G. Lipson, and L. S. Schulman, Phys. Rev. Lett. **45**, 814 (1980).

[41] J. P. van der Eerden, Contributions to the Theory of Crystal Growth, Ph.D. Thesis, University of Nijmegen, 1979.

[42] J. P. van der Eerden, Phys. Rev. **B13**, 4942 (1976).

[43] R. M. Finne and D. L. Klein, J. Electrochem Soc. **114**, 965 (1962).

[44] E. D. Palik, V. M. Bermudez, and O. J. Glembocki, in *Micromachining and Micropackaging of Transducers*, eds. C. D. Fung, P. W. Cheung, W. H. Ko, and D. G. Fleming (1985), p. 135.

[45] D. L. Kendall and G. R. deGuel, in *Micromachining and Micropackaging of Transducers*, eds. C. D. Fung, P. W. Cheung, W. H. Ko, and D. G. Fleming (1985), p. 107.

[46] R. Becker and W. Doering, Ann. Phys. **24**, 719 (1935).

[47] G. H. Gilmer, J. Crystal Growth **42**, 3 (1973).

[48] S. W. H. de Haan, V. J. A. Meussen, B. P. Veltman, P. Bennema, C. van Leeuwen, and G. H. Gilmer, J. Crystal Growth **24/25**, 491 (1974).

[49] S. Balibar, F. Gallet, and E. Rolley, J. Crystal Growth **99**, 46 (1990).

[50] M. Elwenspoek and J. P. van der Eerden, J. Phys. **A20**, 669 (1987).

[51] G. H. Gilmer, J. Crystal Growth **42**, 3 (1973).

[52] I. Sunagawa and P. Bennema, J. Crystal Growth **53**, 490 (1981).

[53] P. Hartman (ed.), *Crystal Growth, An Introduction* (North Holland, Amsterdam, 1973).

[54] H. Hashimoto, S. Tanaka, and K. Sato, Proc. Transducers '91, San Francisco, Calif., June 23–27, 1991, p. 853.

[55] M. Elwenspoek, U. Lindberg, L. Smit, and H. Kock, Proc. Microelectromechanical Systems (MEMS)'94, Oiso, Japan, Jan. 19–28, 1994, p. 223.

[56] L. Ternez, Y. Bäcklund, J. Tirén, and J. O'Connnell, Sensors and Materials **1–6**, 313 (1989).

[57] C. Ju and P. J. Hesketh, Sensors and Act. **A33**, 191 (1992).

[58] W. Choi and J. G. Smits, J. Microelectromechanical Systems **2**, 82 (1993).

[59] P. Hartman, Z. Kristallogr. **121**, 78 (1965).

[60] G. Wulff, Z. Krist. **34**, 449 (1901).

[61] P. Allongue, V. Kosta-Kieling, and H. Gerischer, J. Electrochem. Soc. **140**, 1018 (1993).

[62] P. Bennema, private communication (1993).

[63] P. M. M. C. Bressers, S. A. S. P. Pagano, and J. J. Kelly, J. Electroanal. Chem. **391**, 159 (1995).

[64] J. P. van der Eerden, P. Bennema, and T. A. Cherepanova, Prog. Crystal Growth Charact. **1**, 219 (1978).

[65] M. Elwenspoek, J. Microeng. Micromachining **6**, 405 (1996).

4

Wafer bonding

4.1 Introduction

Wafer bonding is a technology that is of great importance to micromachining. The applications and examples in the next chapter will demonstrate this clearly. Wafer bonding refers to the mechanical fixation of two or more wafers to each other. It is possible to process wafers prior to bonding, in this way circumventing the necessity of assembling tiny structures. By bonding, one can assemble all devices on the wafer in one single step. Waferbonding is indespensible for batch processing. In literature, many different terms are used for bonding processes. The names of the different types of bonding are not consistently used and do not show a systematic classification. The following terms are used:

Direct bonding: This term refers to the bonding of wafers without an intermediate layer. A large number of different materials can be bonded in this way (see Haisma [1]).

Silicon fusion bonding (SFB): This term refers to the bonding of two silicon wafers with or without a thin layer of thermal or native oxide.

Direct silicon bonding: This term refers to bonding of bare silicon wafers (with a native oxide layer).

Indirect bonding: This term refers to the bonding of wafers with an intermediate layer. The use of conventional glue belongs to this type of bonding.

Field-assisted bonding: This term refers to bonding in the presence of an electric field. An important example is anodic bonding of a silicon wafer to a Pyrex wafer.

Field-assisted bonding can be direct or indirect. Direct bonding can be done between different materials; anodic bonding of a Pyrex wafer to a silicon wafer and the so-called silicon fusion bonding (SFB) are examples of direct bonding. In literature the term "direct bonding" is often used for SFB. Note that the term silicon fusion bonding is misleading; the fusion point of silicon at atmospheric pressure is 1410°C, well below the relevant process temperatures. Silicon does not fuse during the process.

4.2 Silicon fusion bonding

Silicon direct bonding is of great interest when fabricating Silicon on Insulator (SOI) structures. In fact, historically, this has been the motivation for the study of SFB [2]. For silicon micromechanics the basic process consists of contacting two wafers and annealing them in a wet or dry oxidation furnace at 900–1100°C. The great advantage of silicon direct bonding is that one obtains practically monolithic structures and one has no problem with residual stress after the bonding process. If one succeeds in proper preparation of the wafers, the process is very simple. The disadvantage is that one has to anneal the stack of wafers at a high temperature. This excludes the use of some materials and processes (such as dopant diffusion) before the bonding step.

4.2.1   Description of the process

Fusion bonding is a critical process. The success of bonding depends strongly on the quality of the wafer surfaces. The wafers must be clean, smooth, and flat. Wafer curvature can be fatal. Special cleaning procedures and even polishing steps are sometimes required prior to bonding [1].

Great care must be taken in process steps that can induce surface roughness and wafer curvature. The former is the result of too many process steps, and the latter of an asymmetry in deposited thin films on either side of the wafer. Cavities etched into the wafer in combination with thin films under intrinsic stress can also cause a deformation of the wafer. Attempts to bond these wafers will fail or result in large voids that do not disappear after heat treatments.

When two silicon wafers are brought into contact, spontaneous bonding occurs. There is a so-called "bonding wave," which can be seen in a transmission infrared video system, a system that easily reveals bonded and unbonded regions (or voids). After the surfaces are brought into contact at room temperature, the bonded region propagates across the wafers within a few seconds.

Direct bonding is possible between, e.g.,

- two bare silicon wafers,
- between one bare wafer and one wafer with a thermal oxide,
- between two wafers with a thermal oxide,
- between one wafer with a thin layer of nitride (100–200 nm) and one bare wafer, and
- between two wafers with a thin layer of nitride.

It appears that hydrophilic wafers bond more easily than hydrophobic ones. In many reports it is claimed that hydrophobic wafers do not bond at all. However, as discussed by Bäcklund et al. [3, 4], the before-mentioned spontaneous bonding can also be observed on hydrophobic surfaces. It has been suggested that the bonding mechanism at low temperature, i.e., without annealing, is due to van der Waals interactions in the case of hydrophobic surfaces and to hydrogen bridging in the case of hydrophilic surfaces. This will be discussed in greater detail below.

When spontaneous bonding at room temperature is successful, the pair of wafers can be handled easily without any danger that the wafers will fall apart. For example, it is quite safe to transport the wafers with a pair of tweezers or to insert them into a wafer boat for an anneal furnace.

The standard process proceeds with hydrophilic wafers that have been cleaned in the standard RCA1 (and sometimes RCA2) process. These processes leave the wafers with a native oxide layer and a high density of OH groups on the surface. The latter gives rise to the before-mentioned hydrogen bonding. When the wafers are treated in an oxygen plasma, the number of OH groups at the surface is considerably increased [5, 6].

When spontaneously bonded wafers are annealed, the bond becomes much stronger. Around 300°C the OH groups form water molecules [7]. A more elaborate description of this process will be given below. The voids that form at this temperature are believed to be due to water vapor formed during the process. These voids tend to disappear after annealing at temperatures around and above 700°C. At such temperatures, the water molecules dissociate, oxygen bonds to silicon, and hydrogen diffuses through the silicon. At still higher temperatures (800–1400°C), the oxygen also diffuses into the silicon lattice.

As reported by Kissinger and Kissinger [8], the application of a small amount of pressure during the bonding process increases the final bond strength. It is also worth noting that Stengl et al. [9] reported that they did not observe the formation of voids if the wafers were contacted at a temperature of 50°C.

The surface energy between the bonded surfaces depends strongly on the temperature at which the wafers have been annealed. The surface energy found is as high as 2.3 J/m^2, close to the fracture surface energy of bulk silicon.

Because of the critical cleaning process, a bonding machine is used in several laboratories. The basic idea is that one should not handle the wafers between the rinse step and the contacting step. For this purpose a bond tool has been developed by A. Söderbärg at the University of Uppsala [10] in which both process steps, namely rinsing and contacting, are combined. In this way the chance of contamination of the wafers with dust before bonding is greatly reduced. A precursor of this bond machine has been described by Stengl et al. [11].

Several authors report on micromachining steps on one or both wafers prior to bonding [12–14]. It is possible to grow an oxide on the wafer, do photolithography, and etch cavities in the wafer. The flat parts of the wafer will still bond, if "not too many steps are performed" [15]. It is, however, difficult to make this statement more quantitative. There are process steps that seem to make bonding impossible. One of them is the formation of a highly boron-doped layer; the author's experience and remarks found in the literature [12] give evidence to this fact.

Harendt et al. [14] investigated the possibilities of micromachining prior to bonding in more detail. They prepared wafers with V grooves and cavities anisotropically etched in KOH using a nitride mask. The size of the cavities ranged from 100 μm to 10 mm, and the grooves had a width of 10–500 μm. The bonding surface was either LPCVD nitride (140 nm), residual oxide after removal of the nitride, silicon, or an oxide grown in a wet oxidation furnace. Most interestingly, wafers containing cavities showed less voids after bonding than wafers without cavities. There seems to be a correlation between the cavity size and the success of bonding (expressed in the percentage of nonbonded area that decreases with increasing cavity size). The bonds are characterized as being satisfactory for all materials present on the surfaces (silicon, nitride, or oxide).

There are several ways to characterise the bond. Of importance are inspection by infrared light, using an IR-sensitive video camera, or ultrasound imaging. The former gives information about possible voids if they are large and thick enough (thicker than one half the wavelength of the light and larger than the numerical aperture of the optical system; the latter is typically small, in the range of 0.05 or less, allowing detection of voids larger than 20–30 μm). Ultrasound imaging gives qualitative information about the bond quality [16]. The bond strength, in terms of interface energy, is most conveniently measured by a method introduced by Maszara et al. [17]: a thin blade is inserted between the bonded wafers and a crack is introduced. The length of the crack is a measure for the surface energy W of the bond. The surface energy is given by:

$$W = \frac{Et^3y^2}{8L^4},$$

(4.1)

where $2y$ is the thickness of the blade, t is the thickness of the wafers (both are assumed to be of equal thickness), L is the crack length, and E is Young's modulus of the wafer material (for silicon, $E = 1.66 \, 10^{11}$ Pa).

The bonding force can be measured directly by pulling the wafers apart. This method has been proposed, e.g., by Muller and Stoffel [18]. The idea is to glue the waferstack to, e.g., stainless steel parts which can be inserted in a conventional aparatus used for pull tests. Unfortunately there is no direct link between the bond strength (or interface energy) obtained with the blade method mentioned above and the bonding force (or pressure) obtained with the latter method.

4.2.2 *Mechanism of fusion bonding*

To understand the mechanism of fusion bonding, it is useful to discuss three temperature ranges [8, 9]. The surface chemistry of silica has been discussed by Iler [19], and what is reported here rests on this work and on what is explained in [8]. Below 200°C, hydrogen bonding seems to be responsible for the bonding of hydrophilic wafers. At 200–400°C, chemical reactions occur and a bond between the hydroxyl groups probably forms directly, which is responsible for the bond strength. At the higher temperature one believes that there is a covalent bond between SiO units. This idea is shown schematically in Fig. 4.1.

The silicon oxide surface is hydroxylated at room temperature and atmospheric pressure. This is the reason why a silicon oxide surface is hydrophilic, and the surface is covered by a thin layer of water molecules at normal atmospheric conditions. For each hydroxyl group,

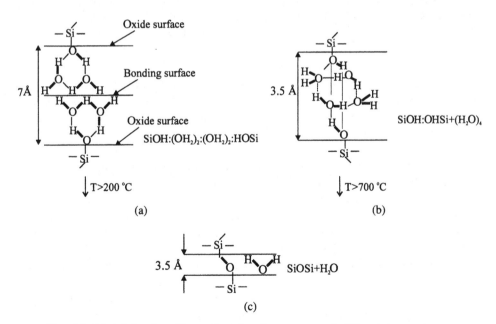

Fig. 4.1. Model for the silicon direct bonding process at different temperatures. (a) Bonding via hydrogen bonds between water clusters. The hydrogen bonds bridging the two wafer interfaces are drawn as dashed lines. (b) Bonding via hydrogen bonds between silanol groups, with a water cluster present at the interface. (c) Bonding via siloxane bonds. Covalent bonds are indicated as solid lines (from Stengl et al. [9]).

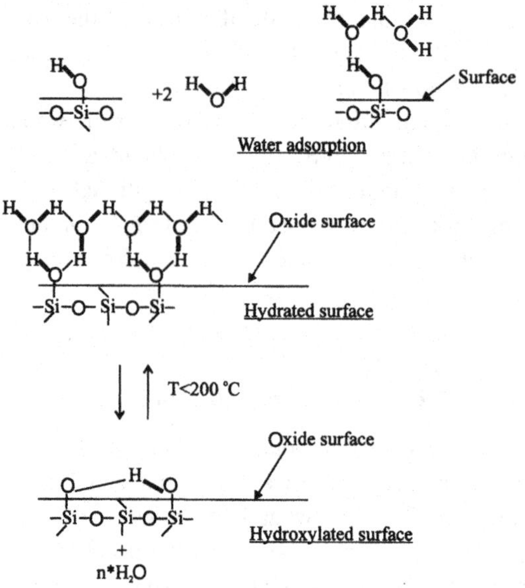

Fig. 4.2. Schematic representation of the interaction between water molecules and a hydrated surface (from Stengl et al. [9]).

two water molecules tend to adsorb, both in different steric positions, one pointing with the O to the hydroxyl group and the other pointing with an H to the wafer (see Fig. 4.2). In this way a network of adsorbed water molecules forms a monomolecular layer on the silica surface. One speaks of a hydrated surface. The interaction with the silica surface is strongly influenced by the presence of tiny amounts of sodium ions. Between 100 and 200°C the water desorps. The surface remains hydroxylated. The number of hydrogen atoms per silica surface is estimated to be between 3 and 5 per 100 Å^2, and the area per silica is approximately 13 Å^2.

Stengl et al. [9] determined the speed of the bonding wave as it proceeds over the wafer surface as a function of the temperature. The most important result is that the speed decreases with increasing temperature and that it approaches zero at a given temperature T_c, which depends on the state of the surface. Above T_c there is no spontaneous bonding anymore. T_c is given for different interfaces in Table 4.1. Qualitatively, the bond wave speed is proportional to the density of adsorbed water molecules, indicating that the state of surface hydration is of essential importance to spontaneous bonding. The presence of the water film on the face therefore seems to be a prerequisite for fusion bonding. This observation is in perfect agreement with observations reported in many papers that a hydrophilic surface is required for SFB (note however Bäcklund et al.'s work [3, 4]). The small T_c for quartz is possibly due to small concentrations of sodium ions.

The bond strength at room temperature can be estimated from the number of hydrogen bonds (two per hydroxyl group, as indicated in the upper part of Fig. 4.1). Hydrogen bonding is not a linear process in the sense that the energy of two bonds is twice the energy of two single bonds. The reason is that hydrogen bridging occurs in a liquid state, and the position and orientation between two neighboring molecules are more strongly correlated than the position and orientation of next-nearest neighbors. Stengl et al. considered this situation similar to hydrogen bonding of a dimer with a monomer, in which case one has a bonding energy of 1 kcal/mole-bond = 4.3 kJ/mole-bond = 45 meV/bond [20]. If one has

Table 4.1. *Maximum spontaneous bonding temperature of
hydrophilic surfaces.*

Interface	$T_c(°C)$
silicon–silicon	330
silicon–thermal oxide	270
quartz–thermal oxide	100
quartz–bare silicon	130–180

two hydrogen bonds per hydroxyl group, and a maximum density of hydrogen atoms bonded
to the silica surface of $0.77/Å^2$, one gets a surface energy of ca. $100 \text{ erg/cm}^2 = 100 \text{ mJ/m}^2$.
This value fits perfectly with experimental results given, e.g., by Bäcklund et al. [4]. Others
report surface energies that are somewhat smaller ($60–85 \text{ mJ/m}^2$ [17]).

Kissinger and Kissinger [8] treated silicon wafers in an oxygen plasma and found a
considerably larger surface energy after spontaneous bonding (280 mJ/m^2). This result
cannot be explained in the framework of Stengl's model, since a surface energy of 100 mJ/m^2
is the maximum that can be accounted for. If indeed the plasma treatment increases the
number of hydrogen atoms adsorbed to the surface by more than a factor of two, the
agreement of Stengl's theory and Bäcklund's results would be fortuitous: there were less
hydrogen atoms available for bonding at the surface than assumed, and the interaction
between the adsorbed water molecules should be stronger than that assumed by Stengl et al.

In the middle-temperature range, indicated in the middle part of Fig. 4.1, we have the
reaction

$$SiOH—(OH_2)_2—(OH_2)_2 - OHSi \leftrightarrow SiOH—OHSi + (OH_2)_4. \qquad (C4.1)$$

The complex of four water atoms becomes free, and the bonding between the silica surfaces is established by direct hydrogen bonds, i.e., not by hydrogen bonds mediated by
water molecules, as is the case in the low-temperature range. Stengl et al. assume that the
bonding energy for the middle-temperature range is close to the hydrogen bond energy
between monomers (i.e., the energy of a water dimer). This energy would correspond to
$2 \times 6.1 \text{ kcal/mol}$ per binding site, leading to a surface energy of 630 mJ/m^2. This implies
from the reaction kinetics, assuming that the conversion from the left- to the right-hand
side of the above equation is complete, that the concentration of the direct hydrogen bonds
becomes

$$[SiOH—OHSi](t) = n_0 \exp(-k_1 t). \qquad (4.2)$$

For the bond strength as a function of annealing time we may write

$$W(t) = (W_{max} - W_0)[1 - \exp(-k_1 t)] + W_0. \qquad (4.3)$$

Here W_{max} is the surface energy for the case that all bonds between the silica surfaces are
described by the right-hand side of reaction (C 4.1), and W_0 refers to the surface energy in

the low temperature phase. k_1 is given by

$$k_1 = \tau^{-1} \exp\left(\frac{-E}{kT}\right), \tag{4.4}$$

where τ is a phenomenological time constant, and E an activation energy. Experimental results [8, 11] give an activation energy of the process of 50 meV and a time constant of 2.35 hours. The agreement between the activation energy and the binding energy of the water dimers to the monomers may be accidental. The long time constant makes it difficult to compare published experimental bond strengths. For example, Bäcklund et al. [4] do not mention the annealing time. They give surface energies of 200 mJ/m^2 and 350 mJ/m^2 at 200 and 400°C, respectively. After a long enough anneal time ($t > 30$ hours!) the results for the surface energy saturate, and Kissinger and Kissinger [8] find 600 mJ/m^2 and 840 mJ/m^2 at 200 and 400°C, respectively. While the low-temperature results fit in with what is expected, the high-temperature results look somewhat high. However, Stengl et al. admit that perhaps doubly linked SiOH—HOSi bonds exist, and that would increase the total surface energy by a factor of 2. The model given in Eq. (4.3) does not allow any variation of the surface energy with temperature, in clear contradiction with experimental evidence [4, 8]. Obviously, the assumption that all sites react according to reaction (C 4.1) is not valid, and an extension of the model appears to be necessary.

The bond strength (surface energy of 2.3 mJ/m^2) observed at the highest annealing temperature fits well with what one would expect for covalent bonds between silicon oxide groups.

4.2.3   Effect of wafer surface imperfections

A typical scan of a silicon wafer surface is shown in Fig. 4.3(a), and Fig. 4.3(b) shows the power spectrum of the surface irregularities (both figures are from Ref. [21]). It can be seen that the overall peak-to-peak amplitude is of the order of 20 nm, and the prominent wavelength is of the order of 1 mm. There is a second peak in the power spectrum close to a wavelength of 0.1 mm, having an amplitude of perhaps 1 nm.

If two surfaces with the roughness as indicated in Fig. 4.3 are bonded, the wafers must deform, since otherwise the effective contact area would be much too small to account for

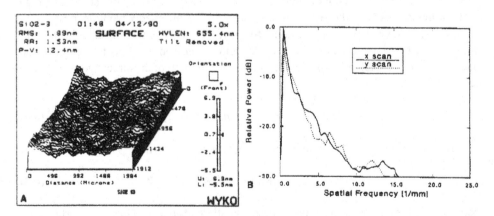

Fig. 4.3. Surface profile from optical profilometry of (a) a 2×2 mm^2 part of a silicon wafer and (b) distribution of the amplitude and the wavelength of its spatial components.

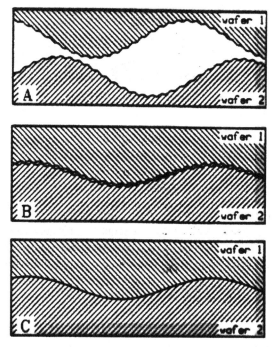

Fig. 4.4. Schematic representation of the deformation by the bonding process. (a) Before bonding, (b) after low-temperature bonding, and (c) after annealing.

the large surface energies found. That the deformation occurs has been demonstrated by Maszara et al. [21]. Direct-bonded wafers are strained (as revealed by X-ray topology) with a wavelength of 1 mm and with an amplitude comparable to that of the long wavelength surface variations. Hence, by bonding the wafers they deform according to the schematic shown in Fig. 4.4. The strain is observed after room temperature bonding and does not change after annealing, indicating that most of the deformation occurs during spontaneous bonding. The resolution of X-ray topography is not good enough to reveal deformations of the small wavelength (0.1 mm). The authors assume that the surface roughness with the small wavelength stays until the wafers are annealed at a higher temperature, and they conclude that the larger bonding energy is (partly) due to more sites that come into contact, if the wafers deform in a way that the surfaces fit smoothly.

This conclusion, however, does not agree with the analysis of the bonding energies given in the previous section. There we saw that we get a satisfactory agreement between the expected and the experimentally observed surface energy, taking all possible bond sites into account. A simple calculation can give us some indication whether the forces caused by hydrogen bonding are strong enough to deform a wafer with a long wavelength and a small wavelength. Following Maszara, we approximate the surface morphology by assuming a spherical (or cylindrical) shape of the hills and valleys on the wafer. One wave is composed of four pieces of a sphere. For the geometry we refer to Fig. 4.5. The radius of curvature R is given by a^2/h. For the long waves, $a = 0.25$ mm and $h = 10$ nm, and for the short waves, $a = 0.025$ mm and $h = 1$ nm. The radii of curvature are therefore 3 m and 30 cm, respectively. The work required to bend a plate of width b and length l is given by

$$W_{\text{elastic}} = \frac{Ed^3bl}{24R^2}.$$ (4.5)

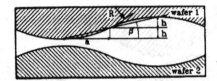

Fig. 4.5. Geometry of the surface waviness.

Here E is Young's modulus of silicon (1.66×10^{11} Pa) and d is the thickness of the wafer. The surface energy is simply

$$W_{\text{surface}} = \gamma bl, \tag{4.6}$$

in which γ denotes the surface tension (and 2γ is thus the bonding energy). Imagine now that two surfaces have a given area of contact, but outside this area the surfaces bend away. In order to bend one plate a length l toward the other one, it costs the energy given in Eq. (4.5), and one gains the energy in Eq. (4.6). Equilibrium will be established if

$$\frac{\partial W_{\text{elastic}}}{\partial l} - \frac{\partial W_{\text{surface}}}{\partial l} = 0, \tag{4.7}$$

which is independent of l. This means that if the surface energy is strong enough, the whole structure collapses, but if it is too weak, the whole structure does not bend at all. From Eq. (4.7) we find the transition at

$$R^2 = \frac{Ed^3}{12\gamma}. \tag{4.8}$$

A factor 2 arises from bending two plates. For $d = 300$ mm and $\gamma = 0.1$ J/m^2 we find $R = 4$ m, which is remarkably close to the 3 m reported by Maszara et al. for the long wavelength roughness and is well above the 30 cm, corresponding to the radius of curvature belonging to small wavelength roughness. The surface energy in the middle-temperature region of 0.6 J/m^2 is equally unable to bend the wafers in order to smooth the short wavelength roughness. It may be noted that the thickness of the wafers enters with the third power in Eq. (4.8). As a consequence, it is much more difficult to bond thicker, i.e., larger, wafers in SFB.

4.2.4 Bonding of hydrophobic wafers

In a number of papers papers it was reported that SFB is also possible with hydrophobic wafers [2, 3, 22, 23]. This is in contrast to earlier claims that the wafers must be hydrophilic for SFB. In fact, it appears that spontaneous bonding of hydrophobic wafers leads to very weak bonding if not annealed. The bonding energy obtained with hydrophobic wafers is as low as 26 mJ/m^2 [4]. However, as reported by Bäcklund et al. [4], a bond wave can also be discerned during the bonding of hydrophobic wafers. During annealing of the wafers, the bonding energy raises quickly when the temperature exceeds 400°C and reaches a value of 2.5 J/m^2 at 600°C. This value is obtained for hydrophilic wafers only when annealed at 900°C or above. Bäcklund's results are reproduced in Fig. 4.6.

More details can be found in the Ph.D. thesis of Karin Ljungberg [38]. Examples of successful applications of SFB will be given in another chapter.

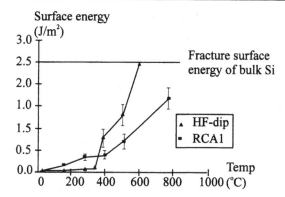

Fig. 4.6. Surface energy of the bond as a function of the anneal temperature for RCA1-cleaned samples and HF-dipped samples (from Ref. [4]).

4.2.5 Silicon fusion bonding after chemical mechanical polishing

Recent experiments in our own labs [39] have demonstrated that direct bonding seems to be possible between all materials that can be made smooth enough. It seems that if the micro roughness is smaller than a few nanometers and the radius of curvature is large enough, that all materials bond to each other, as one could expect simply from adhesion.

Chemical mechanical polishing (CMP) seems to be the key technology for preparing very smooth surfaces. CMP is the standard technology to polish silicon wafers, and in application to silicon it is well-known. CMP can be applied, however, to more materials than silicon alone, and it is even unrestricted to brittle materials: metals can be polished as well with good results. A group at Philips Natlab [40] reported its bonding results for ceramics and metals after careful surface treatment, and we get the impression from this paper that the difficult step is only the flatness of the material, in other words, if the material can be polished, then it bonds, certainly to the same material.

Our studies are in the beginning phase, and the interested reader should watch for Chenqun Gui's Ph.D. thesis in 1999 [42]. After polishing we are able to bond all standard (as tested so far nonmetallic) materials for IC processes, regardless of how the materials have been treated or deposited, and deposited material regardless of its thickness. This applies to polysilicon up to 5-μm thick, doped or not, Si_xN_y tested with thickness up to 3 μm [41], silicon after heavily doping with B, or silicon after etching in KOH (which results in very rough surfaces). Unfortunately, CMP is rather expansive.

4.3 Anodic bonding

The term anodic bonding refers to bonding assisted by an electric field. Anodic bonding can be accomplished between Corning 7740 glass wafers (Pyrex) and virtually any metal [24], but it is also possible to bond two silicon wafers with an intermediate sputtered Pyrex thin film [25]. The advantages of the anodic bonding process are low process temperature (450°C), low residual stress, and less stringent requirements on the surface quality of the wafers as compared to SFB. However, for some applications, even these residual stresses and relatively low process temperatures are not tolerable, and people are still seeking processes at even lower temperatures. We shall first describe anodic bonding of a silicon wafer to a Pyrex wafer.

4.3.1 Anodic bonding of a silicon wafer to a Pyrex wafer

A schematic anodic bonding setup is shown in Fig. 4.7. Generally, one uses a hot plate on which the silicon wafer is placed, and on top of the silicon wafer the Pyrex wafer. A cathode is connected to the uppermost surface of the glass wafer. Typical voltages are between 200 and 1000 V, and typical process temperatures are 180–500°C. At elevated temperature the glass becomes a conductive solid electrolyte, and the positive sodium ions become quite mobile and migrate toward the negative pole, leaving a space charge (depletion) region adjacent to the glass–silicon interface. Most of the voltage drop is situated across this space charge region, and a high electric field strength between glass and silicon results. The wafers are pulled into close contact. The elevated temperature finally results in covalent bonds between the surface atoms of the wafers in Fig. 4.8. A typical plot of the current as a function of time is shown. During the bonding process, temperature and voltage are kept constant. As soon as the voltage is switched on, a current peak occurs, indicating the drift of the sodium ions to the cathode and the buildup of the space charge region. The bonded

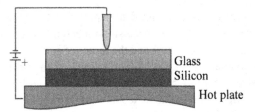

Fig. 4.7. Schematic of an anodic bonding setup (from Hanneborg [25]).

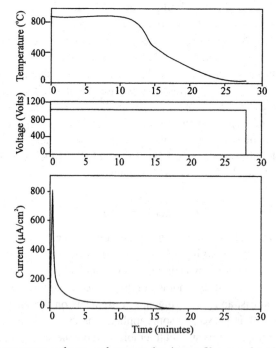

Fig. 4.8. Temperature, voltage, and current density profiles as a function of the time during an anodic bonding process (from Hanneborg [25]).

region can easily be seen through the glass – it looks grey. The bonding front proceeds starting from the spot below the cathode.

Besides Pyrex, Corning #7070, soda lime #0080, potash soda lead #0120, and aluminosilicate #1720 are suitable for anodic bonding. For silicon bonding, Pyrex is the most commonly used.

Anodic bonding is possible under atmospheric conditions as well as in vacuum. The following requirements must be met for the anodic bonding process to occur:

- The glass must be slightly conductive in order to be able to build up a space charge region;
- the temperature must stay well below the softening point;
- the metal must not inject charge carriers into the glass;
- the surface roughness of the wafers must be smaller than 1 μm rms, and the surfaces must be clean and dust free;
- the (native or thermally grown) oxide layer on the silicon must be thinner than 200 nm; and
- the thermal expansion coefficients of the bonded materials must match in the range of temperatures.

In Fig. 4.9 we show the thermal expansion coefficient of silicon and Pyrex as a function of temperature. As can be seen, above 450°C the thermal properties of the materials begin to deviate seriously, therefore, the bonding process with Pyrex glass should be limited to 450°C. Recently, new glass types have been developed, e.g., the types SD1 and SD2 of the Japanese company HOYA (see bottom of Fig. 4.9), which have thermal expansion coefficients very close to that of silicon. For example, for the type SD2 glass the thermal expansion coefficients are 3.20 ppm/°C (30–300°C) and 3.41 ppm/°C-1 (30–450°C) [27].

In a valuable paper, T. R. Anthony [26] gives a quantitative discussion of many important effects in anodic bonding. The most important results are:

- A *DC voltage* is preferable over an *AC voltage*. The electrostatic bonding pressure varies with the inverse frequency of the AC voltage.
- The advantage of using *a point cathode* (instead of an extended planar cathode) is that the electrostatic pressure develops due to the finite conductivity of the glass, starting below the cathode tip, radially propagating across the wafer. The bonding front can be observed by looking through the glass. The velocity of the bonding front is given

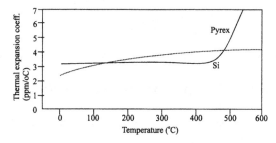

Fig. 4.9. Thermal expansion coefficients of glass wafers compared with silicon. Pyrex and silicon [25]. Glass from HOYA company [27].

by

$$\frac{dR}{dt} = \frac{1}{\rho C(1 + \ln(R/a))},$$

(4.9)

where ρ is the conductivity of the glass, C is the capacitance per unit area of the metal–glass interface, and a is the radius of the cathode tip. For typical values ($\rho = 10^7$ Ω cm, $C = 8.85 \times 10^{-9}$ F/cm^2, $a = 2.5 \times 10^{-4}$ cm) one obtains a velocity of 1.2 cm/s at $R = 1$ cm. Velocities in this range are also observed experimentally [24].

- When using *nonconductive glass*, the bonding pressure is given by

$$P = P_0 \left(\frac{\varepsilon_G D_0}{\varepsilon_G D_0 + \varepsilon_0 D_G} \right)^2,$$

(4.10)

where

$$P_0 = \frac{\varepsilon_0 V^2}{2 D_0^2}$$

(4.11)

is the electrostatic pressure that would be present if the glass was a good conductor. D_0 and D_G are the thickness of the air gap and the glass, respectively, and ε_G and ε_0 are the appropriate dielectric constants. It can be seen that the pressure drops very quickly if the nonconductive glass becomes thick.

- When using *conductive glass*, the pressure is

$$P = \frac{2V^2 \varepsilon_G^2}{\varepsilon_0 D_P^2},$$

(4.12)

where D_p is the thickness of the polarized region. For reasonable values of the parameters in Eq. (4.12), one has $P \sim P_0$, i.e., the electrostatic pressure that would be present if the glass were a metal.

- The presence of *surface roughness* decreases the effective electrostatic pressure considerably. The details depend on the shape of the surface imperfections. Scratches, such as those from polishing, are much worse than a more smooth variation of the surface. In the case of scratches, the electrostatic pressure becomes roughly

$$P = P_0 \left(\frac{D_0}{2h} \right)^2,$$

(4.13)

with $2h$ the peak-to-peak amplitude of the surface roughness, as in Fig. 4.5. In case of a sinusoidal surface profile, one gets

$$P = 2P_0 \sqrt{\frac{D_0}{2h}}.$$

(4.14)

- To get a collapse of the surface so that nonflat surfaces are deformed to join each other closely, the electrostatic pressure must exceed the elastic forces. This is just analogue to the case of SFB, as is easily seen if the surface energy in Eq. (4.6) is replaced by the electrostatic energy,

$$W_{es} = \frac{\varepsilon_0 V^2}{D_0},$$

(4.15)

which leads to a collapse if [27]

$$R^2 = \sqrt{\frac{Ed^3 D}{12\varepsilon_0 V}}, \tag{4.16}$$

with d being the thickness of the wafer. If surface curvatures are larger than the expression given in Eq. (4.16), bonding will not proceed. Equation (4.16) is derived for sinusoidal-like roughness.

Anodic bonding is now a well-developed technology. If care is taken to achieve good cleaning procedures and a dust-free environment, and if polished wafers are used, this process has a high yield. Examples of successful applications will be given in another chapter.

4.3.2 Anodic bonding using a sputtered thin film of Pyrex

It is also possible to use the anodic bonding scheme with an intermediate thin film. Using a thin film for bonding has the advantage that the devices come closer to a monolithic structure. Residual stress and differences in thermal expansion coefficients have only minor effects on the performance of sensors if the thickness of the films used for bonding is very small compared to the thickness (actually the stiffness) of the devices.

Probably the first demonstration of this very promising technique was given in 1972 by Brooks and Donovan [29], who used sputtered borosilicate glass as a bonding layer. Since then, extensive work has been done to develop the technique. Nowadays most researchers use thin films sputtered from a Pyrex target. Usually, argon is used as the sputtering gas; Hanneborg [25] reports a growth rate of the thin film of 100 nm/hr at a pressure 5 mTorr and sputtering power of 1.6 W/cm^2. By adding 10% oxygen to the gas, the growth rate is reduced by a factor of two. Generally a thin film thickness of 1.5–2 μm is used for bonding. Hanneborg's process is as follows [25]:

1. Growth of a thermal oxide layer of 100–500-nm thickness to obtain the required dielectric strength during the anodic bonding.
2. Growth of an LPCVD silicon nitride layer of 100-nm thickness for photolithographic purposes.
3. Sputtering of a Pyrex thin film.
4. Bonding at 400°C with 50–200 V DC, with the negative electrode connected to the wafer that is coated with the sputtered thin film. Bonding time is typically 10 minutes for a 3″ wafer.

Again, clean and dust-free polished wafers must be used. A class 100 clean room environment is recommended. The strength of the bonds was measured by pulling the wafers apart. The bonding pressures found in these experiments scatter between 2 and 3 MPa. The second wafer may be coated with a thin film [30]. Surface layers of oxide (100 nm), nitride (100 nm), polysilicon (500 nm), and aluminium (1 μm) have been bonded successfully. In this case the bond strength seems to be somewhat smaller (0.5–2.5 MPa, with a tendency to larger bond strength when wafers were coated by nitride or polysilicon).

Esashi et al. [31] report on bonding using a sputtered thin film from a Corning glass #7570 target. This type of glass has a softening point at 440°C (as compared to Pyrex: 821°C), and the bonding can be done at room temperature. A thin film with a thickness between 0.5 and 4 μm was sputtered at 30% oxygen in an argon ambient at 6 mTorr. The bonding was completed within 10 minutes, requiring a minimum voltage of 30–60 V. Application of a

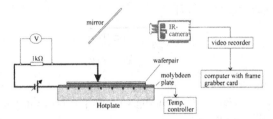

Fig. 4.10. Experimental setup used to follow the bonding process (from Berenschot et al. [32]).

pressure up to 160 kPa on the wafer package apparently supported the bonding process: at a pressure of 160 kPa the minimum voltage to achieve a bond dropped by a factor of two.

The bond strength was measured in a pull test. The results strongly depended on surface characteristics (roughness and contamination by particles), but the strength was found to be larger than 1.5 MPa. Most interestingly, the success of the bonding procedure did not seem to depend on the thickness of the thin film, although wafers with a sputtered glass thickness of less than 0.5 micron could not be bonded.

Recent work at the MESA Research Institute [32] with thin films sputtered from a Pyrex-like target (in fact, Schott 8330 glass was used) led to similar results: wafers could be anodically bonded at a voltage of 12 V at 450°C with thin films of a thickness down to 20 nm, without loss of yield or degrading of the bond strength. It appears as if there is no lower limit for the thickness of the thin film to get a bond. In view of Anthony's calculations [26], this is not too surprising; the bonding pressure becomes large if the layer over which the potential drops is small. Probably, in bonding with very thin Pyrex-like films, the space charge region is of minor importance.

In these bonding experiments an experimental setup, as shown in Fig. 4.10, was used. This setup allows infrared inspection of the bonding process in progress. The heating element is used as the infrared source. To obtain a better contrast in the infrared image, a 3″ molybdeen plate is inserted as a filter between the heater and the waferpair. A tilted silicon wafer acts as an infrared mirror. The bonding current was derived from voltage measurements over a known resistor.

Figure 4.11 shows the progress of bonding of two silicon wafers with a 1-μm thick glass layer sputtered on one of the two wafers. Figure 4.11(a) is the infrared image of the waferpair at 450°C before an electric field was applied. Apparently, two dust particles are present. In the experiment the electric field was gradually increased. In Fig. 4.11(b) it can be seen that bonding starts when the applied voltage reaches 8 V. Figures 4.11(c) and (d) show the situation 22 sec and 37 sec after bonding started. No further bonding is observed to occur after these 37 secs. By increasing the voltage in steps of 100 mV, bonding continues at 10 V (Fig. 4.11(e)). In Figs. 4.11(f), (g), and (h) it can be seen that it is possible to manipulate the bonding process manually, by applying an external force at the bonding front with a pin on the backside of the upper wafer. In this way it is possible to avoid large enclosures of trapped gas. Voids caused by dust particles remain, but their volume can be reduced in this way. At 10 V the whole wafer was bonded. Figure 4.11(i) shows an infrared image of the waferpair after cooling down to room temperature. In this case a halogen lamp was used as the infrared source. No difference can be observed between the infrared image at 450°C and the one at room temperature.

In this work the feasibility of the sputtered glass films as masking layers in anisotropic etching of Si in KOH solutions was also investigated [32]. It turned out that the glass films etch very slowly in KOH, with etch rates of ca. 0.2–0.3 μm/hr, depending on sputter

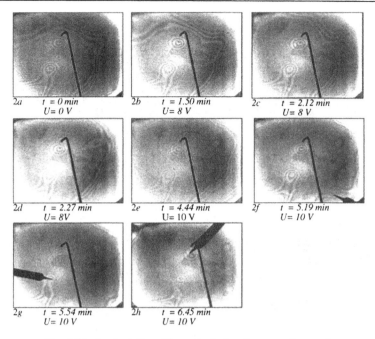

2a   t = 0 min   2b   t = 1.50 min   2c   t = 2.12 min
      U= 0 V   U= 8 V   U= 8 V

2d   t = 2.27 min   2e   t = 4.44 min   2f   t = 5.19 min
      U= 8V   U= 10 V   U= 10 V

2g   t = 5.54 min   2h   t = 6.45 min
      U= 10 V   U= 10 V

Fig. 4.11. In situ infrared images of a bond process (see text).

conditions and annealing procedure. This is comparabale with the etch rate of thermally oxidized silicon, which under the same conditions is 0.28 μm/hr. Hence, the glass films can be used as masking material, and it is possible to bond wafers and etch down to the bond interface with a reliable etch stop at the glass layer. This method has been used to verify the infrared pictures in Figs. 4.11(a) to 4.11(i); the result is shown in Fig. 4.11(j), which shows a picture of the same wafer after etching back one half of it in a KOH solution. It can be seen in the figure that the voids agree with the spots in the infrared image of Fig. 4.11(i). The reason that these voids can be seen after KOH etching is the following: when the bonding process is done at 1 atm, the unbonded regions contain trapped gas under a high pressure. After complete removal of the sputtercoated wafer in KOH, the etch process stops at the glass layer. Due to the gas pressure in the unbonded regions, these regions bulge out. Large unbonded regions crack and can be optically detected very easily because the silicon underneath the film is etched very fast. Fig. 4.12 shows a nice demonstration of these phenomena.

The above process of bonding and etchback was applied to fabricate "active joints" [28]. It was also used in a procedure to planarize wafers containing deep cavities in order to be able to perform additional lithographic steps in a so-called "sacrificial wafer bonding process" (see Fig. 4.13 [33]). Considering its low etch rate in KOH, Pyrex-like glass films can be used both as a bonding and as an etch stop layer. The sacrificial wafer bonding process was used to fabricate metal film electrical contacts across deep grooves that were used as stress-relief structures in a membrane pressure sensor [33, 34] (see Fig. 4.14).

4.4 Low-temperature bonding

Low-temperature bonding schemes work with intermediate thin films of various materials. Here we briefly summarize the results of four papers published on this issue. Esashi et al.'s low-temperature (anodic) bonding process [31] has been described above.

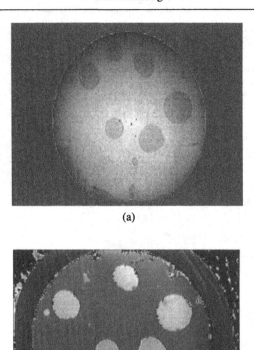

(a)

(b)

Fig. 4.12. (a) Infrared image of bonded wafer; (b) image in normal light, after etching back one of the wafers in a KOH solution (from Berenschot et al. [32]).

Legtenberg et al. [16] used a thin film of boron oxide for bonding at low temperature. Boron oxide becomes quite soft at 450°C. This opens the possibility of bonding a wafer containing features such as grooves, bond pads, and structures protruding out of the wafer surface to a flat wafer covered with boron oxide. When heated to 450–475°C, the thin film becomes soft and will deform, smoothly covering all features on the wafer, as long as they are not too thick (less than the thin film thickness). The films were grown in an atmospheric pressure CVD system. A gas mixture of $N_2:O_2:B_2H_6$ was used to grow the thin film. The effect of glass reflow, smoothly covering a Cr/Au pattern, is shown in Fig. 4.15.

Boron oxide has an important drawback: it is hygroscopic. The bonded wafers fall apart after some weeks if immersed in water, and after one year in atmospheric ambient.

Field et al. used boron-doped silicon dioxide for bonding [35]. This material also becomes soft at a reasonably low temperature (450°C), which is also the typical bonding temperature. First, a thermal oxide of 1 μm was grown and subsequently a thin layer (100 nm) of boron-containing glass was deposited. The latter was achieved in a solid-source drive-in diffusion furnace. Bonding must be done immediately after the deposition of the boron glass, because if allowed to stay in air, some crystalline material (possibly boric acid) forms at the surface of the boron glass. When this happens, bonding becomes impossible. During bonding,

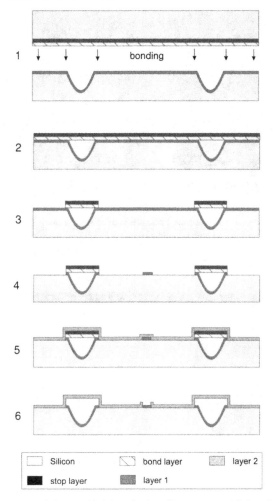

Fig. 4.13. Main steps of the sacrificial wafer-bonding process: (1) bonding of a wafer containing an etch stop layer; (2) etching of the top wafer until the stop layer is reached; (3) patterning of the stop and bond layer with conventional lithography; (4) original layer (e.g., layer 1) can be patterned; (5) new layers (layer 2) can be deposited and patterned; and (6) removal of stop and bonding layer (from Spierling et al. [33]).

a pressure was applied on the waferpair with a quartz piece of 120 g. The main extra requirements are

- The surfaces must be dust free and flat.
- No contamination with phosphorous must be present.

The bond seal appeared to be hermetic. The strength of the bond exceeded that of the material. This was tested by attempts to peel the wafers apart, which resulted in a fracture front right through the material, but not along the bonded surface. The most severe drawback of this technique is the sensitivity to phosphorous contamination.

Quenzer et al. used a thin film of sodium silicate as an intermediate layer [36]. The wafer surface was made hydrophilic by RCA1 cleaning. After drying, a diluted solution of sodium

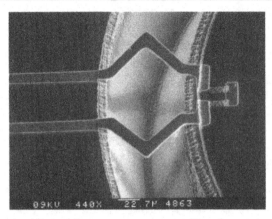

Fig. 4.14. Top view of a membrane pressure sensor with a deep corrugation; the aluminum bridges are electrical contacts. The patterning of the polysilicon strain gauges and the realization of aluminium contact bridges was done with the sacrificial wafer-bonding technique (from Spierling [34]).

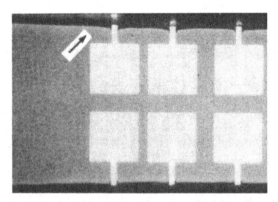

Fig. 4.15. Glass reflow of a 5-μm thick glass layer from the top wafer over SiN and Cr/Au patterns after bonding at 475°C. Dimensions of the bond pads are 150×50 μm^2 (from Legtenberg et al. [16]).

silicate was spun on the wafer surface. The wafers were contacted, and the package was annealed at 200°C. The spinning resulted in film thicknesses between 3 and 100 nm. It turned out that it is possible to bond wafers covered with a native oxide, thermally grown oxide, and silicon nitride. Bonding energies have been determined with the aid of Maszara et al.'s method [17], resulting in surface energies between 1 and 3.2 J/m^2.

Den Besten et al. used thin polymer films for bonding experiments on oxidized silicon substrates [37]. Advantages of this technique are a low bonding temperature (most polymers polymerize at temperatures below 150°C) and relatively high bond strengths. A disadvantage of polymer bonding is that generally no hermetic seal is obtained.

Den Besten et al. tried negative photoresist, polyimide, and epoxy resin, which were applied to the wafers by spin coating. The epoxy layers were spun from a dilute solution of epoxy resin in toluene and NMP (n-methylpyrrolidon). The photoresist layers had a thickness of 1.1 μm, while the other thin films were only 100–200-nm thick. Curing was done at 130°C (resist and polyimide) or 90°C (epoxy resin). The highest bond strengths

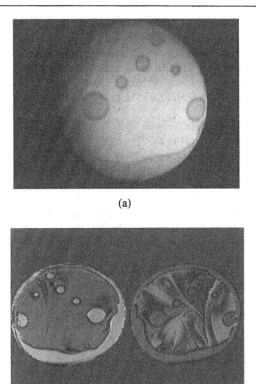

(a)

(b)

Fig. 4.16. (a) Infrared image of two wafers connected by a polymer bond; (b) picture of the two wafers after splitting. Note that splitting occurred in the resist layer (from Spierling [34]).

$(130–170 \text{ kg/cm}^2)$ were obtained with the photoresist layer, as revealed by peel tests. In the bonding experiments with negative photoresist, den Besten et al. observed that the use of HMDS (hexa-methyl disilazane), which is normally used as an adhesion promoter, led to a factor of 2 decrease in bond strength. Poor results were obtained with polyimide; some improvement was obtained when APS (γ-aminopropyl silane) was used as a primer. The results with epoxy were good (bond strength $80–100 \text{ kp/cm}^2$), with and without APS as a primer.

Similar results were reported by Spiering et al. [33], who have tried the before-described "sacrificial wafer bonding process" (Fig. 4.13) with negative photoresist type 3IC (Olin Ciba Geigy) with a prebake (before contacting the wafers) and a postbake step at 150°C in N_2 atmosphere. The prebake step was necessary to evaporate most of the solvent; this step had to be done in a pure N_2 atmosphere in order to prevent hardening of the top layer due to oxidation of the resist. Figure 4.16 shows images of the results.

References

[1] J. Haisma, G. A. C. M. Spierings, U. K. P. Biermann, and A. A. van Gorkum, Appl. Opt. **33**, 1154 (1994).

[2] W. P. Maszara, *Proc. 1st Int. Symp. Semiconductor Wafer Bonding*, eds. U. Gösele et al. (The Electrochemical Soc., Pennington, NJ, 1992), p. 3.

[3] Y. Bäcklund, K. Hermannsson, and L. Smith, *Proc. 1st Int. Symp. Semiconductor Wafer Bonding*, eds. U. Gösele et al. (The Electrochemical Soc., Pennington, NJ, 1992), p. 82.

[4] Y. Bäcklund, K. Ljungberg, and A. Söderbärg, J. Micromech. Microeng. **2**, 158 (1992).

[5] L. G. Sun, J. Zhan, Q. Y. Tong, S. J. Xie, Y. M. Caim, and S. J. Lu, J. Physique Colloq. **C49**, 79 (1988).

[6] M. Shimbo, J. Inst. Electron Inf. Commun. Eng. **70**, 593 (1087).

[7] P. W. Barth, Sensors and Act. **A21–A23**, 919 (1990).

[8] G. Kissinger and W. Kissinger, Sensors and Act. **A36**, 149 (1993).

[9] R. Stengl, T. Tan, and U. Gösele, Jpn. J. Appl. Phys. **28**, 1735 (1989).

[10] A. Söderbärg, personal communication (1992).

[11] R. Stengl, K. Ahn, and U. Gösele, Jpn. J. Appl. Phys. **27**, L2364 (1988).

[12] Y. Bäcklund, L. Rosengren, B. Hök, and B. Svedberg, Sensors and Act. **A21–A23**, 58 (1990).

[13] K. Petersen, P. Barth, J. Poydock, J. Brown, J. Mallon Jr., and J. Bryzek, Techn. Digest IEEE Solid State Sensor and Actuator Workshop, Hilton Head Island, SC, June 6–9, 1988, p. 144.

[14] C. Harendt, B. Höfflinger, H.-G. Graf, and E. Penteker, Sensors and Act. **A25–A27**, 87 (1991).

[15] Y. Bäcklund, personal communication (1992).

[16] R. Legtenberg, S. Bouwstra, and M. Elwenspoek, J. Micromech. Microeng. **1**, 157 (1991).

[17] W. P. Maszara, G. Goetz, A. Caviglia, and J. B. McKitterick, J. Appl. Phys. **64**, 4943 (1988).

[18] B. Muller and A. Stoffel, Techn. Digest Micro Mechanics Europe 1990, Berlin, Germany, Nov. 26–27, 1990, p. 87.

[19] R. K. Iler, *The Chemistry of Silica* (Wiley, New York, 1979).

[20] J. Del Bene and J. A. Pople, J. Chem. Phys. **52**, 4858 (1970).

[21] W. P. Maszara, B.-L. Liang, A. Yamada, G. A. Rozgonyi, H. Baumgart, and A. J. R. de Kock, J. Appl. Phys. **69**, 257 (1991).

[22] S. Bengtson and O. Engström, J. Physique Coll. **49**, C4, 63 (1988).

[23] S. Bengtson and O. Engström, J. Appl. Phys. **66**, 1231 (1989).

[24] G. Wallis and D. I. Pomerantz, J. Appl. Phys. **40**, 3946 (1969).

[25] A. Hanneborg, Proc. IEEE Workshop on MEMS, Nara, Japan, Jan. 30–Feb. 2, 1991, p. 92.

[26] T. R. Anthony, J. Appl. Phys. **54**, 2419 (1983).

[27] Product information brochure of HOYA Europe B.V. (London).

[28] M. Elwenspoek, L. Smith, and B. Hök, J. Micromech. Microeng. **2**, 221 (1992).

[29] A. D. Brooks and R. P. Donovan, J. Electrochem. Soc. **119**, 545 (1972).

[30] A. Hanneborg, M. Nese, H. Jacobsen, and R. Holm, J. Micromech. Microeng. **2**, 117 (1992).

[31] M. Esashi, A. Nakano, S. Shoji, and H. Hebiguchi, Sensors and Act. **A21–A23**, 931 (1990).

[32] J. W. Berenschot, J. G. E. Gardeniers, T. S. J Lammerink, and M. Elwenspoek, Sensors and Act. **A41–A42**, 338 (1994).

[33] V. L. Spiering, J. W. Berenschot, M. Elwenspoek, and J. H. J. Fluitman, J. Micro Electro Mech. Syst. **4**, 151 (1995).

[34] V. L. Spiering, Ph.D. Thesis, University of Twente, The Netherlands (1994).

[35] L. A. Field and R. S. Muller, Sensors and Act. **A21–A23**, 935 (1990).

[36] H. J. Quenzer and W. Benecke, Sensors and Act. **A32**, 340 (1992).

[37] C. den Besten, R. E. G. van Hal, J. Munoz, and P. Bergveld, Proc. IEEE Workshop on MEMS, Travemünde, Germany, Feb. 4–7, 1992, p. 104.

[38] K. Ljunberg, Ph.D. Thesis, University of Uppsala (1995).

[39] C. Gui, H. Albers, J. G. E. Gardeniers, M. Elwenspoek, and P. V. Lambeck, *Microsystems Technologies* (Springer Verlag, Berlin, Germany), p. 122.

[40] J. Haisma, G. A. C. M. Spierings, T. M. Michielsen, and C. L. Adema, Philips J. Res. **49**, 23 (1995).

[41] S. Sanchez, C. Gui, and M. Elwenspoek, J. Micromech. Microeng. **7**, 111 (1997).

[42] Chenqun Gui, Ph.D. Thesis, University Twente, The Netherlands (1999).

5

Examples and applications

5.1 Introduction

This chapter on applications of anisotropic etching and bonding concentrates on the description of the technology required to fabricate the devices shown. How the devices work is only explained qualitatively. Quantitative modeling of the devices is, of course, indispensable for micromechanics. There are several philosophies for the role simulation plays in the design process. Some people prefer to do extensive simulation work before they start the fabrication process. We prefer a different route. We prefer to start with analytical models for the structure taking, as far as possible, only first approximations into account. The next step is the fabrication and extensive characterization of the performance of the device. For optimization, a process that results in a second generation of devices, the simulation is indispensable.

In this chapter, wafer-bonding technology is mentioned as often as anisotropic etching. Probably no device can be fabricated without the use of wafer bonding.

5.2 Membranes

Membranes are used in many types of devices. The most important ones are pressure sensors, microphones, and pumps. In pressure sensors a pressure difference across the membrane causes a deflection which is a measure of the pressure. A microphone can be viewed as a dynamic pressure sensor. In pumps the membrane is deflected to displace a liquid. Some valve designs also make use of membranes. An example for an isotropically etched membrane is given in Fig. 5.1 [1].

5.2.1 Pressure sensors

The design of a pressure sensor depends strongly on the choice of the readout. In the literature there are the four suggestions for readout: capacitive, piezoresistive, resonant, and optic.

Capacitive pressure sensors make use of the fact that the deflection of a membrane changes the capacitance, which is measured. The capacitance of a micromechanical structure is rather small,

$$C = \varepsilon_0 \int \frac{dx\,dy}{d + \delta(x, y)}.$$ (5.1)

Fig. 5.1. View into an etched cavity onto the backside of a membrane [1].

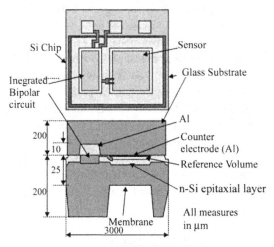

Fig. 5.2. Schematic of an integrated capacitive pressure sensor (from Heuberger [1], original work from Sander et al. [2]).

Here, d is the electrode spacing, and δ is the deflection of the membrane. The integral is evaluated over the entire surface of the capacity. It is clear that C is a nonlinear function of the deflection. But if the deflection is small enough, the function can be linearly approximated. d must not be too large, since otherwise the capacity becomes unstable (this occurs typically if $\delta \gg d/3$). A typical value for a micromechanical capacitance is 10–100 pF, and its change by a deflection is a fraction of this value. It is clear that one has to minimize the gap spacing.

The thickness of membranes (h) used in micromechanical sensors is of the order of 1 μm or more, practically never thinner than the gap spacing. This implies, since $\delta \ll d$, that we automatically have $\delta \ll h$, and we can neglect the cubic term in the pressure dependence of the deflection.

The realization of capacitive pressure sensors requires a wafer-bonding step. A schematic drawing after Sander et al. [2] is given in Fig. 5.2. The fixed counterelectrode (Al) has been deposited on a Pyrex wafer and the deflecting electrode has been realized by an n-doped epitaxially grown silicon layer. The membrane of a thickness of 25 μm has been etched from the backside by anisotropic etching using the electrochemical etch stop. In this work, a bipolar integrated circuit has been integrated on a chip for a first amplification of the small

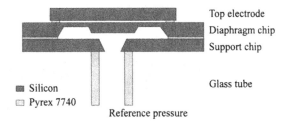

Fig. 5.3. Schematic of a capacitive pressure sensor using a bossed diaphragm and sputtered thin films of Pyrex for anodic bonding.

signals. This particular sensor design has a sensitivity ($[\Delta C/C]/p$) of 0.35 bar^{-1}. A similar design of a capacitive pressure sensor using anodic bonding of a silicon wafer to a Pyrex wafer has been used by Puers et al. [3].

Hanneborg and Ohlckers proposed the use of a bossed diaphragm and anodic bonding using a sputtered thin film of Pyrex, this way minimizing thermal drifts and slow stress relief [4]. A schematic drawing is shown in Fig. 5.3.

Figure 5.4. shows the fabrication sequence of an alternative design using anodic wafer bonding, as proposed by Chau et al. [5]. They fabricated a pressure sensor small enough for use in a cardiovascular catheter, with an outer diameter of 0.5 mm. The chip with the deflecting silicon membrane was bonded on a glass substrate. It was micromachined with concentrated HF solution using a goldmask in order to realize 30-μm deep groves for lead attachment and to realize the metallization. An overall view is given in Figs. 5.4(a) and (b). The process sequence for the fabrication of the membrane is given in Fig. 5.4(c). After etching a shallow cavity (3 μm), in the end providing the cavity that defines the capacity, two boron indiffusion steps are involved: a deep indiffusion (12 μm) to provide the etch stop for the support of the membrane and a shallow indiffusion (1.2 μm) for the etch stop of the membrane itself. A dielectric is deposited on the membrane to protect the membrane against shorts. After this the wafer is bonded using anodic bonding. The last step is etching all undoped silicon in EDP, leaving the membrane and its support that consists entirely of heavily boron-doped silicon and the dielectric layer.

The last example of a capacitive pressure sensor we give here concerns a device that is planned to be implanted in the human eye [6]. It is based on the idea that a change of the capacitance changes the resonance frequency of an LC circuit. The implanted sensor is a passive structure, and the resonance frequency is measured remotely with an external, inductively coupled oscillator. In the fabrication of this device, use was made of direct silicon wafer bonding ("fusion bonding"). The authors report on difficulties combining the boron etch stop with direct bonding. Direct bonding is only possible with perfectly flat silicon wafers, and the indiffusion process seemingly degrades the surface flatness too much. As a consequence, the authors used a time etch stop. A schematic of the device and a microphotograph is shown in Fig. 5.5. The silicon wafer containing the membrane has been etched from both sides and bonded to the second silicon wafer at the narrow 100-μm wide rim. The SEM photo shows the bond region and a part of the membrane.

Piezoresistive pressure sensors use piezoresistive strain gauges for the determination of the deflection of the membrane. The piezoresistive elements can be diffused in the silicon or deposited (as polycrystalline silicon) on the chip. Typical gauge factors are $\Delta R/\varepsilon = 20$. Since it is not necessary to have an electrode very close to the membrane, use of wafer-bonding technology is not a prerequisite for the fabrication of the sensor chip. However,

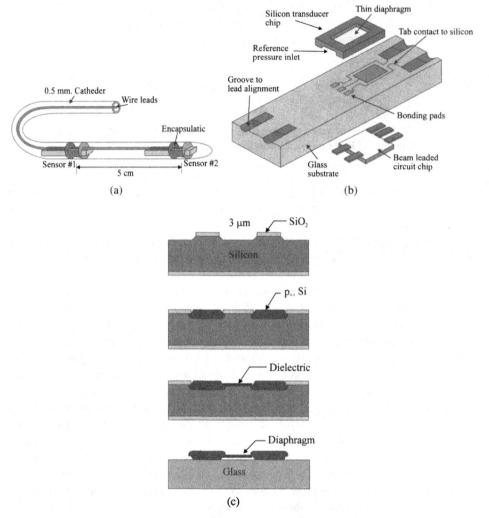

Fig. 5.4. Overall view of (a) the catheter, (b) construction of the pressure sensor, (c) fabrication sequence of the silicon transducer chip (from Chau et al. [5]).

the packaging of the membrane requires some mechanical connection that is often realized using wafer bonding. A schematic of the piezoresistive pressure sensor is shown in Fig. 5.6.

An interesting alternative of the fabrication of single crystalline silicon membranes is suggested by Petersen et al. [7]. The process also involves anisotropic etching of a cavity and direct silicon bonding.

In this process, cavities in the form of negative square pyramids are etched in one wafer in an anisotropic etchant. Wafers 525-μm thick are used, and the pyramidal holes are 250-μm square and 175-μm deep. The other wafer consisted of a p-type substrate and an n-type epitaxial layer. The thickness of this layer corresponds to the thickness of the membrane. After bonding, the wafer is thinned down to the electrochemical etch stop provided by the n-epi layer. Piezoresistors were implanted into the membrane. The process sequence is shown in Fig. 5.7, together with a comparison of conventionally etched membranes. The

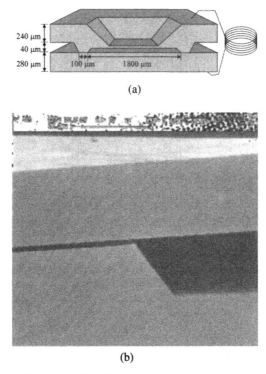

(a)

(b)

Fig. 5.5. (a) Schematic cross section of the intraocular pressure sensor. (b) SEM photo of the membrane showing the bonded region. (The region marked with 100 μm in (a) upside down.) (From Bäcklund [6])

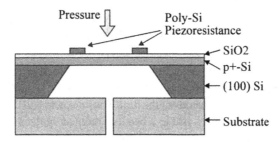

Fig. 5.6. Schematic cross section of a pressure sensor with polycrystalline silicon piezoresistive elements (from Heuberger [1]).

great advantage of this technology is that the support of the membranes can be much smaller, which can be critical, e.g., in medical applications.

Two types of *resonating pressure sensors* have been described in the literature. One type consists of a vibrating membrane, the resonance frequencies of which depend on a distributed load such as applied by a pressure difference across the membrane, and the other uses resonating microbridges, integrated in the membrane, as strain gauge.

The *vibrating membrane* pressure sensor is described in various versions. The common element is the anisotropically etched membrane. This is either fabricated using a time-controlled etch stop or a boron etch stop. Variations concern the way that it is used to excite and to detect the vibrations. Smits et al. [8] proposed the use of a ZnO thin film on top of

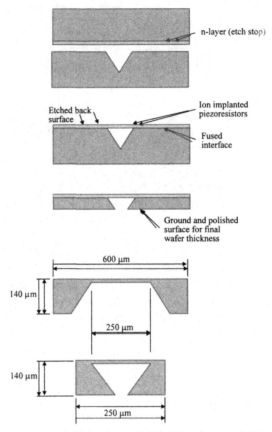

Fig. 5.7. Fabrication sequence of silicon fusion bonded membrane suitable for ultra miniature pressure piezoresistive sensors (from Petersen et al. [7]).

the membrane. ZnO can be sputtered in a magnetron sputterclock in an oxygen atmosphere using a Zn target (reactive sputtering) [8]. ZnO is piezoelectric. A voltage across the thin film leads to a change of the dimensions of the film. The thickness and the lateral dimensions of the film change. The latter gives rise to a bending moment in the membrane that can be used to excite bending vibrations of the membrane.

Lammerink [9] used a thermal excitation principle. Heat dissipated in a resistor on top of the membrane introduces a temperature gradient across the membrane and, consequently, a stress gradient due to the thermal expansion. If the frequency of an AC current superimposed on a DC current matches the resonance frequency of the membrane, it will be excited with a sufficiently large amplitude to detect the vibration. For the detection, optical means and piezoresistive strain gauges have been used.

A similar structure has been described by Bouwstra et al. [10] in a thermally excited resonating membrane mass flow sensor. Here the heat dissipated to excite the vibration is also used for a thermo anemometer, thereby turning the disadvantage of the thermal excitation of the structure (stress, change of resonance frequency) into an advantage. Gas streaming along the heated membrane carries away some heat, lowering the temperature of the structure and the compressive stress and thereby increasing the resonance frequency. A schematic and a microphotograph of the sensor membrane are given in Fig. 5.8. The structure can be used as a mass flow sensor and as a differential pressure sensor.

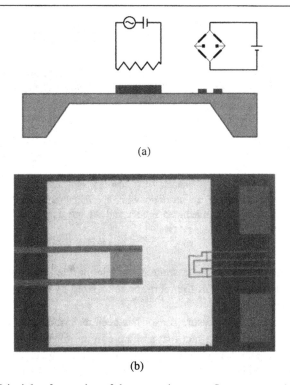

(a)

(b)

Fig. 5.8. (a) Principle of operation of the resonating mass flow sensor and resonating membrane pressure sensor: Indicated are the center resistor for thermal excitation of a bending mode vibration and piezoresistors at the edge of the membrane for detection. (b) Photograph of the 2×2 mm $\times 2$ μm p$^+$ silicon membrane, showing the center resistor for excitation and the piezoresistors for detection of the vibration (from Bouwstra et al. [10]).

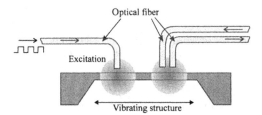

Fig. 5.9. Optical proximity sensors for the use of detectors of the vibration of a membrane (from Lammerink and Gerritsen [11]).

The concept of the vibrating membrane pressure sensor has also been realized as an all-optic sensor. Thermal excitation can be accomplished also by absorption of chopped light. The detection of the vibration can be realized by optical proximity sensors, which are just pairs of optical fibers (see Fig. 5.9) [11].

A more elegant way is the integration of the optical fibers into the chip, as shown in Fig. 5.10 [11]. Here the membrane has been etched anisotropically from both sides of the silicon wafer using a time-controlled etch stop. The center of the membrane has also been masked, and the etch process results in a mesa-type structure. On either side of the membrane, V grooves have been etched that serve as a guide for the optical fibers; the mesa

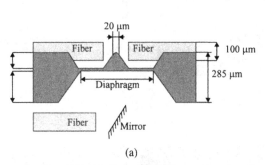

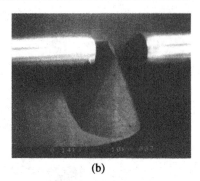

(a)   (b)

Fig. 5.10. (a) Schematic of an all-optic vibrating membrane pressure transducer. (b) SEM photo of the micromachined shutter and the integrated optical fibers (from Lammerink [10]).

on the vibrating membrane acts as a shutter, modulating the intensity of the transmitted light (see Fig. 5.10). Activation of the vibration is realized by chopped light impinging at the rear side of the membrane.

A very elegant resonating pressure sensor has been described by Stemme and Stemme [12]. The design is based on the fact that a change in the form of a body will change its moment of inertia and therefore its resonance frequencies. The general design idea is shown in Fig. 5.11(a). The resonator consists of two diaphragms bonded together so that they form a cavity. Any change in the difference of the pressure inside and outside the cavity will change the resonance frequency. The cavity vibrates in a quadrupole-like mode so that the node lines pass the four support beams. The center of inertia does not move, and there is only a very small coupling of a moment to the support. The damping is dominated by viscous flow of air around the structure.

In Fig. 5.10(b) we show the basic fabrication sequence of the sensor. The wafers are etched in two steps before they are bonded. In the first step the membrane and part of the trench separating the cavity from the support were etched in EDP. Then an oxide was grown in order to protect the membranes from the second etch step, in which the etching of the trenches was completed. Finally, the wafers were bonded using direct silicon bonding. A photograph of the completed sensor is shown in Fig. 5.11(c). The holes at the edges are introduced to minimize air damping.

Vibrating membrane (pressure) sensors have an important disadvantage; the resonance frequency is also dependent on the density of the surrounding medium. Furthermore, the medium is allowed to interact directly with the resonator. Absorption of chemicals and dust, as well as corrosive effects, change the mass of the resonator and therefore cause a drift in the readout of the sensor.

The *integration of resonating microbridges in a membrane* is one of the most exciting developments in micromechanics of the past few years. Two ways of realizing such devices have been described; one of them uses surface micromachining and is described in detail in Chapter 7. The other technology makes use of epitaxial growth of doped silicon and the dopant selectivity of anisotropic etching in electrochemical etching. This pioneering work is due to Ikeda et al. at Yokogawa Electric [13–15].

The fabrication process is shown in Fig. 5.12. The process starts with an HCl dry etch at 1050°C in an epi reactor through a hole in an oxide mask (Figs. 5.12 (a) and (b)). Selective

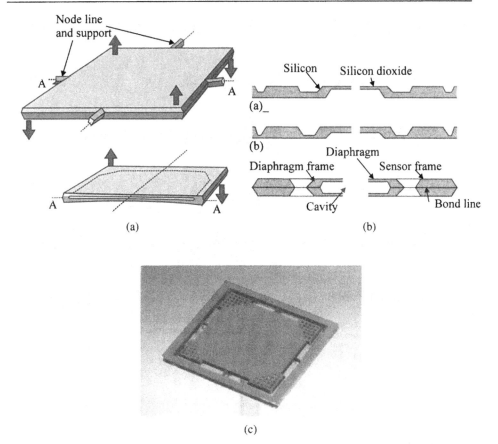

Fig. 5.11. (a) Design of a balanced resonant pressure sensor. Heavy arrows indicate the mode of vibration. (b) Fabrication sequence of the balanced pressure sensor. (c) Photograph of the complete resonant pressure sensor.

epitaxial growth of silicon of p^+, $p^{++\ddagger}$, and p^+, p^{++} doping follows, as indicated in Figs. 5.12 (c), (d), (e), and (f). These steps are carried out in the same epireactor, just by changing the concentration of B_2H_6. The next step is stripping of the oxide in HF (Figs. 5.12(g)), followed by selective electrochemical etching in hydrazine (5.12(h)). In this etching step, the n substrate is pacified against the etchant by the electrochemical potential and the B^{++} structures by the boron etch stop. After this process one has a microbridge covered by a cap, both from single crystalline heavily B-doped silicon. The cap is sealed by growth of an epitaxial layer (5.12(i)), and annealing in a nitrogen atmosphere leads to a pressure inside the cavity of 1 mTorr (5.12(j)).

In order to obtain a structure that works as a pressure sensor, one has to etch a membrane from the backside of the wafer, a step rather trivial after the process just described.

In this particular device, the resonance of the encapsulated beam is activated by a Lorenz force and detected by measuring the inductance.

$\ddagger$ Here p^+ and p^{++} refer to medium and high doping levels, respectively.

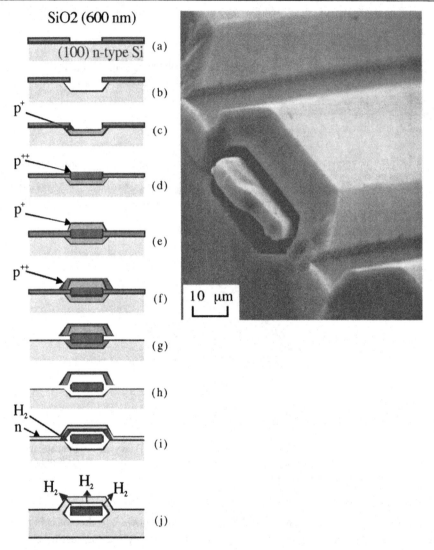

Fig. 5.12. Fabrication process of vacuum sealed resonating microbridges (from Ikeda et al. [13]). The photograph shows the final product. (a) Photo lithography. (b) HCL etching. (c) Selective epi (p^+). (d) Selective epi (p^{++}). (e) Selective epi (p^+). (f) Selective epi (p^{++}). (g) Removing SiO_2. (h) Selective etching. (i) Epi (n). (j) Annealing in N_2.

5.2.2 Miniature microphones

A lot of work on silicon technology-based miniature microphones is done by Bergveld's group at the University of Twente. The essential parts are a charged capacity with a membrane that is deflected by the fluctuating pressure at audio frequencies and a back-plate with apertures to allow for flow of air in order to reduce the viscous damping of the system. The charge of the capacity can be provided by a built-in charge (a so-called electret) or by an external voltage. First designs of silicon microphones used a mylar foil to provide the flexural membrane and Teflon as an electret material.

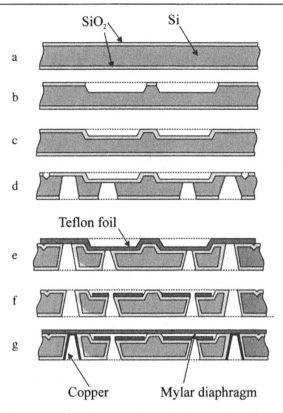

Fig. 5.13. Process sequence for fabrication of a silicon electret microphone (after Sprenkels [16]).

The processing steps for the silicon electret microphone [16] are shown in Fig. 5.13. Since the backplate of the device has to be processed from both sides, a double-sided polished wafer ($\langle 100 \rangle$) is preferable, and the process must start with the definition of alignment marks on either side of the wafer. Using a double-sided mask aligner is the most convenient method, but in this particular work there was no such instrument available. Therefore a specially designed holder was used containing the masks with the alignment marks. Wafers placed in this holder can be illuminated from both sides, providing there are alignment marks on both sides.

In Fig. 5.13(b), the 32.5-μm deep cavities are etched in KOH. After growth of thermal oxide (leading to an oxide of 1 μm in the cavities and of 1.5 μm elsewhere), holes are etched from the backside, providing holes for the airflow out of the cavity (D) and to attach the mylar foil during one of the last steps (G). This etch stops at the oxide layers. Simultaneously, V grooves are etched in the frontside of the wafer in order to break the individual chips out of the wafer after processing. The depth of this etch is defined by the opening in the oxide mask. After this all oxide is stripped and regrown. In step E, the Teflon foil is attached to the wafer. For this procedure, which is by no means easy due to the bad adhesion of Teflon to practically all materials, we refer to Sprenkels's thesis. The Teflon foil is patterned in a dry etch (oxygen plasma, see Chapter 9) and charged. For the charging procedure and the attachment of the mylar foil, we again refer to Sprenkels [16].

A variation of the process scheme is shown in Fig. 5.14, leading to a condenser microphone with electromechanical feedback [17].

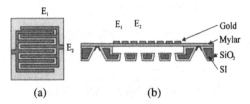

(a)   (b)

Fig. 5.14. Design of a silicon condenser microphone which can be used for feedback. (a) Top view showing the electrode configuration for actuating the membrane. (b) Cross section of the microphone (from van der Donk [17]).

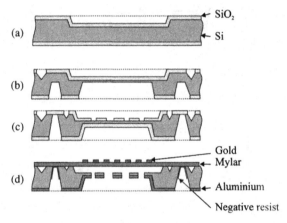

Fig. 5.15. Process sequence for the fabrication of a silicon condenser microphone with a thin airgab and a backplate with many holes (64 in van der Donk's work) (from van der Donk [17]).

The electromechanical feedback is used to increase the bandwidth of the microphone. This requires an actuator that counteracts the deformation of the membrane, keeping the capacity constant. The required voltage at the actuator serves as a signal. In Fig. 5.14 we show the design of the microphone. The combstructure in Fig. 5.14(a) serves as the actuator.

In Fig. 5.15 we show the key process steps to fabricate a condenser microphone with a larger number of holes in the backplate of the microphone. The number of holes is limited rather by the inclined (111) faces during anisotropic etching than by the size of the holes ($40 \times 40 \ \mu m^2$ in this case). If one etches into a thin backplate (in this case 30-μm thick) the patterns on the etchmask can be much smaller, leading to a much higher hole density. After etching the membrane from the backside of the wafer (Fig. 5.15(b)) an oxide is grown for the protection of the backside against KOH. The holes are etched from the frontside. This is necessary because photolithography becomes impossible if the wafer is not flat enough; the backside contains features that are too deep, making the deposition of a homogeneous layer of photoresist by spinning impossible.

5.2.3 Application of membranes in micro liquid handling devices

Micro liquid handling devices are probably the first complex microsystems that will enter the market. We believe that this field is the most important one for microactuators.

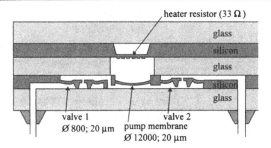

Fig. 5.16. Design of a micropump with two checkvalves. The actuation is thermopneumatic (from van de Pol et al. [18]).

Membranes are inevitable structures in micropumps. An example is given in Fig. 5.16. This design [18], based on an earlier design [19], even contains four membranes. One of them is just a carrier of a resistor for heating a gas cavity in order to actuate the pump membrane (in the center of Fig. 5.16). The other two membranes form constructive elements of checkvalves, which are opened and closed by deflecting the pump membrane. The pump membrane and the valves are all etched in one wafer (use of double-sided polished wafers is essential in this design because the wafer has to be bonded to Pyrex wafers on both sides). The pump membrane and the flexible elements are located in the center of the wafer; they are etched in KOH from both sides. The thickness of the membranes is defined by the etching time.

The membrane carrying the heater resistor is also anisotropically etched. First the membrane is etched from the backside of the wafer until a thickness twice the desired thickness (20–30 μm) is reached, then trenches are etched from the frontside. Since the backside is not protected, it etches simultaneously with the trenches. If the trenches break through the membrane, the latter has been etched to approximately half of its initial thickness. The process results in a membrane suspended by four thin beams. In this way the heat transfer from the heating element to the wafer is minimized.

The sharp edges that result from the anisotropic etching form a problem for the handling of the device. At sharp edges one always gets a stress concentration, and cracks easily nucleate at these places. The sharp edges could be avoided by using an isotropic etchant or by using a dry etch process for the trenches.

The valves consist of circular diaphragms comprising a flexural outer ring and a rigid inner sealing ring. Because of the flexibility of the outer ring, a pressure acting on one side of the valve will deflect the sealing ring, opening or closing the valve. On the inner ring there is a thin oxide layer, giving rise to a small pretension of the valve. This oxide layer is of vital importance in the production process because it prevents bonding to the Pyrex wafer in the final assembly step.

Note that ring, valve diaphragm, and pump membrane are circular. This is essential for the good sealing of the valves if the pressure difference is reversed. Yet, the structures have been fabricated using anisotropic etching (KOH). The circular structures result because of the relatively large size of the lateral structures (ca. 1 cm diameter). The lateral features are much larger than the thickness of the whole structure (300 μm), and the etched membrane more or less follows the shape of the circular opening in the etch mask.

Other membrane pumps have been described by Smits [20] and Shoji et al. [21]. The size of the pumps described so far is defined by the size of the pump membrane plus

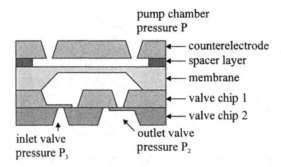

Fig. 5.17. Pump design of Zengele et al. (from Zengele et al. [21]).

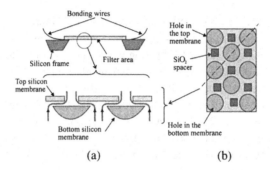

Fig. 5.18. Filter design and principle of operation. (a) Schematic cross-sectional view of the two membranes. (b) Top view showing the apertures and spacers.

the size of the valves. Since they are placed in one wafer, the devices are certainly not of μm size, but of several cm^2 × 1 mm.

A design leading to a more compact micropump has been proposed by Zengerle et al. [21]. The central idea is bonding the wafer containing the valves on the pump chamber (see Fig. 5.17). In Zengerle's design the actuation was first to be electrostatic, but in the paper the authors report only results with external hydrostatic actuation. The valves used by Zengerle et al. were first introduced by Tiren et al. [22].

An alternative could be a combination of van de Pol's thermopneumatic actuator [24] and L. Smith's [25] valves. Those designs should allow a downscaling of micropumps to 1 mm^3 devices.

Perforated membranes such as those we encountered in van der Donk's microphone are suggested as filters and in micromixers. All designs of micropumps suggested so far are such that particles (e.g., dust or coagulation) are fatal for the pumps, mainly because the valves will be jammed.

In all micro liquid handling systems, filters play a vital role. A clever design of a filter has been presented by Kittilsland et al. [26]. The general idea is shown in Fig. 5.18. The filter comprises two perforated membranes sandwiching a perforated spacer layer, the thickness of which defines the filter size.

The process sequence is as follows (Fig. 5.19(a)): on a ⟨100⟩ wafer an oxide layer is deposited, then a polysilicon layer, then again an oxide layer. The oxide layer is patterned with circular holes, through which the other thin films are etched. The next step is deep boron doping of the silicon wafer through the holes for a boron etch stop. The concentration profile is such that the high-doped region overlaps the holes. The undoped silicon is etched

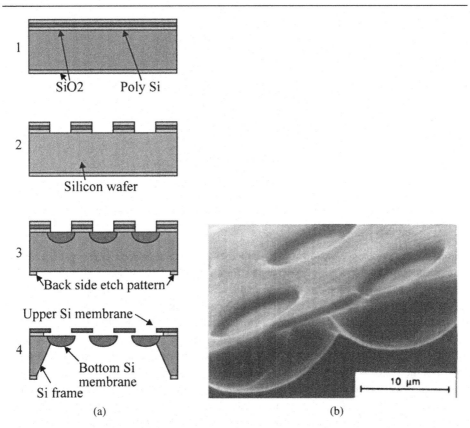

Fig. 5.19. (a) Fabrication sequence of the filter structure. (b) SEM photo of a cleaved part of a 50-nm filter membrane showing the perforated polysilicon membrane and the heavily B-doped hemispheres of the bottom membrane.

away in EDP. After that the oxide is etched so far that in the space between the holes in the thin film stack and the holes in the remaining silicon, there are oxide spacers left.

The last example of a perforated membrane deals with a micromixer.

Microfluidics typically have Reynolds numbers of the order of 1 or smaller; this means that there are no chances for vortices or any turbulence. Mixing of liquids is only possible via diffusion. Typical diffusion times over a distance of 300 μm is still of the order of 1 min or more, so this process is too slow in micro liquid handling with flow velocities of the order of mm/s or more. These considerations lead us to think of the use of perforated membranes for the injection of one liquid into the other [27]. In this way the required diffusion length is of the order of the distance between the apertures, and the overall mixing time is sped up considerably. A schematic drawing of the device is shown in Fig. 5.20.

5.3 Beams

Beams are used in several ways in micromechanical devices, for example, as structural elements for suspending a membrane, this way minimizing transfer of heat generated on the membrane to the substrate (Fig. 5.16, Ref. [18]). They are also used as suspension beams for masses in accelerometers, as resonating strain gauges in mechanical sensors, and as carriers for heating elements and temperature sensors in flow sensors. Finally, they can

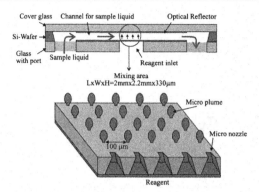

Fig. 5.20. Design of a micromixer with a perforated membrane (from Miyake et al. [27]).

be used as actuators, e.g., as flow directive elements in valves, and as joints and members in microrobot limbs. We shall give examples for all these applications.

A design of an *accelerometer* is given in Fig. 5.21 [27]. Similar to the pressure sensor of Petersen et al. ([7], Fig. 5.7), besides anisotropic etching, fusion bonding is an essential fabrication step. This technique permits great freedom in the production process. In the cross section shown in Fig. 5.21(a), the cavities described as "gap 2" are etched before bonding. After bonding the following fabrication steps can be performed: etching the seismic mass into the lower wafer; etching a membrane by thinning the upper wafer; etching the beams out of the membrane for suspension, bending, and overload protection; fabrication of the piezoresistors and the metallization. This device has two special aspects. One is the overload protection. The central seismic mass is surrounded by the supporting frame (see Fig. 5.21(b)). The interdigited beams for overload protection are present around three sides of the mesa, and cantilever beams suspending the mass are located at the fourth side. This can be seen in the SEM photograph in Fig. 5.21(c).

The other special aspect is the fact that this device can be fabricated with controlled damping. The damping is viscous, and the main contributions come from the narrow gaps indicated in Fig. 5.2(a). The size of these gaps is such that the damping is critical.

The overload protection of the device is better than 1000 G.

An alternative design [29] of an accelerometer is shown in Fig. 5.22. Again, use is made of fusion bonding and anisotropic etching. The readout principle is capacitive, similar to other proposals of accelerometers [30]. The seismic mass is supported by eight beams located at the edges of the mass. For the etching of the beams, the boron etch stop technique has been used. The high symmetry of the design serves as a small cross sensitivity, so that acceleration in the lateral direction of the chip is nearly invisible.

An interesting improvement of the mechanical strength of the suspension beams in accelerometers has been proposed by Koide et al. of Hitachi [31]. One always has stress concentration at sharp corners, and anisotropic etching easily leads to sharp corners. Koide et al. proposed using a layered mask technique to resolve this problem. By growing an oxide layer of varying thickness it is possible to perform different etch steps one after the other without the need of photolithographic steps (which are difficult to realize or impossible if there are structures on the wafer). An example is shown in Fig. 5.23. The technique has been developed for the fabrication of mesas supported by beams in accelerometers. The etching steps are performed in succession after removing the thinnest mask layers. In this etch

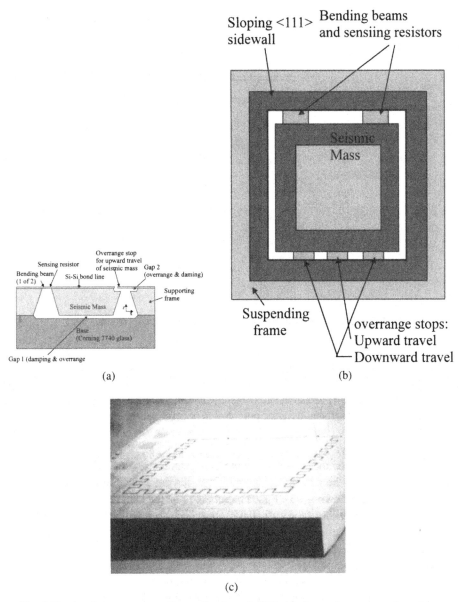

Fig. 5.21. Accelerometer according to Barth et al. [28]. (a) shows the cross section of the design, and (b) the plane view. (c) is an SEM photo of the device before attaching the chip to a glass base. The interdigited fingers for overload protection are clearly visible, and the supporting beams can be seen in the far end (from Barth et al. [28]).

process, etch stop techniques are not used; the wafers are removed from the etch bath after a given time. The etch time has been calculated with an etch profile simulation program [32].

Beams are also used in *resonating sensors*. The resonance frequency of a beam clamped at both sides ("microbridge") is strongly dependent on the axial stress applied to the beam. These beams can be used as force sensors or, as part of a sensor system, as strain gauges in a manner similar to piezoresistive strain gauges. Similar to the latter, a gauge factor can be

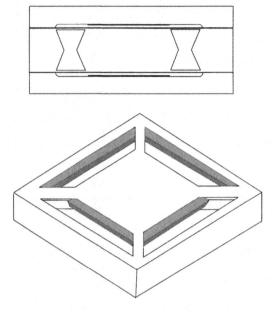

Fig. 5.22. Schematic drawing of a design for an accelerometer [29].

defined (relative change of the output–frequency in the case of resonant sensors, resistance in the other case – over applied strain) which depends on the geometry of the microbridges. Microbridges can be designed such that the gauge factor is several thousand (see, e.g., [33]). Accordingly, resonant sensors have the potential of becoming the most sensitive and precise mechanical sensors.

The first example we give is work done by to Blom [34, 35] and van Mullem et al. [36, 37]. The outline of the sensor is shown in Fig. 5.24. It consists of a silicon frame with a microbridge in the center. The microbridge, having a thickness of ca. 25 μm, has been realized in the following way. First a membrane of a thickness twice the desired beam thickness from the backside of the wafer has been etched. Then the etch mask at the frontside has been opened in order to etch through the membrane. During this step the membrane is etched from both sides, therefore the last etch step results in a microbridge of half the thickness of the membrane.

In later designs the second step is done in a dry etch process, mainly to avoid the sharp corners that result from anisotropic etching. It has been found that the structures etched in the second way are stronger by a factor of nearly 10. The sensor is loaded in a cell as shown in Fig. 5.25. Resonance frequencies of the fundamental mode are in the 10-kHz range. By applying a load of 0.4 Newton the resonance frequency changes from 7.5 kHz (unloaded) to 20 kHz.

A special aspect of this particular sensor is the method of excitation and detection of the vibration. While there are many methods for excitation and detection, c.f. [38–40], here use is made of a piezoelectric thin film, ZnO, which has, as has been shown by Prak et al. [40], in the dimensions of van Mullem's design, the greatest effectivity.

A microbridge fabricated in a different way is presented in Fig. 5.26. The beam bridges an etched cavity diagonally. This design allows etching from one side only, which is much simpler than the procedure described in the foregoing paragraph. The example is from

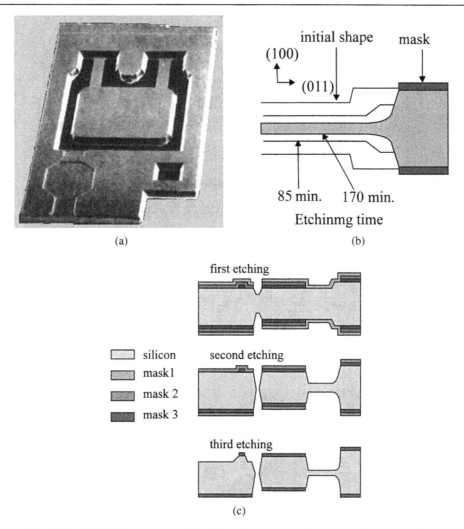

(a)   (b)

(c)

Fig. 5.23. (a) SEM photograph of the silicon chip showing the mesa and the suspending beams. (b) Two-step anisotropic etching process. The emergence of a rounded etch profile. The silicon chip showing the mesa and the suspending beams. (c) How to etch rounded profiles of beams. Cross section of a multistep anisotropic etching process (from Koide et al. [31]).

Zhang et al. [41], but similar structures have also been presented by the group of Baltes from the ETH Zürich (see, e.g., [42]). The trick is, as we have explained in Section 2.3, that etching under a microbridge on a (100) wafer is possible only if it is inclined with respect to the ⟨110⟩ direction of the wafer.

These structures can be used as force sensors in a similar way as Blom suggests [34]. But they are also proposed as carriers for thermal flow sensors [43] where heat is dissipated in a resistor carried by the microbridge. A flow of gas or liquid along the microbridge carries away some heat, thus cooling the microbridge. The temperature of the microbridge can be measured conveniently by measuring the electrical resistance of the dissipating element [44, 45].

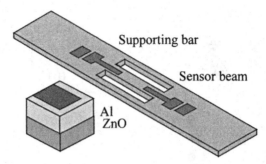

Fig. 5.24. Schematic of a force sensor using a vibrating beam. The inlet shows the thin
film sandwich.

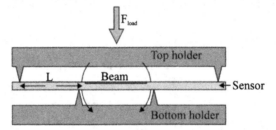

Fig. 5.25. Load cell for force sensors.

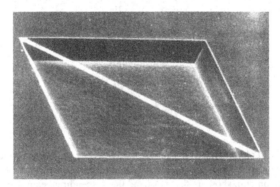

Fig. 5.26. Microbridge crossing a cavity along the diagonal (from Zhang et al. [41]).

A more flexible principle of thermal flow measurement makes use of three separate
elements: a central heat source and two temperature sensors up- and downstream. Any flow
will imbalance the temperature distribution. A measurement of the temperature difference
between the temperature sensors is a sensitive measure for the flow. Flows as small as nl/min
can be measured [45].

An example for a realization of a *flow sensor* is shown in Fig. 5.27. A perforated
membrane carries the resistors for heating and temperature measurement. The resistors are
made from thin films of chromium and gold; chromium is necessary because gold does not
adhere well to silicon, silicon dioxide, or, as is relevant in this case, to silicon nitride. Both
Cr and Au are inert with respect to KOH. The membrane itself is made from a 1-μm thick
LPCVD silicon nitride layer. It merely consists of beams directed along the $\langle 100 \rangle$ direction,

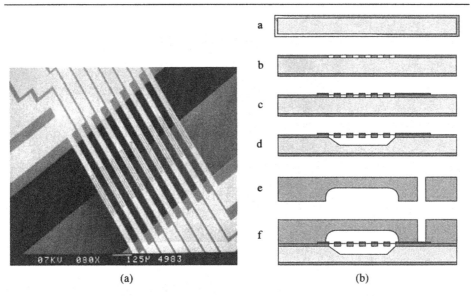

(a)   (b)

Fig. 5.27. (a) SEM-photograph of a thermal flow sensor, showing the perforated carrier membrane (LPCVD silicon nitride) with electrical resistors (Cr-Au) as a heat source (center of photograph) and temperature sensors. (b) Process for the fabrication of the flow sensor (from Lammerink et al. [45]).

while the channel is directed along the ⟨110⟩ direction. This way, the membrane is readily underetched. The process sequence is shown in Fig. 5.27(b). The perforated membrane and the metallization can be made before etching; etching is from the frontside only. The cap wafer is made from Pyrex into which a channel is etched in HF with an LPCVD polysilicon mask. Feedthroughs for the electrical connection ($\phi = 400$ mm) are drilled through the Pyrex. The final step is anodic bonding of the wafers.

Since the fabrication of the micropump designed by van de Pol et al. [18] requires only anisotropic etching, bonding, and metallization, it can be fabricated using the same process steps as the flow sensor. Therefore it is not so difficult to fabricate a flow sensor and micropump integrated in one chip. The combination of pump and flow sensor can be used as a dosage system. The dosage system has been presented by Lammerink et al. [46]. A microphotograph is shown in Fig. 5.28. The next step would be the integration of dosage systems, Miyake's mixer [27], and a chemical sensor for an integrated fluid injection analysis system.

Another alternative for gas flow sensing is shown in Fig. 5.29 [47, 48]. Here the micro-bridge crosses a shallow V groove in a (110) wafer. Therefore the bridge can cross the wafer in a straight line. The microbridge is used as a *resonating force gauge*. The resistors serve for thermal excitation and piezoresistive detection of the vibration.

Again the mechanical carrier is made of LPCVD silicon nitride, but the resistors are made of LPCVD polysilicon. Metallization is by Cr-Au to make the anisotropic etch possible after deposition of the metal wires. To finish the channel, a second (110) wafer with etched V grooves is bonded over the V grooves. Here a special bonding technology had to be developed, which must be a low-temperature bonding with the possibility to bond over thin layers with a step height of 1–3 μm. The bonding could be achieved with a thin film of boron oxide glass [49]. This material becomes soft at 400°C. If wafers containing steps not

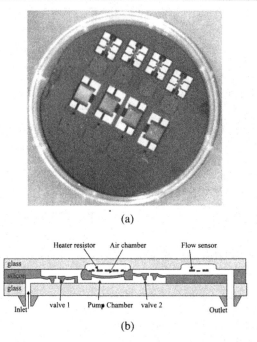

(a)

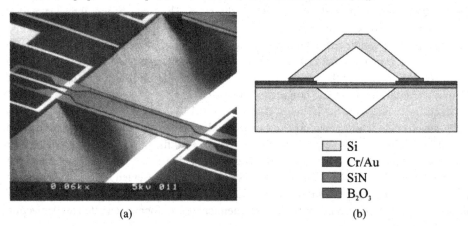

(b)

Fig. 5.28. A dosage system with a micropump and a flow sensor integrated on one chip. (a) Photograph of the chip. (b) Cross section (from Lammerink et al. [46]).

Fig. 5.29. (a) SEM photograph of a resonating microbridge mass flow sensor. The bridge crosses a V groove etched into a ⟨110⟩ silicon wafer (from Bouwstra et al. [46]). (b) Schematic of the cross section of an assembled resonating microbridge mass flow sensor. For bonding, use was made by a boron oxide thin film (from Legtenberg et al. [48]).

higher than the thickness of the film are bonded at 450°C, the flow of the glass is sufficient to fill all shallow cavities without filling the flow channel, and one obtains a tight strong bond. The disadvantage of boron oxide is its reactivity with water. The cross section of the bonded device is also shown in Fig. 5.29.

A special beam structure has been presented by Buser and de Rooij from the University of NeuChâtel [50]. It is a microbridge that contains a mesa to increase the total mass of the

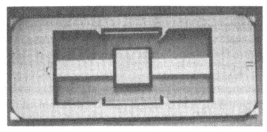

Fig. 5.30. Clamped-clamped beam with a central mesa (from Buser et al. [51]).

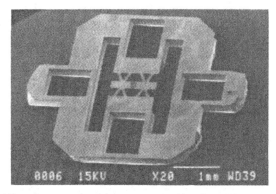

Fig. 5.31. All-optic resonating pressure sensor chip (from Kristeroy et al. [56]).

beam (see Fig. 5.30). The aim of this work was to design a resonator with a high quality factor. The quality factor of a resonant system is equal to

$$Q = (\omega\tau), \tag{5.2}$$

where ω is the resonance frequency, and τ is the decay time of the free oscillation. Analysis shows that one expects that Q is proportional to $\sqrt{m}$, with m the mass of the oscillator. The system containing mesa, beam, and frame has been etched anisotropically using a CAD program for corner compensation. The quality factor found in this system was 25,000 at a pressure below 10 μbar. This value, by the way, is typical for microbridges at low pressure [13, 14, 53, 54]. Buser and de Rooij succeeded in designing structures with even higher quality factors. In these structures, torsional oscillators were coupled in a way that the moment coupling to the support was minimized. The highest Q factors obtained were above 600,000 [55].

A more complex structure for a *resonating sensor* has been proposed by Greenwood [56]. The structure is shown in Fig. 5.31. The structure has been fabricated using anisotropic etching and the B^{++} etch stop. The complexity of the structure has to do with underetching. In Greenwood's original design the resonator is mounted on ridges on a membrane, which deforms under a pressure difference. The ridges couple the deformation to the resonator, it is strained, and the resonance frequencies change.

This structure is now used as the first resonating sensor that is actually applied in industry. The sensor is driven and interrogated thermally by using a light source, therefore, the whole system can be used under circumstances where electric currents must be avoided or that are fatal for integrated circuits, such as high temperature. For the latter application the sensor has been developed. The sensor will be applied as a pressure sensor downhole at a depth of ca. 5 km in oil bubbles. It must operate at 200°C and at a pressure of 1000 bar. The SEM photograph of the silicon chip in Fig. 5.31 is the design for this application. The chip is mounted in an evacuated stainless-steel tube which deforms under the pressure. The deformation is coupled to the silicon frame shown in Fig. 5.31. For more details of the sensor we refer to [57, 58].

Cantilever beams for actuation have been proposed by Benecke and Riethmüller [59, 60]. The actuation is based on the thermal bimorph effect. Two layers of material with different coefficients of thermal expansion attached to each other form a bimorph that bends if heated or cooled. The bimorphs can be fabricated using a combination of thin film deposition and anisotropic etching. Thin films of different material are deposited to realize the bimorph and the heater resistor. In the fabrication scheme proposed by Benecke and Riethmüller [58] (Fig. 5.32), the heater resistor is made by LPCVD polysilicon, and the bimorph effect

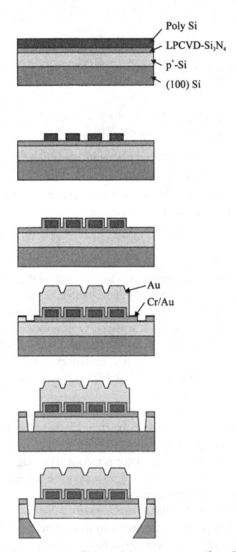

Fig. 5.32. Fabrication sequence of bimorph beam actuators (from Benecke [58]).

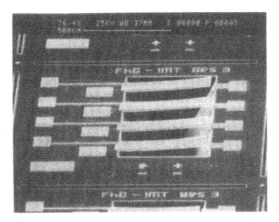

Fig. 5.33. SEM photograph of micromechanical bimorph actuators (from Benecke [60]).

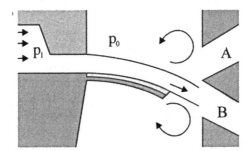

Fig. 5.34. Schematic view of a bimorph cantilever beam for fluid jet deflection (from Döring et al. [62]).

is realized by a heavily boron-doped epilayer of silicon and an electroplated gold layer. The latter was grown on top of an evaporated Cr-Au layer. The resulting structure is shown in Fig. 5.33. The tips of 500-μm long beams deflect up to 100 μm.

Benecke et al. suggested using the beams in valves (in order to close or open a channel), as micromechanical mirrors, for micromanipulation by coupling several actuators, or as micromechanical motors [61].

A new suggestion for an application of these actuators is provided by Döring et al. [62] from Robert Bosch GmbH in Stuttgart (Germany). They used the beam to direct a jet of a liquid (fuel in car engines), this way switching from one channel to the other (see Fig. 5.34).

Since in this application the beam is cooled by the streaming liquid, the device is very fast in contrast to the thermal actuators proposed earlier. The operating frequency is as large as 1000 Hz. At a temperature rise of 100°C, a deflection of the liquid jet of 4° could be reached. In the fabrication sequence, Döring used aluminum instead of gold and the electrochemical etch stop.

An alternative principle to deflect structures (beams [63, 64] and membranes [65]) makes use of electrostatic forces. The great difficulty with electrostatic forces is that they act only over very small distances. Therefore the electrostatic motors work only if one manages to fabricate the gap spacing between rotor and stator of less than 2 μm (see Chapter 8).

An electrostatically activated movement over a distance of more than a few micrometers seems to be difficult to achieve, but it is possible to use the following trick (c.f., Fig. 5.35):

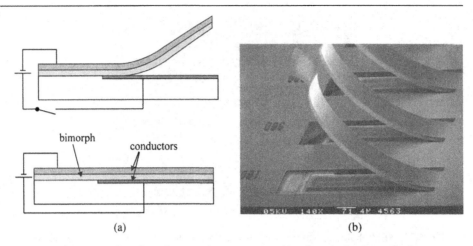

Fig. 5.35. (a) Design of an electrostatic actuator capable of producing large displacements at large forces (from Elwenspoek et al. [64]). (b) SEM photo of the electrostatic actuator.

if a structure (e.g., a cantilever beam), which is deflected by some means, is clamped at one edge so that a capacity is formed at this very edge, the electric field in the capacity at the edge can be large enough – since it is very easy to reduce the distance between capacity plates – to pull the whole structure down. This way the electrostatic force is always acting over a short distance. Furthermore, the force can be increased considerably by using an appropriate dielectric, such as PZT. PZT can be grown in a sol gel process, and one gets a piezoelectric material with a relative dielectric constant of more than 1,000 [66]. The deflection of the beam can be achieved by a hydrostatic pressure [67], by viscous drag of the streaming liquid, or by a vertical stress gradient in the structure. Proposals exist to have an electrostatic counterforce, but ways of realizing microversions have not been put forward so far.

This design possibly leads to the strongest micromechanical actuators that can be realized. Depending on the electrode design, the force can be independent of the deflection (in this case, one has a switch without the possibility of controlling the deflection with the help of an external voltage) or dependent on the deflection. In the latter case the deflection of the tip of the cantilever beam is a function of the applied voltage. This is achieved if one uses, e.g., triangular electrodes [64].

References

[1] A. Heuberger, *Mikromechanik* (in German) (Springer-Verlag, Heidelberg, 1989).
[2] C. Sander, J. Knütti, and J. Meindl, IEEE Trans. Electron Devices **27**, 927 (1980).
[3] B. Puers, E. Peeters, A. van den Bossche, and W. Sansen, Sensors and Act. **A21–23**, 108 (1990).
[4] A. Hanneborg and P. Ohlckers, Sensors and Act. **A21–23**, 151 (1990).
[5] H.-L. Chau and K. D. Wise, IEEE Trans. Electron. Devices **35**, 2355 (1988).
[6] Y. Bäcklund, L. Rosengren, B. Hök, and B. Svedberg, Sensors and Act. **A21–A23**, 58 (1990).
[7] K. Petersen, P. Barth, J. Poydock, J. Brown, J. Mallon Jr., and J. Bryzek, Tech. Digest IEEE Solid-State Sensor and Actuator Workshop, Hilton Head Island, SC, June 6–9, 1988, p. 144.

[8] J. G. Smits, H. A. C. Tilmans, K. Hoen, H. Mulder, J. van Vuuren, and G. Boom, Sensors and Act. **A4**, 565 (1983).

[9] T. S. J. Lammerink Ph.D. Thesis, University of Twente (1990).

[10] S. Bouwstra, P. Kemna, and R. Legtenberg, Sensors and Act. **A20**, 213 (1989).

[11] T. S. J. Lammerink and S. J. Gerritsen, Proc. SPIE **798**, Fiber Optical Sensors II (1987), p. 67.

[12] G. Stemme and S. Stemme, Sensors and Act. **A21–A23**, 336 (1990).

[13] K. Ikeda and T. Watanabe, Proc. SICE '86, Tokyo, Japan, 1986, p. 651.

[14] K. Ikeda, H. Kuwayama, T. Kobayashi, T. Watanabe, T. Nishikawa, T. Yoshida, and K. Harada, Sensors and Act. **A21–A23**, 1007 (1990).

[15] K. Ikeda, H. Kuwayama, T. Kobayashi, T. Watanabe, T. Nishikawa, and T. Yoshida, Sensors and Act. **A21–A23**, 193 (1990).

[16] A. J. Sprenkels, Ph.D. Thesis, University of Twente (1988).

[17] A. G. H. van der Donk, Ph.D. Thesis, University of Twente (1992).

[18] F. C. M. van de Pol, H. T. G. van Lintel, M. Elwenspoek, and J. H. J. Fluitman, Sensors and Act. **A21–A23**, 198 (1990).

[19] H. T. G. van Lintel, F. C. M. van de Pol, and S. Bouwstra, Sensors and Act. **A15**, 153 (1988).

[20] J. G. Smits, Sensors and Act. **A21–A23**, 203 (1990).

[21] S. Shoji, S. Nakagawa, and M. Esashi, Sensors and Act. **A21–23**, 189 (1990).

[22] R. Zengerle, A. Richter, and H. Sandmayer, Proc. MEMS '92, Travemünde, Ger., Feb. 4–7, (1992), p. 19.

[23] J. Tiren, L. Ternez, and B. Hök, Sensors and Act. **A18** (1989).

[24] F. C. M. van de Pol, D. G. J. Wonnink, and M. Elwenspoek, Sensors and Act. **17**, 139 (1989).

[25] L. Smith and B. Hök, Transducers '91, San Francisco, CA, USA, June 24–27, 1991, p. 1049.

[26] G. Kittilsland, G. Stemme, and B. Norden, Sensors and Act. **A21–A23**, 904 (1990).

[27] R. Miyake, T. S. J. Lammerink, M. Elwenspoek, and J. H. J. Fluitman, Proc. MEMS '93, Ft. Lauderdale, FL, USA, Feb. 7–10, 1993, p. 248.

[28] P. W. Barth, F. Pourahmadi, R. Mayer, J. Poydock, and K. Petersen, Tech. Digest IEEE Solid State Sensor and Actuator Workshop, Hilton Head Island, SC, June 6–9, 1988, p. 35.

[29] H. Seidel, H. Riedel, R. Kolbeck, G. Mück, and M. Königer, AMA Seminar, Heidelberg, October 19–20, 1989, p. 177.

[30] A. Jornod, F. Rudolf, Sensors and Act. **A17**, 415 (1989).

[31] A. Koide, K. Sato, A. Suzuki, and M. Miki, Technical Digest 11th Sensor Symposium, Japan, 1992, p. 23.

[32] A. Koide, K. Sato, and S. Tanaka, Proc. MEMS '91, Nara, Japan, Jan. 30–Feb. 2, 1991, p. 216.

[33] H. A. C. Tilmans, M. Elwenspoek, and J. H. J. Fluitman, Sensors and Act. **A30**, 35 (1992).

[34] F. R. Blom, Ph.D. Thesis, University of Twente (1990).

[35] F. R. Blom, S. Bouwstra, J. H. J. Fluitman, and M. Elwenspoek, Sensors and Act. **A17**, 513 (1989).

[36] C. J. van Mullem, F. R. Blom, J. H. J. Fluitman, and M. Elwenspoek, Sensors and Act. **A25–A27**, 379 (1991).

[37] C. J. van Mullem, H. A. C. Tilmans, A. J. Mouthaan, and J. H. J. Fluitman, Sensors and Act. **A31**, 168 (1992).

[38] M. Elwenspoek, *Journal A* **A32**, 15 (1991).

[39] G. Stemme, J. Micromechan. Microeng. **1**, 113 (1991).

[40] A. Prak, T. S. J. Lammerink, and J. H. J. Fluitman, Sensor and Materials, **5**, 143 (1993).

[41] L. M. Zhang, D. Walsh, D. Uttamchandani, and B. Culshaw, Sensors and Act. **A29**, 73 (1991).

[42] D. Moser, H. Baltes, M. Parameswaran, S. Linder, and S. Rudin, AMA Seminar, Heidelberg, October 19–20, 1989, p. 259.

[43] G. Wachutka, R. Leggenhager, D. Moser, and H. Baltes, Proc. Transducers '91, San Francisco, CA, June 24–27, 1991, p. 22.

[44] K. Petersen and J. Brown, Proc. Transducers'85 (1985), p. 361.

[45] T. S. J. Lammerink, N. R. Tas, M. Elwenspoek, and J. H. J. Fluitman, Sensors and Act. **A37–A38**, 45 (1993).

[46] T. S. J. Lammerink, M. Elwenspoek, and J. H. J. Fluitman, Proc. MEMS'93, Ft. Lauderdale, FL, Feb. 7–10, 1993, p. 254.

[47] S. Bouwstra, R. Legtenberg, H. A. C. Tilmans, and M. Elwenspoek, Sensors and Act. **A21–A23**, 332 (1990).

[48] R. Legtenberg, S. Bouwstra, and J. H. J. Fluitman, Sensors and Act. **A25–A27**, 723 (1991).

[49] R. Legtenberg, S. Bouwstra, and M. Elwenspoek, J. Micromechan. Microeng. **1**, 157 (1991).

[50] R. A. Buser and N. F. de Rooij, Sensors and Act. **17**, 145 (1989).

[51] R. A. Buser, B. Stauffer, and N. F. de Rooij, Ext. Abstr. Fall Meet. Electrochem. Soc., San Diego, CA, Oct. 1986, p. 879.

[52] R. A. Buser, AMA Seminar, Heidelberg, October 19–20, 1989, p. 231.

[53] F. R. Blom, S. Bouwstra, M. Elwenspoek, and J. H. J. Fluitman, J. Vac. Sci. Technol. **B10**, 19 (1992).

[54] H. Guckel, J. J. Sniegowski, T. R. Christenson, and F. Raissi, Sensors and Act. **A21–A23**, 346 (1990).

[55] R. A. Buser and N. F. de Rooij, Sensors and Act. **A21–A23**, 323 (1989).

[56] J. C. Greenwood, J. Phys. E **17**, 561 (1984).

[57] T. Kvisteroy, O. H. Gusland, B. Stark, H. Nakstad, M. Ericsrud, and B. Bjornstad, Sensors and Act. **A31**, 164 (1992).

[58] B. S. Douma, P. Eigenraam, and P. Hatlem, Sensors and Act. **A31**, 215 (1992).

[59] W. Benecke and W. Riethmüller, IEEE Trans. Electron Devices **35**, 758 (1988).

[60] W. Benecke, AMA Seminar, Heidelberg, October 19–20, 1989, p. 217.

[61] W. Benecke and W. Riethmüller, Proc. MEMS '89, Salt Lake City, Utah, Feb. 20–22, 1989, p. 20.

[62] C. Döring, T. Grauer, J. Marek, M. S. Mettner, H. -P. Trah, and M. Willman, Proc. MEMS '92 (1992), p. 12.

[63] T. Ohnstein, T. Fukiura, J. Ridley, and U. Bonne, Proc. MEMS'90, Napa Valley, CA, Feb. 11–14, 1990, p. 95.

[64] M. Elwenspoek, L. Smith, and B. Hök, Proc. MME'92, Leuven, Belgium, June 1–2, 1992, p. 211, J. Micromech. Microeng. **2**, 221 (1993).

[65] J. Branebjerg and P. Gravese, Proc. MEMS '92, Travemünde, Ger., Feb. 4–7, 1992, p. 6.

[66] A. M. Flinn, L. S. Tavrow, S. F. Bart, R. A. Brooks, D. J. Ehrlich, K. R. Udayakumar, and L. E. Cross, J. Microelectromechanical Systems **1**, 44 (1992).

[67] K. Sato and M. Shikida, Proc. MEMS '92 Travemünde, Germany, Feb. 4–7, 1992, p. 1.

6

Surface micromachining

6.1 Introduction

After R. Howe's and Muller's work on the surface micromachined resonating vapor sensor [1], an explosive development in surface micromachined sensors and (mainly) actuators began. A culmination of papers on surface micromachined electrostatic micromotors, grippers, accelerometers, pressure sensors, flow sensors, vacuum encapsulated resonating strain gauges, gears, x-y stages, and other structures was observed at Transducers 1989 in Montreux, Switzerland, and in the period after this conference many groups started work in this field.

The great advantage of the surface micromachining (or "sacrificial layer etch" technique) lies in the fact that, compared to the "conventional" bulk micromachining methods (as described in previous chapters), mechanical structures with larger freedom in design can be built. The biggest disadvantage of surface micromachining is the restriction to thin films. It must be said, however, that during the last few years more and more work has been published on surface machining of thicker films, with the help of alternative ways for photolithography, so that surface micromachining using thicker structures becomes possible with the same precision as machining of thin films using conventional photolithography. The most-developed alternative technology capable of surface micromachining is the so-called LIGA process[‡]. Originally synchrotron radiation was needed for the LIGA process, but alternative, cheaper ways are now being developed for deep lithography.

6.2 Basic fabrication issues for surface micromachining

The "sacrificial layer etch" technique was first demonstrated by Nathanson et al. [2] in 1967 with the fabrication of resonant gate transistors that employed free-standing gold beams. The fabrication of free-standing structures from polycrystalline silicon using a silicon dioxide sacrificial layer that was removed in hydrofluoric acid was introduced by Howe and Muller [1]. Since then, many new surface micromachining techniques have been developed, leading to a great variety of possible structural materials and sacrificial layer combinations.

Many combinations of structural material and sacrificial layer can be, and have been, used; examples are combinations of silicon nitride and polysilicon, gold and titanium, nickel and

---

[‡] LIGA refers to the German "Lithographie, Galvanoformung, Abformung." In this process, a thick resist layer (up to 1-mm thick) is illuminated by a strong X-ray source (from a synchrotron). After developing the resist, one has structures with very high aspect ratios, into which metal is grown in an electroplating process. The resist is then removed, and a metal mold remains that can be used to fabricate microstructures from plastics and other materials.

titanium, polyimide and aluminium, tungsten and silicon dioxide, and aluminium and polymer. The majority of the surface micromachining work has, however, focused on the combination of polysilicon as the structural material and silicon dioxide or related glasses as the sacrificial material that is etched in HF solutions. This combination is the best-documented surface micromachining technique, which has as an additional advantage that it most closely meets many of the requirements for micro-electromechanical systems (MEMS), while the materials and the etchants are IC compatible and can be used in combination with conventional IC processes. This chapter will therefore almost exclusively discuss the polysilicon surface micromachining technique.

6.2.1 The basic idea

The basic fabrication steps are shown in Fig. 6.1. The first step is the deposition and subsequent patterning of the silicon dioxide sacrificial layer. This is followed by the deposition and patterning of the polysilicon structural layer. Next, the sacrificial layer is selectively

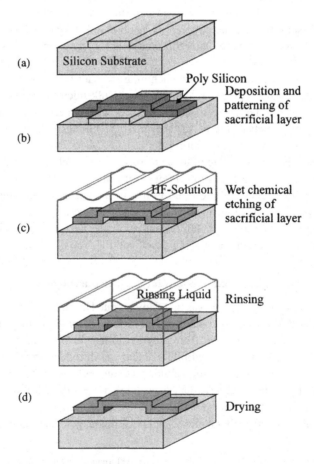

Fig. 6.1. Basic process scheme of surface micromachining: (a) deposition and patterning of the silicon dioxide sacrificial layer; (b) deposition and patterning of the polysilicon structural layer; (c) sacrificial layer etching in a hydrofluoric acid solution; (d) rinsing and drying procedures.

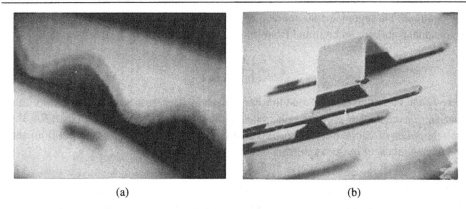

(a)   (b)

Fig. 6.2. Buckled polysilicon structures demonstrating compressive stress of unannealed thin films [3].

etched in a fluoric acid solution. The last steps are rinsing procedures to remove the fluoric acid and drying from the rinse liquid.

For the growth of the thin films a good step coverage is important. This is achieved in LPCVD (low-pressure chemical vapor deposition) processes. Furthermore, a sacrificial layer that etches fast is advantageous. Phosphorous-doped silicon oxide (PSG) meets this requirement. It follows that besides the usual equipment for micromachining, LPCVD equipment for polysilicon and PSG is required. For the fabrication of electrostatic micromotors, one also needs advanced lithographic facilities, because these motors operate only if the gap spacing is small enough (typically 2 μm).

6.2.2   Properties of thin film materials

The material of which the structures are made are thin films, mainly polysilicon. The mechanical properties of thin films strongly depend on process parameters and thermal treatment. As deposited, the polysilicon films are under a strong compressive stress; when such films are released by sacrificial layer etching, they may buckle. Examples of buckled polysilicon structures are shown in Fig. 6.2 [3]. The compressive stress disappears after annealing [4].

One has to distinguish between two kinds of stress that deform the released thin films. Compressive axial stresses lead to buckling of the structures above a critical lateral size. Compressive and tensile stresses are important, e.g., for microbridges. Vertical stress gradients become visible in structures such as cantilever beams (beams that are clamped at one side): the stress gradients cause the beams to curl up or down, depending on the sign of the stress gradient. Stress gradients are invisible in microbridges, while on the other hand compressive axial stresses are invisible in cantilever beams. The latter should be taken into account in the design of test structures for the determination of film stresses.

A variety of different techniques and approaches has been used in order to determine the residual stress and Young's modulus of thin films.

Wafer curvature

One of the easiest approaches to measure the residual strain or stress, is the measurement of the curvature of a substrate wafer covered at one side with a layer of the thin film under investigation. This approach does not require patterning and etching of the film or substrate. The residual stress σ_0 can be found from [6]

$$\sigma_0 = \frac{E}{(1-\nu)} \frac{t_s^2}{6t_f R},$$ (6.1)

where E and ν are the Young's modulus and Poisson's ratio, respectively, of the substrate material, t_s is the thickness of the substrate, t_f is the thickness of the thin film, and R is the substrate radius of curvature. The wafer curvature can be measured using a profilometer, optical techniques, or by X-ray diffraction.

Bulge test

Another method to determine both the Young's modulus and the residual stress is the bulge test [7, 8]. This technique is based on the measurement of the load–deflection characteristics of a membrane. The membrane is constructed from the thin film material or a composition of a mechanical carrier and the thin film material under investigation. Circular, square, and rectangular membranes have been used for this technique. The deflection shape can be measured optically, by interference, by focusing of a microscope, or by an optical or mechanical profilometer. The load–deflection relationship is given by

$$P = \frac{C_1 h \sigma_0}{a^2} \delta + \frac{C_2 h E}{a^4 (1-\nu)} \delta^3,$$ (6.2)

where P is the load (a pressure, e.g.), h is the membrane thickness, a is the length or radius of the membrane, and d is the deflection at the center of the membrane. The constants C_1 and C_2 in Eq. (6.2) depend on the shape of the membrane and have been verified by Pan et al. [9] by FEM (finite element method) analysis for circular and square membranes.

Surface profiler techniques

Concentrated load–deflection measurements of micromechanical structures – like loading of microfabricated structures by a stylus-type surface profiler [10], cantilever beam bending experiments using a nanointender [11], and loading of membranes by a surface profiler [12] – have been used to extract thin film mechanical properties.

Extraction from resonance data

Resonant microstructures can also be used to determine the residual strain and the Young's modulus by using the resonant frequency data of beams [12–14]. The expression for the mode frequencies ω_n of beams with rectangular cross sections and a large width ($b > 5h$) can be written as

$$\omega_n = \frac{\alpha_n^2}{\sqrt{12}} \sqrt{\frac{E}{\rho(1-\nu^2)}} \frac{h}{l^2} \sqrt{1 + \gamma_n \varepsilon_0 (1-\nu^2)\left(\frac{l}{h}\right)^2},$$ (6.3)

where α_n and γ_n are mode shape constants, ρ is the density of the beam material, h is the beam thickness, l is the beam length, and ε_0 is the residual strain. A linear fit of $(\omega_n l)^2$ versus $(h/l)^2$ gives the Young's modulus from the slope of the fit and the residual strain from the zero offset. The residual strain can also be found from the ratio of higher overtones to the fundamental frequency [13].

Diagnostic test structures

A number of micromechanical diagnostic test structures have been developed in order to determine thin film mechanical properties. The advantage of such test structures is that their fabrication and development resemble that of conventional device microstructures. This is therefore suitable in order to create standard "drop-in" mechanical test structures analogous to transistor test structures used for extraction of electrical device parameters.

An array of microbridges can be used to measure *compressive strain*. A doubly clamped beam will buckle if its length exceeds a critical length L_{cr} (see also Fig. 6.2):

$$L_{cr} = \sqrt{\frac{\pi^2 h^2}{3\varepsilon_0}}. \tag{6.4}$$

Under similar conditions a square plate, clamped at its edges, will buckle if the length exceeds the critical length a_{cr}:

$$a_{cr} = \sqrt{\frac{4\pi^2 h^2}{9(1+\nu)\varepsilon_0}}. \tag{6.5}$$

The strain is deduced by determining the smallest geometry for which buckling occurs. Buckling can be readily determined using a phase contrast microscope.

It is possible to measure *tensile strains*, if a ring-shaped structure is used to convert tensile strain into compressive strain and measure the critical buckling length of the center part of the structure (Fig. 6.3).

Cantilever beams can be used to measure *strain gradients* along the beam thickness (see, e.g., Fig. 6.4). Variation in film stress with thickness results in an internal moment per unit width M_0 equal to

$$M_0 = \int_{-\frac{h}{2}}^{\frac{h}{2}} \sigma_x(z) z \, dz, \tag{6.6}$$

where z is taken as the distance from the center of the film. After release, the internal stress relaxes and the structural film will curl with a radius of curvature:

$$R = \frac{Eh^3}{12M_0}. \tag{6.7}$$

Other structures have also been used to determine the residual strain, e.g., strain diagnostics with a bent beam deformation multiplier using a vernier structure [15]. Furthermore,

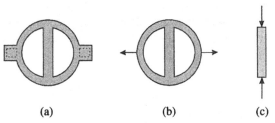

(a)　　　　　　　　(b)　　　　　　　　(c)

Fig. 6.3. Ring and beam diagnostic structure: (a) top view; (b) free-body diagram of the ring for a film in tension; (c) free-body diagram of the cross beam for a film in tension [4].

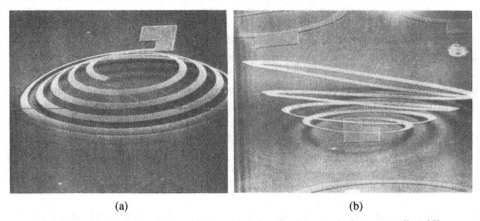

(a)   (b)

Fig. 6.4. Polysilicon spirals that, if released, deform under a vertical stress gradient [5].

active micromechanical test structures have been fabricated, like combdrive structures, to
determine Young's modulus and residual strain [16, 17]. Also, the pull-in voltage of cantilever beams, microbridges, and membranes can be used to determine the residual stress,
Poisson's ratio, and Young's modulus [18, 19]. The pull-in voltage of cantilevers, bridges,
and circular membranes can be found from

$$V_{PI}^2 = \frac{B_n h^3 d^3 E}{\varepsilon_0 L^4} + \frac{S_n h d^3 \sigma_0}{\varepsilon_0 L^2}, \tag{6.8}$$

where B_n is a bending term constant, S_n is a stress term constant, d is the capacitor gap
spacing, and b is the width of the beam.

Strain control

The residual strain in polysilicon can be controlled by varying the deposition variables or
by annealing procedures [4]. Polysilicon is generally deposited by LPCVD from pyrolysis
of silane (SiH_4) at temperatures around 600°C and at pressures of several hundred mTorr.
Both compressive and tensile stresses of large magnitude are known to exist in polysilicon.
Deposition temperatures below the silicon crystallization temperature (ca. 605°C) are generally found to produce tensile films, while higher deposition temperatures generally result
in compressive films. Polysilicon films deposited near 600°C have been shown to grow
initially in the amorphous state and subsequently recrystallize during further deposition.
As the density of amorphous silicon is smaller than the density of polysilicon, it contracts
against boundary constraints, inducing a tensile stress. Tension occurs only in films that are
deposited in the amorphous state and subsequently recrystallize. Annealing at high temperatures induces recrystallization which allows intrinsic stresses in the silicon films to relax.
In Fig. 6.5 the strain in polysilicon as a function of the annealing temperature and time is
shown for undoped and boron-doped samples [4].

Fracture strain

Polysilicon has good mechanical properties. Although it is polycrystalline and in principle
can be plastically deformed more easily than single-crystalline silicon, it appears that at
least at room temperature polycrystalline silicon is comparable to single-crystalline silicon

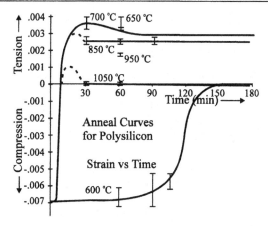

Fig. 6.5. Internal strain for LPCVD polysilicon films deposited on oxidized silicon substrates as a function of postdeposition anneal time and temperature. These films were deposited using the pyrolysis of 100% silane at a temperature of 580°C and at a pressure of 300 mTorr [4].

regarding plasticity. The fracture strain has been measured by Tai and Muller [20], who used a polysilicon bridge-slider structure; a mechanical probe was used to push against a plate end, compressing and forcing the bridge to buckle until it breaks. Unannealed samples were found to be mechanically stronger than annealed ones: the fracture strain was found to be $1.72 \pm 0.09\%$ for unannealed samples and $0.93 \pm 0.04\%$ for samples annealed at 1000°C for one hour. If Young's modulus would be the same in both samples and equal to the bulk value (averaged over crystallographic orientations), these fracture strains would correspond to fracture stresses of 3.2×10^9 N/m^2 and 1.8×10^9 N/m^2. These values must be compared with 7×10^9 N/m^2, which is the yield strength of single-crystalline silicon.

6.2.3   Sacrificial layer etching in HF solutions

An important process step in surface micromachining is the etching of the sacrificial layer to release the microstructure. For the polysilicon/silicon dioxide combination this is generally done by wet, hydrofluoric acid etching. This process exploits the isotropic nature and high selectivity of wet etching to release the overlying structural polysilicon films (see Fig. 6.1). It is important to model and understand this sacrificial layer etching process in order to be able to predict etching times a priori instead of simply extrapolating initial etch rates and using trial and error to fully release microstructures.

A heterogeneous reaction model for the etching of sacrificial silicon dioxide layers, as presented by D. J. Monk et al. [21–23], is shown in Fig. 6.6. The etching can be divided into seven steps:

(1)   Mass transfer of the reactant (species A) by diffusion from the bulk to the external etch opening.
(2)   Diffusion of the reactant from the etch opening through the etch channels to the immediate vicinity of the internal catalytic surface.
(3)   Adsorption of the reactant onto the catalyst.
(4)   The reaction on the surface of the catalyst.

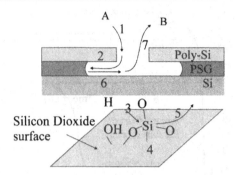

Fig. 6.6. Model for etching of sacrificial silicon dioxide layers.

(5)   Desorption of the products from the surface.
(6)   Diffusion of the products from the interior of the etch channels.
(7)   Subsequent mass transfer of the products from the etch opening to the bulk fluid.

Etching is believed to occur in two elementary steps. First, the protons break up the siloxane bonds to form silanol species on the surface, and second, this reaction is followed by an attack of fluoride ions on the silicon in the silanol, leading to the formation of SiF_4, which dissolves in aqueous solutions as H_2SiF_6. Silicon dioxide etching is therefore much more complicated than the overall reaction that is usually mentioned in literature may suggest:

$$SiO_2 + 6HF \rightarrow H_2SiF_6 + 2H_2O. \tag{C6.1}$$

Sacrificial layer etching has been modeled by Monk et al. [21–23] and Liu et al. [24] as a combination of the diffusion of reactant to the silicon dioxide surface at the etching front underneath a nonetching layer and the subsequent reaction at that surface. The main difference in the models of Monk and Liu lies in the assumed surface reaction model: Monk uses a model consisting of a first-order chemical reaction at the surface in combination with steady state diffusion [25]. This model failed to fit the experimental data of the underetch distance as a function of time for different geometrical situations [21–24]. An excellent fit was obtained by Liu, who used a combined first- and second-order reaction model, following from a Freundlich adsorption isotherm [24]. Figure 6.7 shows a plot of the calculated underetch length as a function of time for HF solutions of different concentrations and for structures of varying dimensions.

At low HF concentrations the etching is dominated by a first-order chemical reaction, and at high concentrations by a second-order reaction. Diffusion limitations are observed when the etch length reaches ca. 200 μm. This is in general more than the total etch length needed to free micromechanical structures.

The overall etch rate in HF solutions is higher for phosphosilicate glass (PSG) films, with higher etch rates for films with a higher phosphorous content; furthermore, it was found that the addition of HCl to the HF solutions increases the overall etch rate.

6.2.4   Stiction

A notorious problem in surface micromachining is stiction of the thin films after underetching by wet chemical agents. There is a large amount of literature on this problem. It has

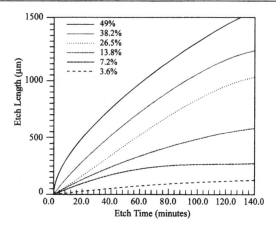

Fig. 6.7. Modeled sacrificial layer etch length as a function of etch time.

been found that if the thin film structures come into contact with the substrate, they stick to it and there is no chance of freeing them. Basically, the latter effect is the same phenomenon as wafer bonding.

The mechanisms and the nature of the forces that attract the thin film to the substrate are controversial. In our opinion, two different mechanisms play a key role [26]: the force that pulls the structure down to the surface probably comes from surface tension between the liquid in which the wafers are rinsed. When the liquid evaporates (during drying) a liquid film between the thin film and the substrate is bounded by a meniscus. The latter causes an attractive force between substrate and thin film. If the thin film structure is too weak, i.e., too thin, or too large, one finds an instability of the structure quite similar to the instability of charged plate capacitors. As a result, the structure collapses. By this collapse the thin film and the substrate approach each other so close that other forces like van der Waals forces, electrostatic forces, hydrogen bridging, and chemical reactions come into play, resulting in permanent attachment of the structures to the substrate.

Stiction can be prevented or reduced in several ways, which can roughly be divided into two categories:

(1)   *Methods which prevent the physical contact between the structures and the substrate during fabrication.* This is done by avoiding the attractive meniscus forces that bring the structures in contact with the substrate during the drying process, e.g., by freeze drying [27–29], critical point drying [30], or dry etching techniques [31, 32]. Temporarily increasing the mechanical stiffness of the structures with the aid of breakaway supports [33] has also been used. These methods have the disadvantage that stiction can still occur if the structures are afterward brought into contact with the substrate by external forces like large accelerations or static charges [29].

(2)   *Methods based on the reduction of the adhesional forces.* The latter can be done by minimizing the surface energy, for instance, by using hydrophobic surfaces or by surface treatments [34], by reduction of the contact area with the aid of stand-off bumps [35, 36], or by increasing the surface roughness [37]. All these methods result in a permanent reduction of the adhesional forces.

An elegant way to prevent the sticking problem is "freeze drying." The freeze-drying method has been developed as a drying method for biological specimens by Boyde and

Wood in 1969 [38]. Guckel et al. [39] and later Takeshima et al. [28] implemented this technique for sacrificial layer etching of surface micromachined structures. They describe a process whereby a final rinse agent is frozen and subsequently sublimated in a few hours under vacuum conditions.

A faster freeze-drying technique has also been presented, which does not require any vacuum equipment but instead can be performed under atmospheric conditions [29]. Cyclohexane, which freezes at about 7°C, is used as the final rinsing agent. Freezing and subsequent sublimation are readily accomplished by placing the substrate under a nitrogen flow on a regulated Peltier element with a temperature below the freezing point. The total time for the freeze-sublimation process depends on the geometry of the sample and typically takes 5–15 minutes for the structures described in Legtenberg and Tilmans [29].

Details of the rinse-freeze-sublimation procedure are as follows: after sacrificial layer etching in an HF solution a dilution rinse in deionized water is performed to remove the etchant. Isopropyl alcohol (IPA) is added to keep the hydrophobic wafer surface wet after removal from the solution. The wafers are placed in another beaker containing IPA and next are placed in cyclohexane, the final rinse agent. The IPA serves as an intermediate mixing agent. After rinsing the wafer in cyclohexane, it is placed upon a Peltier element, which has already been cooled to −10°C. A nitrogen flow aids in the sublimation process by removing cyclohexane vapors and by preventing condensation of water. After sublimation is complete, the Peltier element is raised to room temperature, which completes the fabrication of the free-standing structures. No residues have been observed. Microbridges with aspect ratios as high as 2000, at a gap spacing of 1 mm, were obtained by this freeze-drying procedure [29].

6.3   Applications

The technology of surface micromachining is now so advanced that devices fabricated by this technology are entering the market. This concerns mainly sensors for pressure, acceleration, and angular frequency, fabricated by large companies such as Bosch in Germany or Analog Devices (Norwood, MA). For a number of structures, in particular for the famous micromotors, no real application has been anticipated so far. The potential applications, however, are enormous. This is not restricted to surface micromachining alone but to the microtechnology as described in this book as a whole. A survey of possible applications has been described by Gabriel et al. [40]. In the following we give a few examples of microstructures and micromachines realized with the aid of surface micromachining.

6.3.1   Combdrives

Combdrives are the most simple electrostatic actuators that can be fabricated by surface micromachining and related technologies. Their working relies on the fact that opposite charged capacitor plates not only attract each other but an additional force drives the plates to overlap as far as possible. This is related to the energy of a charged capacitor with charge Q:

$$W = \frac{Q^2 d}{2\varepsilon bl},\tag{6.9}$$

with d being the plate spacing, ε the permittivity, and d, b, and l the plate spacing, width, and length, respectively. A force can be derived by differentiating with respect to d, b, or l, and it is easily seen that the attractive force between the plates has a tangential component. The use of the normal component is quite limited. First of all, if this force is voltage-controlled, one gets an instability. Second, the range of the deflection is very much limited, since appreciable forces at tolerable voltages require small gap spacing of the order of a few μm, so the normal deflection cannot exceed this figure.

An interesting aspect is that the tangential force is independent on the overlap when the actuator is voltage-controlled. To see this, we perform a Legendre transformation of W, $W' = W - uQ$ (with u the voltage), and the force is then given by

$$F = -\left(\frac{\partial W'}{dl}\right)_u = \frac{u^2 \varepsilon b}{d}. \tag{6.10}$$

Typically, in micromachined structures d and b in Eq. (6.9) (here actually the height of the capacitor plate) are of the same order, accordingly the force is of order $u^2 \varepsilon = 10^{-7}$ N for 100 V. These actuators are weak.

In Figs. 6.8 and 6.9 we show an example for a combdrive actuator [41, 42]. This actuator has fabricated from a 5-μm thick LPCVD polysilicon layer grown at 590°C, 250 mTorr,

Fig. 6.8. Example of a combdrive actuator (R. Legtenberg [41]).

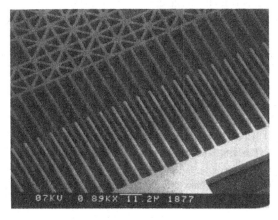

Fig. 6.9. Close-up of the combdrive actuator in Fig. 6.8 (R. Legtenberg [41]).

and silane flow of 250 sccm[‡]. The deposition was done on a silicon wafer covered with 2-μm thick thermal oxide. The polysilicon has been doped with boron at 1100°C for 3 hours to obtain a sheet resistivity of 4.5 Ω/cm. This procedure results in relatively small stress and stress gradient in the structure. During indiffusion, a boron silicate glass (BSG) layer grows on top of the polysilicon which has been stripped in buffered HF. The etching of the BSG is much faster if the wafers are put into a wet oxidation furnace for half an hour at app. 1000°C, as we have mentioned in Section 2.8. On top of the doped polysilicon a 0.6-mm thick PECVD (plasma-enhanced chemical vapor deposition) silicon oxide layer has been deposition and patterned. This layer served as an etch mask for reactive ion etching (RIE, see Chapter 9) to etch the structures in the polysilicon. The structures are released after standard cleaning for 50 min in BHF. For drying, use has been made of a freeze-drying procedure to prevent sticking. Finally, for backside contact a 1-mm thick aluminum layer has been evaporated.

The design of the mask is such that the anchors of the suspension beams holding the shuttle and the movable part of the combdrive (the so-called rotor) are wider than the parts that must be released. These parts are readily underetched while the anchors and the bond pads are still fixed to the substrate by the thermal oxide. This way the actuators can be fabricated in a single-mask process.

Electrostatic combdrive x-y stages were first described by Zhang and MacDonald [45]. A different design of an electrostatic drive, not based on comb actuators, has been proposed by Takeshima et al. from the University of Tokyo [28]. It is based on the deformation of tips of a parallelogram, as illustrated in Fig. 6.10. One point of the parallelogram is fixed and two points are attached to capacitor plates that move if the capacitor is charged. The remaining point is displaced. The displacement at the last point is much greater than that of the capacitor plates. First prototypes demonstrated a displacement of 5 μm.

6.3.2 Manipulators and x-y stages

Manipulators can be used in biology for manipulating cells, in microoptics for adjusting optical components, and for the interface between the microworld and the macroworld. Manipulators could be needed for industrial robot arms in microfabrication. The latter application is very intriguing, because a significant part of production costs of ICs (and other microsystems) stems from the huge clean rooms. Why are they so huge? Because the fabrication machines are so large. A future dream would be a fully automatic fabrication unit the size of a table, with inside microrobots that handle the wafers.

Micromanipulators were one of the first demonstrators of freely moving polysilicon thin film structures (published at Transducers 1987 in Tokyo). A review paper on the early surface micromachined structures was written by Fan et al. [43]. The first planar microarm is a four-joint crank.

Later work by Behi et al. [44] resulted in more complex structures, an example of which is shown in Fig. 6.11. It has three degrees of freedom (x, y, and orientation) and a joint clearance of 0.3 μm. The length of the arms in the example shown in Fig. 6.11 is of the order of 50 μm. The path motion of the structure in Fig. 6.11 is shown in Fig. 6.12. In Fig. 6.13, dashed and solid lines show two possible positions of the central stage.

---

‡ Standard cubic centimeter per minute.

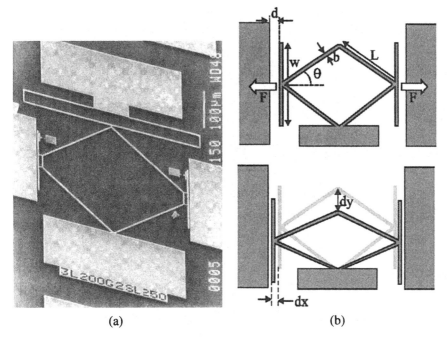

(a)   (b)

Fig. 6.10. Design of an electrostatic actuator [28]. (a) is a SEM image, (b) is a schematic drawing to illustrate the mechanical transmission of the device.

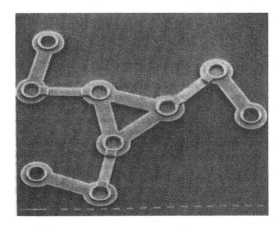

Fig. 6.11. SEM photograph of a released link mechanism [44].

The fabrication process of the above micromanipulator is shown in Fig. 6.14. First, an oxide is deposited on a silicon wafer in a low-temperature LPCVD process. Next, a heavily phosphorous-doped LPCVD polysilicon layer is deposited and patterned in a reactive ion etcher (RIE, using thermal oxide as a mask) to form the outer links and the platform. Where the pin joints and the bearings are to be formed, the oxide is isotropically etched. Again, a thermal oxide of 0.3-μm thickness is grown that defines the clearance of the joints, an anchor is etched to the substrate for the fixed joints, and a second polysilicon layer is deposited and patterned to form the inner links and the joints. The final thickness of the first polysilicon

Fig. 6.12. A close-up of the link mechanism showing the nonplanarity of the joint [44].

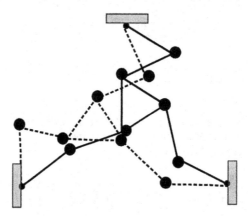

Fig. 6.13. Schematic drawing of the possible movement of the micromanipulator of Fig. 6.12 [44].

layer is 2.2 μm, and of the second is 1.3 μm. The structure is carried by the fixed joints and slides on the substrate on the free joints.

To move the structure, a minimum torque to overcome inertia and dry friction is estimated to be 50×10^{-12} Nm. As we shall see below, this torque tends to exceed the torque delivered now by electrostatic micromotors. Accordingly, the structures are passive structures and will become active structures only if a new generation of electrostatic micromotors arrives.

6.3.3 Microgrippers

Impressive work on microgrippers has been published by Kim et al. from the University of California, Berkeley (see Ref. [46] and references therein). A schematic drawing is shown in Fig. 6.15. For the actuation an electrostatic combdrive is used, which was first proposed by Tang et al. [47, 48]. The basic idea of combdrives is to increase the electric energy stored in the capacitor.

The polysilicon microgripper structure is protected by a cantilever beam located underneath the gripper structure. Only a 400-μm-long tip of the gripper protrudes over the

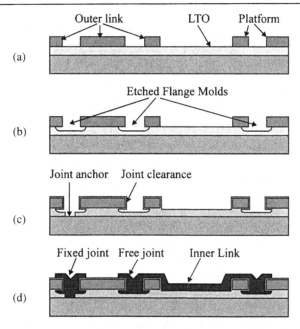

Fig. 6.14. Fabrication process of the micromanipulator: (a) after the first polysilicon step, (b) after isotropic flange mold etch, (c) after joint clearance oxide growth and joint anchor openings, (d) after second polysilicon step [44].

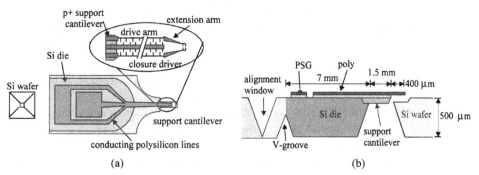

Fig. 6.15. Schematic of the overhanging microgripper (not to scale) [46]. (a) is the top view and (b) is a cross section.

cantilever beam of 1.5 mm. The gripper has a working range of ca. 6 μm, producing gripper forces in the range of 50 nN at a drive voltage of 50 V.

The fabrication process of the gripper involves a combination of anisotropic bulk etching (using EDP type F) and surface micromachining. The basic features of the process are shown in Fig. 6.15. First, the supporting cantilever beam is defined by a deep (12 μm) boron diffusion that serves as an etch stop for EDP. Then a PSG and a polysilicon layer are deposited and patterned.

The polysilicon layer will form the gripper and the electrical connections. An essential step in the process is the following deposition of a second PSG layer of 6-μm thickness. This layer is deposited in three steps with annealing steps in between to planarize the layer,

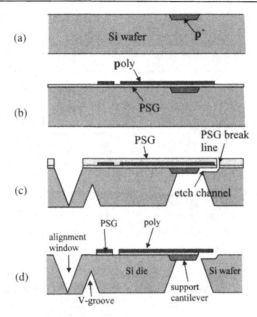

Fig. 6.16. Process sequence for the overhanging microgripper (redrawn from [46]). The process sequence (a) through (d) is described in the text.

filling possible pinholes, and for diffusion of the phosphorous into the polysilicon. This PSG layer has to serve as an etch mask in the following anisotropic etch steps. After etching an alignment window, break lines are defined in the frontside of the wafer and etch windows in the rearside. Then a second EDP etch step follows, in which the cantilever beam and the overhanging gripper structure are etched free. At this stage, the polysilicon structure is imbedded and protected in PSG. To work as an etch stop in EDP, the PSG layer had to be thick and pinhole-free.

The large stress in the heavily boron-doped region that forms the cantilever beam causes cracks in the PSG membrane. A progression of the cracks to the gripper structure could be prevented by fabrication of break lines in the PSG layer. The etch channels indicated in Fig. 6.16(d) are important to make final rinsing in water easier; in first process schemes the authors etched right through the substrate with much larger etch windows, but it turned out that nearly all grippers broke during rinsing.

The last step is the etching of the PSG layer, therewith releasing the polysilicon structure. This etch step is time-controlled. Microphotographs of the resulting microgripper are shown in Fig. 6.17.

The authors also report on the use of the microgripper: it was found that if the device looked right under an optical microscope, it inevitably worked properly. It was possible to handle polysterin spheres (2.7-μm diameter), dried red blood cells, and various protozoae. One problem that still needs to be addressed is the sticking of the objects to the gripper. Generally, the reproducibility of sticking was poor. It appeared that freshly processed grippers have a lower tendency to stick to objects and that the sticking problem was not less pronounced for "heavy" objects (as expected from dimensional analysis, weight does not play a role in the microworld as compared to surface forces). An obvious further problem is the fragility of the gripper tip. Great care in packaging and handling of the device is required.

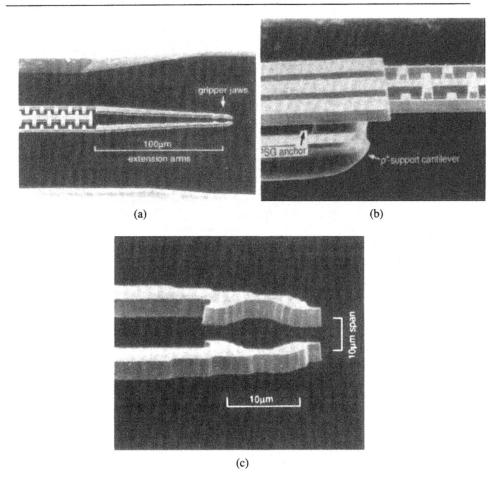

Fig. 6.17. (a) SEM photograph showing the flexible combdrive structures, extension arm, and the gripper jaws. (b) SEM photograph showing the three conduction lines and the first combdrive teeth and the tip of the support cantilever. (c) Close-up of the microgripper jaws [46].

6.3.4 Rotating micromotors

Although the micromotor is much farther away from any application that makes direct sense at this time, the results of the attempts to realize the first rotation micromotors were most spectacular. The groups first active in development of micromotors are at MIT, AT&T Bell Labs, and Berkeley. Later developments at the University of Wisconsin in Guckel's group resulted in a magnetic micromotor using the LIGA process [49]. Also, at Georgia Tech [50], a surface micromachined magnetic motor has been presented using polysilicon as the mechanical material. A different technology using electroplating has been described by a Japanese group [51]. In Europe, de Rooij's group at the University of Neuchâtel has developed an electrostatic micromotor. For more details, we refer to a recent excellent review article [52] on electrostatic micromotors.

The first polysilicon micromotors were quite small due to residual stress problems in the thin film, from which the films tend to take the shape of a potato chip. Larger motors can be fabricated by increasing the thickness of the structural layer.

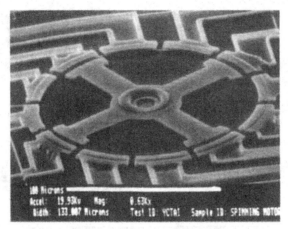

Fig. 6.18. Side-drive polysilicon variable capacitance motor. The rotor diameter is 120 μm, and air gaps are 2 μm wide [53].

Fig. 6.19. Top-drive polysilicon variable capacitance motor. The rotor diameter is 300 μm, and air gaps are 2 μm wide [54].

Basically, there are three types of variable capacitance micromotors. Side-drive motors and top-drive motors are shown in Figs. 6.18 (work of Tai from Berkeley [53]) and 6.19 (work of Mehregany from MIT [54]). Top-drive motors have the potential of producing a larger torque, but they are considerably less stable than side-drive motors due to large vertical forces. The torque in these motors is produced by the electric field between the activated stator poles and the rotor poles. The gap in top-drive motors is defined by the thickness of a PSG thin film, which is comparably easy. In side-drive motors the gap can be defined by photolithography, or by the lateral thickness of a PSG sacrificial layer. Since for successful operation the gap spacing must be below 2 μm, this type of motor puts high demands on the fabrication equipment and care of fabrication. In Fig. 6.20 we show a typical process sequence.

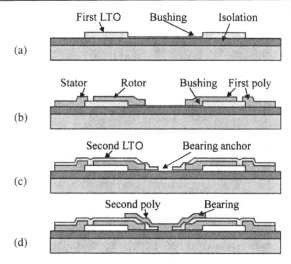

Fig. 6.20. Basic process sequence for a side-drive motor. (a) After patterning the first PSG layer ("LTO," low-temperature oxide); (b) after patterning rotor and stator; (c) after patterning second sacrificial layer; (d) completed device after sacrificial layer etching [54].

The third type of variable capacitance motors is the harmonic side-drive motor ("wobble motor"). In this motor, the rotor rolls ("wobbles") on the stator; the motor operation relies on normal electric forces attracting the rotor, which is electrically grounded. As the excitation of the poles rotate, the rotor will follow. Since the material for micromotors (polysilicon) is conductive, the rotor must not touch the excited stator poles. Therefore the design of micro wobble motors requires a bearing on which the rotor rolls. An example is shown in Fig. 6.21.

The voltage needed to drive the motors lies in the range of 10–100 V. This voltage corresponds to an electrical field of more than 10^8 V/m, well in agreement with the expectations of the breakdown extrapolated from the Paschen relationship [56]. The motive torque of the side-drive motors is typically 10 pNm (10^{-11} Nm). This is five times smaller than the torque that is needed to move the micromanipulator that is discussed in Section 6.3.1.

Typical revolution speeds are in the range of 10,000–20,000 rpm [57]. The revolution times are so small that the rotors do not move smoothly, rather they tend to flip from one energy minimum position to the next. For wobble motors the maximum speed is ca. 700 rpm [58].

The performance and lifetime of the micromotors described so far vary widely. With the first motors, just a few slow-stepping motions could be achieved at driving voltages of more than 100 V. One of the problems was a strong electrostatic attractive force between the rotor and the underlying substrate. This problem is now solved by changing the fabrication process, which now includes a conductive plate (polysilicon) on the substrate to shield the rotor from the substrate. The performance and lifetime have now been improved significantly. Motors rotating for many days have been fabricated, some have even rotated for so long that the actual lifetime is not yet known. In a recent *Scientific American* article [59] it is claimed that friction is no longer an issue for micromotors.

In order to reduce the friction between rotor and substrate, the contact surface between rotor and substrate must be reduced as much as possible. Two fabrication schemes were proposed to achieve this: the so-called "center-pin process" (Fig. 6.22, left), in which the

Fig. 6.21. A typical flange-bearing wobble motor. The rotor diameter is 100 μm, and air gap is 2.5 μm wide [55].

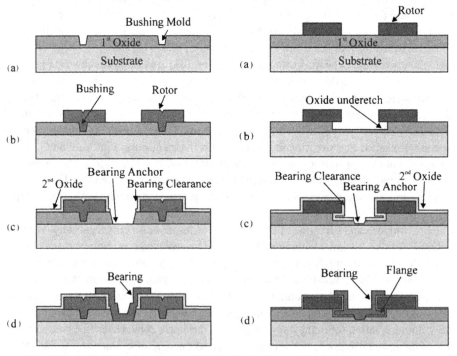

Fig. 6.22. (left) Center-pin process: (a) after the bushing mold is patterned; (b) after the rotor is patterned; (c) after the bearing anchor is patterned; (d) device before sacrificial layer etching. (right) Flange fabrication process: (a) after the rotor is patterned; (b) after isotropic oxide etching; (c) after the bearing anchor is patterned; (d) device before sacrificial layer etching [52].

Fig. 6.23. A released gear train fabricated in the center-pin process. The smallest gear is 120 μm wide [52].

rotor slides on the bushings which are formed during deposition of the first polysilicon layer, and the "flange process" (Fig. 6.22, right), in which a flange is formed by isotropic underetching of the oxide after deposition of the first polysilicon layer; during operation, the rotor slides on the flange. An example for a device fabricated using the center-pin process is given in Fig. 6.23, which shows a freely rotating (passive) gear train [52].

Wear limits the lifetime of micromotors and other sliding micromachines in at least two ways. The first obvious way is the wearing down of structures to dimensions where they fail to function. The other way may be more serious: dust and particles produced by wear can contaminate the structures, causing additional (perhaps fatal) friction and short circuits. Little is known about wear of microstructures.

Recently, Mehregany and Duhler demonstrated that polysilicon wobble motors can also be operated in demiwater and in silicone oil [60].

A discussion on design rules for side-drive motors can be found in [61, 62]. The main issues are

- the number of rotor poles to stator poles should be 2:3;
- there should be as many poles as possible;
- stator pole width over stator pole pitch should equal 0.6;
- the radius of the bearing should be as small as possible (the radius is limited by photolithography);
- the radius of the rotor should be as large as possible (limited by warpage);
- the gap spacing should be as small as possible, certainly below 2 μm;
- the bearing clearance must be small compared to the gap spacing; and
- the distance between the rotor and the shield plate (i.e., the height of the bushing in center-pin motors) should be as large as possible.

Generally, the mass of the rotor is so small that inertia forces can be completely ignored, since they are small compared to surface forces such as dry friction and sticking. This makes quite a difference with ordinary motors. In order to reduce the dry friction in motors and other rotating devices, such as the gear train shown in Fig. 6.21, the possibility of levitation of the rotors was considered using electrostatic and fluid flow forces. The feasibility of these ideas is demonstrated in [63–65]. However, combination of motors and the levitation technology has not been demonstrated yet.

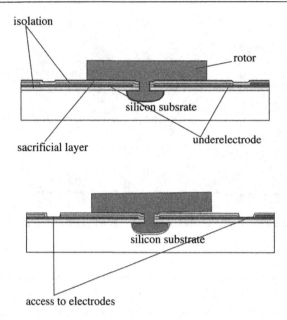

Fig. 6.24. Schematic of an axial gab wobble motor (redrawn from R. Legtenberg [41]). Above is the prior sacrificial layer etching, below is after release and opening of the bond pads.

In our own labs we succeeded in the fabrication of an electostratically driven axial gab wobble motor which is capable of driving micromechanical structures [41, 66]. The special aspect of this work is the anchoring of the axis. This is shown schematically in Fig. 6.24. The process starts with the deposition of a 1-μm thick low-stress silicon nitride layer. Low-stress Si_xN_y is obtained if one grows this material under excess of silicon; too much silicon leads to a compressive stress. Legtenberg used in an LPCVD furnace 10 sccm $SiCl_2H_2$ and 18 sccm NH_3 at 850°C and at a pressure of 200 mTorr. Then the underelectrode material is deposited (polysilicon, heavily doped with boron). The process parameters are (1) deposition: LPCVD at 590°C, 50 sccm silane flow at 250 mTorr, and (2) doping: solid source drive in for 1 hour at 1150°C. After stripping the boron silcate glass in BHF a second low-stress LPCVD nitride is deposited to isolate the rotor from the stator. Contact windows are opened by RIE using CHF_3/O_2 plasma. A first part of the sacrificial layer is deposited now (PECVD oxide). In a photolithographic step the ball bearing is defined by etching through the layer package (RIE using CHF_3/O_2), and the photoresist is stripped in an oxygen plasma. At the location of the bearing the silicon surface is now open, and the ball bearing is etched in an SF_6/N_2 plasma by isotropic RIE. A second sacrificial layer is deposited by LPCVD from tetra ethyl ortho siliate (TEOS). This layer conformally covers the ball bearing.

The next step is deposition of the structural layer and its patterning by photolithography. The structural layer consists of a heavily doped polysilicon layer, deposited and doped similarly to the electrode layer described above, except with a 2-μm thickness and a 6-μm thick amorphous silicon layer sputtered in argon. This layer is annealed at 450°C for one hour to reduce the residual stress in the amorphous layer. Patterning is done by anisotropic RIE in an SF_6/O_2 plasma. The process is finished by sacrificial layer etching and by opening

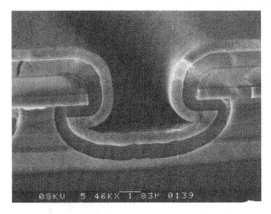

Fig. 6.25. Cross section of a slider bearing, similar to the ball bearing used for the wobble micromotor [41].

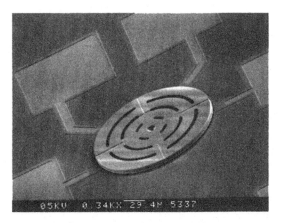

Fig. 6.26. SEM image of an axial gab wobble motor, showing the rotor and bond pads to four stator electrodes. The rotor diameter is 200 μm, and its thickness is 8 μm. [41].

the contact windows. Results of the process are seen in Figs. 6.25 (ball bearing[‡]) and 6.26 (the top view of the device).

By introducing teeth to the rotor, other structures can be driven, such as gears or sliders. These structures have been fabricated in a process similar to the one described above, but it is more complicated because more masks are required. A few examples are shown in Figs. 6.27 through 6.30.

6.3.5 *Resonators and sensors*

Perhaps except for the microgripper and the micropumps, all actuators described so far are far away from any industrial application. The situation may be different for surface micromachined sensors (Guckel and Burns [67]).

[‡] Actually, this is not the ball bearing but a slider bearing. One needs very good luck to break the wafer right through a bearing with a radius of 10 μm.

Fig. 6.27. A gear train with ball bearings and an axial gab wobble motor [41, 66].

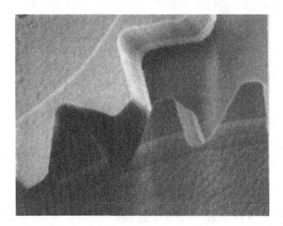

Fig. 6.28. Close-up of the gear teeth [66].

Fig. 6.29. An axial gab wobble motor driving a linear slider [66].

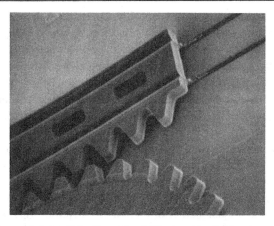

Fig. 6.30. Close-up of the slider from Fig. 6.29, showing the bearing at the upper left of the image [66].

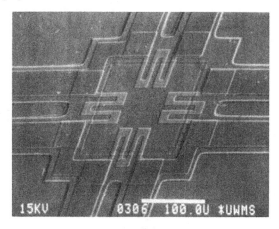

Fig. 6.31. Micromachined polysilicon membrane pressure sensor. The contours of the membrane can be seen as well as the polysilicon piezoresistive strain gauges [68].

A very simple but ingenious pressure sensor has been described by Guckel and Burns [67, 68]. The fabrication process is based on a local oxidation over which a polysilicon layer was deposited. Via openings in the membrane, the oxide is etched away, leaving a polysilicon membrane over a shallow cavity. The openings are sealed by growing an LPCVD silicon nitride layer. This step leaves the cavity under a low pressure (of the order of 1 mbar [65]). A second polysilicon layer is grown for piezoresistive elements. Figure 6.31 gives an SEM photograph of the completed device.

A lot of work has been published on resonating sensors. The first surface micromachined polysilicon resonator is due to Howe and Muller [1]. In fact, the very first underetched resonator is due to Nathanson et al. [2] and was published as early as 1967. However, in this work, use was made of quite different materials. The great achievement of Howe was the use of IC-compatible materials and processes (LPCVD polysilicon and PSG), and he succeeded in integrating an NMOS[‡] amplifier on a chip.

[‡] n-metal oxide silicon doped.

Howe used electrostatic excitation and capacitive detection of the vibration. The capacitive detection results in signals that are too small for off-chip detection. Consequently, on-chip amplification was the only way to obtain a measurable signal. Later it was claimed by Linder, also working on resonating polysilicon microbridges, that the integration of ICs on-chip was not necessary to obtain sufficiently large signals [70]. Howe's sensor was designed to measure the change of mass of the resonator after absorption of chemicals from surrounding vapor; therefore, the bridge had to be exposed to the ambient. It turned out however that the microbridge was overdamped due to squeezing of the gas film in the narrow gap between the microbridge and the substrate. In a second generation, apertures were etched into the bridge, facilitating the flow of gas out of the gap. This measure resulted in a quality factor of 4, which is still far too small for the use in a sensor system.

Other disadvantages are that the mass of the microbridge, being exposed to the ambient, can change due to contamination or corrosion. Therefore for sensor applications, microbridges should be sealed under vacuum. This was pioneered by Ikeda et al. [71–73] as reported in a previous chapter. Guckel demonstrated that it is also possible to fabricate

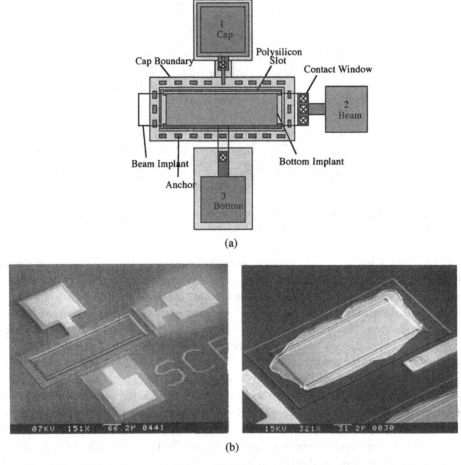

(a)

(b)

Fig. 6.32. (a) Typical resonator layout. (b) SEM photograph of a single encapsulated polysilicon resonator showing the cap and the bond pads (left) and the beam after the cap has been removed manually (right) [13].

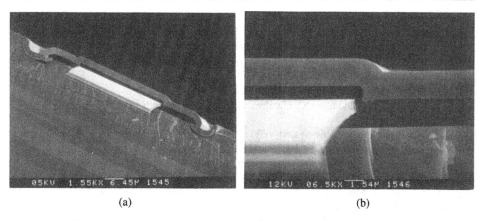

Fig. 6.33. (a) Cross-sectional view of the cavity showing the released microbridge. (b) A close-up [13].

sealed resonating microbridges by surface micromachining [74–77]. Related work has been published by Tilmans and Legtenberg [13, 29]. Excitation and detection in this case are electrostatic and capacitive, respectively. The boron-implanted region in the first step serves as the bottom electrode, which is electrically isolated from the rest of the structure by a silicon nitride film. Deposition and patterning of the sacrificial layer (PECVD SiO_2) and deposition of a polysilicon layer follow. This polysilicon layer will form the resonator. The polysilicon is boron doped (to increase the electrical conductivity), annealed, and patterned by RIE. A second deposition of a sacrificial oxide follows. After patterning the oxide, an oxide layer is thermally grown onto the polysilicon and a thin silicon nitride layer is deposited. Anchors are opened, and a polysilicon layer is deposited, boron doped, and patterned. The sacrificial layer is etched in HF through the etch channels.

The freeze-drying procedure with cyclohexane, described in a previous paragraph, was used to prevent stiction. After this step, the cavity is reactively sealed by growth of LPCVD silicon nitride. This deposition method leads to a good step coverage so that silicon nitride grows in the cavity and in the etch channels. Because the latter are much narrower than the cavity, they will be filled by growth of a thin film only, and the cavity is hermetically sealed. Yet, the pressure in the cavity is so high that the quality factor of the beams is only about 600. SEM photographs of the resulting device are shown in Figs. 6.32 and 6.33.

The resonant frequency of Tilmans's and Guckel's resonators are in the range of 100 kHz. Tests of the devices show that the resonance frequency is very sensitive to loads on the chip, in agreement with the theory. The small signal problem with capacitive excitation has been solved by Tilmans by using dummy bridges that will not vibrate. In a compensation circuit the difference signal is large enough for off-chip detection.

A problem of both designs is associated with large-amplitude effects that lead to a dependence of the resonance frequency on the amplitude and ultimately to instability of the sensor [13, 78, 79].

References

[1] R. T. Howe and R. S. Muller, IEEE Trans. Electron Devices **ED-33**, 499 (1986).
[2] H. C. Nathanson, W. E. Newell, R. A. Wickstrom, and J. R. Davies Jr., IEEE Trans. Electron. Dev. **ED-14**, 117 (1967).

[3] R. T. Howe, unpublished results (1988).

[4] H. Guckel, D. W. Burns, H. A. C. Tilmans, D. W. DeRoo, and C. R. Rutigliano, Techn. Dig. IEEE Solid State Sensor and Actuator Workshop, Hilton Head, SC, June 1988, p. 96.

[5] L.-S. Fan, R. S. Muller, W. Yun, R. T. Howe, and J. Huang, Proc. IEEE Workshop on Micro Electro Mechanical Systems, Napa Valley, CA, Feb. 11–14, 1990, p. 177.

[6] G. G. Stoney, Proc. R. Soc. London Ser A **82**, 172 (1909).

[7] E. I. Bromley, J. N. Randall, D. C. Flanders, and R. W. Mountain, J. Vac. Sci. Technol. **B1**, 1364 (1983).

[8] O. Tabata, K. Kawahata, S. Sugiyama, and I. Igarashi, Proc. IEEE Workshop on Micro Electro Mechanical Systems, Salt Lake City, Utah, Feb. 20–22, 1989, p. 152.

[9] J. Y. Pan, P. Lin, F. Maseeh, and S. D. Senturia, Techn. Dig. IEEE Solid-State Sensor and Actuary Workshop, Hilton Head, SC, June 4–7, 1990, p. 70.

[10] Y. C. Tai and R. S. Muller, Proc. IEEE Workshop on Micro Electro Mechanical Systems, Napa Valley, CA, Feb. 11–14, 1990, p. 147.

[11] K. E. Crowe and R. L. Smith, J. Electrochem. Soc. **136**, 1566 (1989).

[12] K. E. Petersen and C. R. Guarnieri, J. Appl. Phys. **50**, 6761 (1979).

[13] H. A. C. Tilmans, Ph.D. Thesis, University of Twente, Enschede, The Netherlands (1993).

[14] L. M. Zhang, D. Uttamchandani, B. Culshaw, and P. Dobson, Meas. Sci. Technol. **1**, 1343 (1990).

[15] Y. B. Gianchandani and K. Najafi, Techn. Dig. IEEE Solid-State Sensor and Actuator Workshop, Hilton Head, SC, June 13–16, 1994, p. 116.

[16] R. I. Pratt, G. C. Johnson, R. T. Howe, and J. C. Chang, Proc. Transducers, San Francisco, CA, June 24–27, 1991, p. 205.

[17] M. Biebl, G. Brandl, and R. T. Howe, Proc. Transducers Stockholm, Sweden, June 25–29, 1995, p. 80.

[18] P. M. Osterberg, R. K. Gupta, J. R. Gilbert, and S. D. Senturia, Techn. Dig. Solid-State Sensor and Actuator Workshop, Hilton Head SC, June 13–16, 1994, p. 184.

[19] K. Najafi and K. Suzuki, Proc. IEEE Workshop on Micro Electro Mechanical Systems, Salt Lake City, Utah, Feb. 20–22, 1989, p. 96.

[20] Y. C. Tai and R. S. Muller, Techn. Dig. IEEE Solid-State Sensor and Actuator Workshop, Hilton Head SC, June 1988, p. 88.

[21] D. J. Monk and D. S. Soane, J. Electrochem. Soc. **140**, 2339 (1993).

[22] D. J. Monk, D. S. Soane, and R. T. Howe, J. Electrochem. Soc. **141**, 264 (1994).

[23] D. J. Monk, D. S. Soane, and R. T. Howe, J. Electrochem. Soc. **141**, 270 (1994).

[24] J. Liu, Y. C. Tai, J. Lee, K.-C. Pong, Y. Zohar, and C.-M. Ho, Proc. IEEE Workshop on Micro Electro Mechanical Systems, Ft. Lauderdale, FL, Feb. 7–10, 1993, p. 71.

[25] B. E. Deal and A. S. Grove, J. Appl. Phys. **36**, 3770 (1965).

[26] R. Legtenberg, H. A. C. Tilmans, J. Elders, and M. Elwenspoek, Sensors and Act. **A43**, 230 (1994).

[27] H. Guckel, J. J. Sniegowski, and T. R. Christenson, Proc. IEEE Micro Electro Mechanical Systems Workshop, Salt Lake City, Utah, Feb. 20–22, 1989, p. 71.

[28] N. Takeshima, K. J. Gabriel, M. Ozaki, J. Takahasji, H. Horiguchi, and H. Fujita, Proc. 6th Int. Conf. Solid-State Sensors and Actuators (Transducers'91), San Fransisco, CA, June 24–27, 1991, p. 63.

[29] R. Legtenberg and H. A. C. Tilmans, Sensors and Act. **A45**, 57 (1994).

[30] G. T. Mulhern, D. S. Soane, and R. T. Howe, Proc. 7th Int. Conf. Solid-State Sensors and Actuators (Transducers'93), Yokohama, Japan, June 1993, p. 296.

[31] M. Orpana and A. O. Korhonen, Proc. 6th Int. Conf. Solid-State Sensors and Actuators (Transducers'91), San Francisco, CA, June 1991, p. 957.

[32] D. Kobayashi, C. J. Kim, and H. Fujita, Proc. 7th Int. Conf. Solid-State Sensors and Actuators (Transducers'93), Yokohama, Japan, June 1993, p. 14.

[33] R. A. Brennen, M. G. Lim, A. P. Pisano, and A. T. Chou, Proc. IEEE Solid-State Sensors and Actuators Workshop, Hilton Head Island, SC, June 1990, p. 135.

[34] R. L. Alley, G. J. Cuan, R. T. Howe, and K. Komvopoulos, Proc. IEEE Solid-State Sensors and Actuators Workshop, Hilton Head, SC, June 1992, p. 288.

[35] W. C. Tang, T. C. H. Nguyen, and R. T. Howe, Proc. IEEE Workshop on Micro Electro Mechanical Systems, Salt Lake City, Utah, Feb. 20–22, 1989, p. 53.

[36] F. S. A. Sandejas, R. B. Apte, W. C. Banyai, and D. M. Bloom, Proc. 7th Int. Conf. Solid-State Sensors and Actuators (Transducers'93), Yokohama, June 1993, p. 6.

[37] R. L. Alley, P. Mai, K. Komvopoulos, and R. T. Howe, Proc. 7th Int. Conf. Solid-State Sensors and Actuators (Transducers'93), Yokohama, June 1993, p. 288.

[38] A. Boyde and C. Wood, J. Microsc. **90**, 221 (1969).

[39] H. Guckel, J. J. Sniegowski, T. R. Christenson, S. Mohney, and T. F. Kelly, Sensors and Act. **A20**, 117 (1989).

[40] K. Gabriel, J. Jarvis, and W. Trimmer, (eds.), *Small Machines, Large Opportunities: A Report on the Emerging Field of Microdynamics*, National Science Foundation (1989).

[41] R. Legtenberg, Ph.D. Thesis, University of Twente, The Netherlands, 1996.

[42] R. Legtenberg, A. W. Groeneveld, and M. Elwenspoek, J. Micromachining Microeng. **6**, 320 (1966).

[43] L.-S. Fan, Y. C. Tai, and R. S. Muller, IEEE Trans. Electron Devices **35**, 724 (1988).

[44] F. Behi, M. Mehregany, and K. G. Gabriel, Proc. IEEE Workshop on Micro Electro Mechanical Systems, 1990, p. 159.

[45] Z. Liza Zhang and Noel C. MacDonald, Proc. 6th Int. Conf. Solid-State Sensors and Actuators (Transducers'91), San Francisco, CA, June 1991, p. 520.

[46] C.-J. Kim, A. P. Pisano, and R. S. Muller, J. Micro Electro Mech. Syst. **1**, 31 (1992).

[47] W. C. Tang, T.-C. H. Nguyen, and R. T. Howe, Sensors and Act. **A20**, 25 (1989).

[48] W. C. Tang, T.-C. H. Nguyen, and R. T. Howe, Sensors and Act. **A21–A23**, 328 (1990).

[49] H. Guckel, K. J. Skrobis, T. R. Christenson, J. Klein, S. Han, B. Choi, E. G. Lovell, and T. W. Chaplman, J. Micromech. Microeng. **1**, 135 (1991).

[50] C. H. Ahn, Y. J. Kim, and M. G. Allan, Proc. MEMS, Ft. Lauderdale, FL, February 7–10, 1993, p. 1.

[51] T. Furuhata, T. Hirano, L. H. Lane, R. E. Fontana, L. S. Fan, and H. Fujita, Proc. MEMS, Ft. Lauderdale, FL, February 7–10, 1993, p. 161.

[52] M. Mehregany and Y.-C. Tai, J. Micromechanics Microeng. **1**, 73 (1991).

[53] Y.-C. Tai and R. S. Muller, Sensors and Act. **A21–A23**, 180 (1990).

[54] M. Mehregany, S. F. Bart, L. S. Tavrow, J. H. Lang, S. D. Senturia, and M. F. Schlecht, Sensors and Act. **A21–A23**, 173 (1990).

[55] V. R. Dhuler, M. Mehregani, S. M. Phillips, and J. H. Lang, Proc. IEEE Workshop on Micro Electro Mechanical Systems, Travemünde, Germany, Feb. 4–7, 1992, p. 171.

[56] F. Paschen, Ann. Phys. Lpz. **37**, 69 (1889).

[57] M. Mehregany, P. Nagarkar, S. D. Senturia, and J. H. Lang, Proc. IEEE Workshop on Micro Electro Mechanical Systems, Napa Valley, CA, Feb. 11–14, 1990, p. 1.

[58] M. Mehregany, S. D. Senturia, and J. H. Lang, Proc. IEEE Solid-State Sensors and Actuators Workshop, Hilton Head Island, SC, June 1990, p. 17.

[59] G. Stix, Sci. Amer. November, p. 72 (1992).

[60] M. Mehregany and V. R. Duhler, J. Micromech. Microeng. **2**, 1 (1992).

[61] S. Tavrow, S. B. Bart, and J. H. Lang, Sensors and Act. **35**, 33 (1992).

[62] S. Tavrow, Ph.D. Thesis, MIT Cambridge (1991).

[63] W. C. Tang, M. G. Lim, and R. T. Howe, Proc. IEEE Workshop on Micro Electro Mechanical Systems, Napa Valley, CA, June 11–14, 1990, p. 24.

[64] S. Kumar and D. Cho, Proc. 6th Int. Conf. Solid-State Sensors and Actuators (Transducers'91), San Francisco, CA, June 1991, p. 882.

[65] J.-B. Huang, Q.-Y. Tong, and P.-S. Mao, Proc. 6th Int. Conf. Solid-State Sensors and Actuators (Transducers'91), San Francisco, CA, June 1991, p. 894.

[66] R. Legtenberg, E. Berenschot, M. Elwenspoek, and J. H. J. Fluitman, Proc. MEMS, San Diego, CA, February 11–15, 1996, p. 204.

[67] H. Guckel and D. W. Burns, Technical Digest IEEE IEDM Conf., 1984, p. 233.

[68] H. Guckel and D. W. Burns, and C. R. Rutigliano, Proc. IEEE Solid-State Sensors and Actuators Workshop, Hilton Head Island, SC, June 1986.

[69] H. Guckel, Sensors and Act. A28, 133 (1991).

[70] C. Linder, E. Zimmermann, and N. F. de Rooij, Sensors and Act. A25–A27, 591 (1991).

[71] K. Ikeda and T. Watanabe, Proc. SICE'86, Tokyo, Japan, 1984, p. 651.

[72] K. Ikeda, H. Kuwayama, T. Kobayashi, T. Watanabe, T. Nishikawa, T. Yoshida, and K. Harada, Sensors and Act. A21–A23, 1007 (1990).

[73] K. Ikeda, H. Kuwayama, T. Kobayashi, T. Watanabe, T. Nishikawa, and T. Yoshida, Sensors and Act. A21–A23, 193 (1990).

[74] H. Guckel, J. J. Sniegowski, T. R. Christenson, and F. Raissi, Sensors and Act. A21–A23, 346 (1990).

[75] J. J. Sniegowski, H. Guckel, and T. R. Christenson, Proc. IEEE Solid-State Sensors and Actuators Workshop, Hilton Head Island, SC, June 1990, p. 9.

[76] J. J Sniegowski, Ph.D. Thesis, University of Wisconsin, Madison (1991).

[77] J. D. Zook, B. W. Burns, H. Guckel, J. J. Sniegowski, R. L. Engelstad, and Z. Feng, Sensors and Act. A35, 51 (1992).

[78] Chengqun Gui, R. Legtenberg, M. Elwenspoek, and J. H. J. Fluitman, J. Micromechanics Microeng. 5, 183 (1995).

[79] Chengqun Gui, R. Legtenberg, H. A. C. Tilmans, J. H. J. Fluitman, and M. Elwenspoek, Proc. IEEE Workshop on Micro Electro Mechanical Systems, Amsterdam, The Netherlands, Jan. 29–Feb. 2, 1995, p. 157.

7

Isotropic wet chemical etching of silicon

7.1 Etchants and etching diagrams

Isotropic wet chemical etching is much less relevant for micromachining than anisotropic etching. This is primarily because the shape of the etched structures is controlled by diffusion and convection in the etchant, and these processes are much less reproducible than the great reduction of the etch rate of a particular crystallographic direction.

However, isotropic etching is useful in several respects. For example, anisotropic etched structures have sharp corners. Mechanical engineers avoid sharp corners leading to stress concentration. The structure turns out to be much more fragile than expected from estimates of the yield strength; they tend to break at sharp corners. Sharp corners might also cause problems in liquid handling devices such as pumps, valves, flow sensors, channels, and so on. Problems can arise as sharp corners trap gas bubbles more easily than curved structures.

Compared to anisotropic etching, the process of isotropic etching has some attractive features. It is a room temperature process. It is much faster (probably due to the greater undersaturation), and photoresist is a suitable mask material. We shall return to these points below.

The most well-known isotropic etchant is a mixture of HNO_3 and HF. The first is an oxidizing agent and the latter dissolves the oxide. More details regarding the chemical reaction will be given below. The HF–HNO_3 mixture is dissolved with water or acetic acid (CH_3COOH). Pioneering work has been done by Robbins and Schwartz [1–4].

In Fig. 7.1(a) we give in a ternary diagram etch rates as a function of the composition of the etchants (water, HNO_3, and HF) at 25°C. The highest etch rate is observed around a weight ratio HF–HNO_3 of 2:1. Note the maximum etch rate of 800 μm/min, nearly three orders larger than anisotropic etch rates. The etch rate becomes smaller by adding water, and the maximum shifts to higher HNO_3 concentrations. The etch rate depends exponentially on the concentration. As a consequence, the etch results reproduce rather badly, since small perturbations in the concentration, caused, e.g., by instabilities in the convection in the etch vessel, cause great changes in the etch rate. See also the discussion in [5].

Dissolving the HF–HNO_3 mixture with acetic acid gives qualitatively similar results (see Fig. 7.1(b)). The most noticeable difference is that water has a much greater effect on the etch rate in the HF-rich corner than formic acid.

The topology of the surfaces [4] depends strongly on the composition of the etch solution. Around the maximum etch rates the surfaces appear quite flat with rounded edges, and the slower etching solutions lead to very rough surfaces. The concentration dependence of the surface topology in solutions of the HF–HNO_3 mixture with water and formic acid is given in Fig. 7.2.

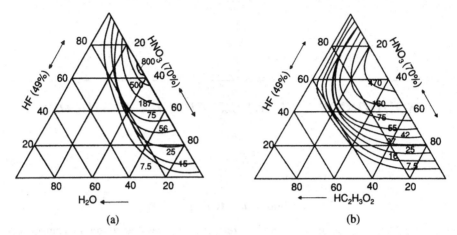

Fig. 7.1. Silicon isotropic etch rate in μm/min. (a) HF—HNO$_3$—H$_2$O; (b) HF—HNO$_3$—CH$_3$COOH [5].

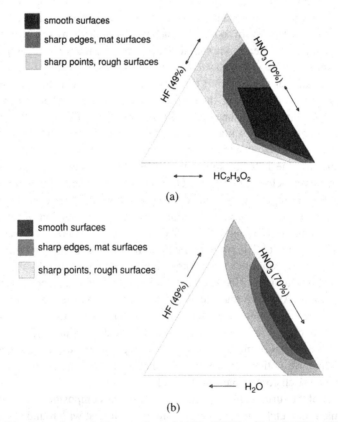

Fig. 7.2. Topology of etched silicon surfaces. (a) HF—HNO$_3$—H$_2$O; (b) HF—HNO$_3$—CH$_3$COOH [5]. Dark: smooth surfaces; medium: sharp edges, nonreflecting surfaces; light: rough surfaces.

Fig. 7.3. Cross section of a hole etched into the silicon through a mask opening using isotropic etching. (left) Without agitation; (right) with agitation.

Although doping seems to change the etch rate in these etch systems, attempts to exploit these differences for micromachining have failed so far [5]. This situation is different in anodic isotropic etching.

7.2   Diffusion and stirring

Diffusion in the solution to and from the crystal surface contributes to the etch rate. Stirring therefore has a considerable effect on the etch rate.

If one etches holes, stirring influences the shape of the hole. Without stirring the hole becomes flat, while with stirring the hole approaches the shape of a hemisphere. This is illustrated schematically in Fig. 7.3.

7.3   Mask materials

We already mentioned the possibility of positive photoresist as a convenient mask material. An alternative is Si_3N_4, which etches very slowly in the isotropic solutions. Within limits, SiO_2 can also be used as a mask material, since the etch rate of thermal oxide is 100 times smaller than the etch rate of silicon. Gold is also resistant against HF–HNO_3-based etchants and can be used as a mask material and as metallization before the isotropic etch step. Aluminium is etched very quickly and is therefore not suitable as a mask material.

7.4   Anodic HF etching

7.4.1   Introduction

Anodic etching of silicon with the aid of aqueous HF solutions is a technique which has been used for many years to thin or polish silicon wafers [6, 7] and to manufacture thick porous silicon layers or SOI structures [8, 9]. Although not many practical applications have been reported so far, HF anodic etching can also be used as a micromachining tool. This can be achieved by exploiting the selectivity of the method for silicon of certain dopant types and dopant concentrations [10–12]. Some researchers have reported on micromechanical applications of this technique, exploiting the etching selectivity between low-doped n-type epitaxial layers and highly doped n-type substrates [13–15].

In this chapter we comment on the etch mechanism, as far as it is understood, and demonstrate the possibilities of anodic HF etching for micromachining. Finally, we report on the use of the sharp etching selectivity between n-type and p-type material, both low-doped, which is very appropriate for micromachining purposes. We will demonstrate that the latter method can be used to realize monocrystalline silicon microstructures.

7.4.2   Electrochemical etching mechanism

The electrochemical behavior of silicon in fluoride-containing solutions has been treated at length in the literature of the last decennial [6, 7, 10, 12, 16–19], mostly with an emphasis on anodic dissolution. In this section we present a short resumé of the current state of understanding of this matter. It has to be mentioned, however, that up to now no thoroughly convincing and consistent description of the reaction mechanism has been given, despite the availability of a significant amount of experimental data and the large interest in the subject.

The current-voltage characteristics of lightly doped silicon samples in contact with concentrated HF solutions roughly resemble that of diodes, where the rectifying action is in opposite directions for n-type and p-type silicon: in the anodic region the current increases exponentially with the electrode potential in the case of p-type silicon, whereas for n-type the current saturates at a certain current density [16]. The anodic saturation current in the case of n-type silicon increases when the sample is illuminated, indicating that the current is limited by hole diffusion.

Similar behavior has been found for p-type material in the cathodic regime, i.e., in the cathodic regime the current is limited by electron diffusion. The fact that simultaneously hydrogen evolution is observed at the silicon electrodes indicates that the cathodic current is due to the discharge of protons, a process which proceeds almost entirely over the conduction band [16].

A closer look at the anodic regime of the current-voltage curves for p-type silicon leads to the conclusion that there exists a critical current density, observable as a peak in the curves, below which, at least at high HF concentrations (20 wt% HF), the formation of an amorphous film, composed largely of silicon, and evolution of hydrogen gas are observed. For low HF concentrations (a few wt% HF), no formation of amorphous silicon is observed. It was observed that in all cases for currents above the critical current density, electropolishing occurred [6, 7, 12, 16, 17]. The current peak normally occurs at anodic voltages of about 0.2 to 1 V, dependent on the HF dilution (the peak occurs at a higher voltage if the HF concentration is higher) and the p-type dopant concentration (the peak shifts to higher voltages if the dopant concentration is lower).

For the low-voltage regime, i.e., below the current peak, experimental evidence exists that silicon dissolution is largely divalent and leads to an amorphous film, which is consistent with the following scheme of reactions [6, 16, 17].

(a) Charge transfer:

$$Si + 2HF + 2h^+ = SiF_2 + 2H^+ \tag{7.1}$$

(b) Disproportionation:

$$2SiF_2 = Si\ (amorphous) + SiF_4 \tag{7.2}$$

$$SiF_4 + 2HF = H_2SiF_6 \tag{7.3}$$

(c) Evolution of hydrogen results mainly from the reactions

$$SiF_2 + 4HF = H_2SiF_6 + H2 \tag{7.4}$$

$$SiF_2 + 2H_2O = SiO_2 + 2HF + H_2 \tag{7.5}$$

(d) The silicon dioxide may dissolve by the reaction

$$SiO_2 + 6HF = H_2SiF_6 + 2H_2O \tag{7.6}$$

At potentials positive to the current peak, silicon is dissolved in the tetravalent state [6, 17], and no hydrogen is evolved. This change of the dissolution mechanism is also observed in concentrated HF at high currents [16]. Since the dissolution reaction in this case also proceeds entirely over the valence band [16], the corresponding reaction will probably be of the following kind:

$$Si + 2H_2O + 4h^+ = SiO_2 + 4H^+. \tag{7.7}$$

SiO_2 is not soluble in water and forms a highly resistant film on the electrode surface, which can be dissolved by a chemical reaction with the fluoride in the solution, i.e., Reaction (7.6). Very recently experimental evidence for the presence of such an oxide film was derived from the impedance response of silicon during electropolishing in HF solutions, although it was also found that the film does not behave like bulk SiO_2 [18].

While p-type specimens can very easily be etched electrochemically by HF, low-doped n-type electrodes need the generation of holes either by illumination or by strong oxidants which can extract electrons from the valence band [19]. Thus, practically no oxidation current is observed for n-type silicon at anodic polarizations in HF in the dark, not even at rather high voltages.

However, if in a 5 wt% HF solution an anodic polarization is applied to n-type silicon specimens with dopant concentrations higher than $2 \times 10^{16} \, \text{cm}^{-3}$, dissolution does occur [10–12], although the etching process is very inhomogeneous and leads to porous (but monocrystalline) surfaces. An explanation of this phenomenon is proposed in Ref. [10]: because of the positive bias of an n-type silicon anode in an aqueous HF solution, a positively charged depletion layer forms at its surface, which supports the almost complete voltage drop in the system. Besides the before-mentioned ways of injecting holes in the n-type silicon surface, excess holes can also be generated by Zener breakdown at this depletion layer. The essential point in the explanation is that this breakdown will occur locally at the surface, possibly at positions where crystal defects or dopant fluctuations occur, leading to etching channels and porous layer formation. The density of the channels was found to increase with the donor concentration.

At relatively high donor concentrations, however, a bright polished surface was obtained for etching at 10 V [10] (at lower voltages the formation of a brownish layer was observed). It seems that at the high dopant concentrations tunneling is the most probable breakdown mechanism.

7.4.3 Geometry control of etched structures

For the implementation of HF anodic etching as a tool to realize monocrystalline silicon microstructures, several methods are possible which exploit the characteristic properties of the etching technique. We shall discuss these methods in more detail below.

(a) *Masking the surface*

This can be done by conventional photolithography and layer deposition. The method takes advantage of the fact that silicon only etches in HF and NH$_4$F if it is anodically polarized. This has the important advantage that etching can be stopped by simply turning off the voltage supply.

Advantage of the isotropic character of the etching could be made in liquid handling systems to etch smooth channels. It is obvious that geometry control is easily accomplished by switching off the voltage after a given process time rather than by removing the wafer from the etch bath. Without use of etch stop mechanisms the applicability of

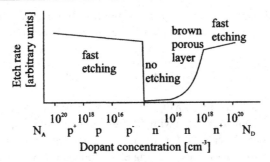

Fig. 7.4. Etch rate dependence on dopant concentration for HF anodic etching of silicon [20].

just masking is rather restricted, since the shape of the etched structures is completely
determined by diffusion fields in the etchants.

(b)   *Changing the conductivity type of Si via dopant incorporation methods*
This method takes advantage of the anodic etching selectivity of the technique, as
depicted in Fig. 7.4 (this figure was derived from the experimental data in Refs.
[10–12]).
Ion implantation or dopant diffusion through masks can be used to create local changes
in the concentration and type of dopants in the silicon crystal. Another technique is
the deposition of epitaxial Si layers which are doped differently than the substrate.
The latter method has been used exploiting the selectivity of HF anodic etching
between n+-silicon and n-silicon. The main features of this method have been reported
by Theunissen et al. [12], who investigated several combinations of epitaxial layer
and substrate doping concentrations and types. Other usable combinations are n-epi
on p-type substrates and p-epi on n-type or p-type substrates.
Esashi et al. [14] used n-doped epitaxial layers (10 Ω cm) on n+-doped silicon substrates (0.01 Ω cm) in order to etch circular diaphragms for pressure sensors. An etch
selectivity in a 5 wt% aqueous HF solution of 1:1000 is reported.
In Fig. 7.5, we show a schematic drawing of the etch apparatus and a microphotograph
of an isotropically etched membrane. This photo is from Shengliang et al. [21] who
used an NH₄F solution (5 wt%) instead of HF. Shengliang reports a selectivity of
n-silicon to n+-silicon (0.001 Ω cm) of 300.
In the etching apparatus used by Esashi et al. the electrodes were placed as close to
each other as possible (500 μm) in order to realize a sufficiently large electrical field
strength with reasonable low voltages. Typically, an electrical field of 50–100 V/cm
is required.
The application of this method was also discussed by Benjamin [22]; this author uses
the dopant-selective anodic formation of porous silicon which is dissolved chemically
afterwards.

Figures 7.6(a) and (b) give examples of the inactivation of p-type Si by n-doping and the
activation of n-type by p-doping, respectively. The use of the selectivity between n- and
p-silicon is described in detail below.

(c)   *Optical activation of unetchable silicon*
When n-type silicon is exposed to light, electron-hole pairs are generated in the
crystal. Because of the applied anodic potential, the electrons are drawn away from

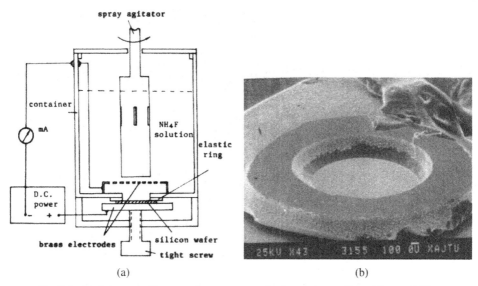

Fig. 7.5. (a) Schematic diagram of an apparatus for isotropic anodic etching, and (b) a microphotograph of an etched circular membrane [21].

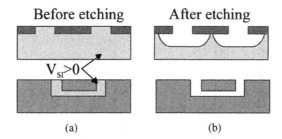

Fig. 7.6. Geometry control in HF anodic etching. (a) p-silicon with implanted n-type regions; (b) n-doped wafers with a combination of diffused and implanted p-type regions.

the surface, and the remaining free holes can participate in the etching process. Local light exposure can thus be used to etch the n-type silicon locally. This method has been tested by Ternez and Hök [23], who demonstrated the etching on a hole into a silicon wafer just by local illumination with a 5-mW HeNe laser. The etch selectivity between illuminated and nonilluminated silicon was more than 250, with an etch rate of 2.5 μm/min of the illuminated silicon. The structure obtained is shown in Fig. 7.7. For other methods using illumination for anodic HF etching we refer to [24].

7.4.4   Low-doped selective anodic HF etching

7.4.4.1   Experimental details

In the following we demonstrate the possibility of realizing structures using method b, which emphasizes the possibilities of the HF anodic etching technique. To this purpose, some exploratory experiments were performed according to method b.

All of the silicon wafers used were (100) oriented with an acceptor (boron) concentration of 1.4×10^{15} cm^{-3}. In some of these wafers, regions were implanted with phosphorous

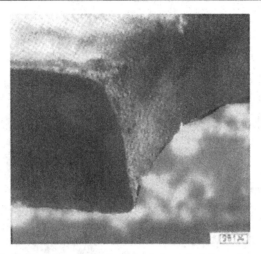

Fig. 7.7. SEM showing a cross section of a hole etched through a silicon wafer by photoelectrochemical etching [23].

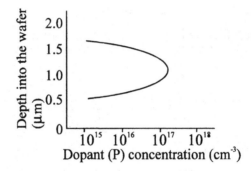

Fig. 7.8. Doping profile of the P-implanted regions after annealing.

using positive photoresist as a mask. Figure 7.8 shows the doping profile of the implanted regions.

After removing the resist and annealing (in the case of wafers with implanted regions), an aluminium layer was deposited on the reverse side to serve as an ohmic contact to the p-substrate. This aluminium layer was covered with a protective resist layer, except for a small contact area.

The anodic etching process was performed as follows: the wafer to be etched is immersed in a 5% aqueous HF solution at room temperature (see Fig. 7.9). A positive voltage of 1.5 V with respect to a platinum electrode in contact with the etchant is applied to the aluminium; corresponding current densities at the silicon–solution interface during etching range from 60 to 80 mA/cm. The surface area being etched is approximately 0.25 cm, and the size of the platinum electrode is 4×5 cm. Apart from these experiments investigating the feasibility of the etching technique for micromachining purposes, some experiments were also performed to characterize the etching process. In those experiments, other voltages and HF concentrations were used.

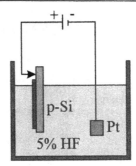

Fig. 7.9. Experimental setup for HF anodic etching of silicon.

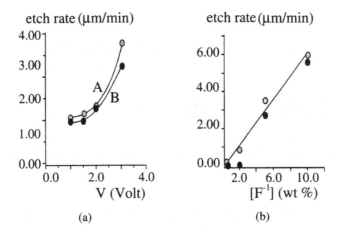

Fig. 7.10. The etching rate as a function of (a) the anodic potential, for etching in large holes and (b) the F^- concentration, for etching in small holes. Arrows indicate experiments with NH_4F–HF mixtures.

7.4.4.2 Results and discussion

(i) *Influence of process parameters*

Figure 7.10(a) shows the etching rate of p-type Si covered with a silicon nitride mask, as a function of the anodic potential for two different structures: structure A has relatively large mask openings, while structure B has small ones. Structure A has a somewhat higher etching rate. Furthermore, it was observed that the etching rate increases if the solution is stirred.

Figure 7.10(b) demonstrates that the etching rate of p-type Si, at an anodic potential of 3 V for the structures A and B mentioned above, is linearly dependent on the fluoride concentration. It has to be mentioned that the results from buffered HF solutions, i.e., solutions containing NH_4F, are not consistent with the HF results (see figure). We think that this has to do with the difference in pH of these solutions.

The above observations indicate that the rate-determining step in the etching process is a first-order reaction in the fluoride concentration (Fig. 7.10(b)) and that this step is (at least partially [16]) diffusion limited. As the voltage in our experiments is higher than the voltage at the current peak mentioned in a foregoing paragraph, this step is

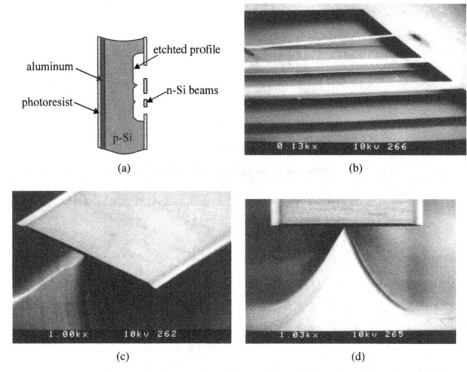

Fig. 7.11. (a) Schematic cross-sectional view of the etched wafer with undercut n-Si beams. (b) SEM micrograph of low-doped n-silicon underetched beams with varying width. (c) Close-up of a broken almost completely underetched beam, showing the very uniform thickness caused by the strong etching selectivity between n-silicon and p-silicon. (d) Front view of the beam in (c), showing the rim remaining after etching.

most probably one of the subreactions leading to the dissolution of the passivating oxide (see Reaction (7.6)); the exact nature of the rate-limiting step is, however, still unknown [16, 17].

(ii) *Etching underneath implanted phosphorous doped layers*
      A striking example of the passivation of active silicon by patterned n-doping of p-Si is shown in Fig. 7.11. Figure 7.11(a) shows a schematic drawing of the sample structure; Fig. 7.11(b) shows some etched beams, 1.0-μm thick (the thickness of the implanted n-layer as calculated from doping conditions equals approx. 1.1 μm) and 2-mm long, with varying widths (10, 20, and 30 μm). The thickness of the beams is very well defined and uniform (Fig. 7.11c).

In the experiments leading to these beams, the applied voltage was much lower (1.5 V) than the voltages to which Fig. 7.4 applies. Consequently, in our case, formation of porous silicon will occur at higher doping levels, probably above 10^{18} cm^{-3}. The maximum doping level in the implanted layer is approx. 2.2×10^{17}/cm^2. Etching stops as the etch front reaches the edge of the implanted layer where the doping level is low (Fig. 7.10). Figure 7.11(d) demonstrates the isotropic character of the anodic etching process. The photograph shows a front view of the underetched beam of Fig. 7.11(c). A standing rim of unetched p-type Si can be seen very clearly. The etched surfaces are very smooth (polishing effect): from the surface profile measured at the bottom of an etched hole of 8 μm depth and 200 μm width, it was derived that the average roughness is about 7 nm.

The results indicate that the etch selectivity is due entirely to the doping dependence of etching and not to some effect of the p-n diode: even if the n-layer is contacted, it will not be etched. So the etch stop mechanism involved differs from that of the frequently used "p-n stop" in KOH etching of silicon (see, e.g., Ref. [25]).

The process described above has several advantages. The etching process is IC compatible, since no metal ions as in alkaline (e.g., KOH) solutions are present. It is a room-temperature process. Photoresist can be employed as an etching mask since the etchant in this electrochemical process is much milder than in more degrading solutions used for chemical etching of silicon. Unlike electroless etching, the etching time can be defined very accurately since the starting and stopping points of time can be controlled by switching the voltage on and off, respectively.

Conventional IC technology can be used to define the microstructures (photolithography, metal and silicon nitride deposition, dopant diffusion, ion implantation). The reproducibility and accuracy of etching are thus mainly determined by the properties of these well-established IC processes.

The technology is very young, but it offers new opportunities in the field of micromachining of silicon for micromechanical applications because it is complementary to the commonly used techniques of bulk micromachining (anisotropic etching with KOH or EDP or isotropic etching with HF—HNO_3—H_2O mixtures) and surface micromachining with sacrificial layer techniques.

References

[1] H. Robbins and B. Schwartz, J. Electrochem. Soc. **106**, 505 (1959).

[2] H. Robbins and B. Schwartz, J. Electrochem. Soc. **107**, 108 (1960).

[3] B. Schwartz and H. Robbins, J. Electrochem. Soc. **108**, 365 (1961).

[4] B. Schwartz and H. Robbins, J. Electrochem. Soc. **123**, 1903 (1976).

[5] H. Seidel, in *Mikromechanik*, ed. A. Heuberger (Springer, Heidelberg, 1989), p. 162.

[6] D. R. Turner, J. Electrochem. Soc. **105**, 402 (1958).

[7] D. R. Turner, in *The Electrochemistry of Semiconductors*, ed. P. J. Holmes (Academic Press, London, 1962), p. 155.

[8] Y. Watanabe, Y. Arita, T. Tokoyama, and Y. Igarashi, J. Electrochem. Soc. **122**, 1351 (1975).

[9] S. S. Tsao, IEEE Circuits and Devices Mag., p. 3 (1987).

[10] M. J. J. Theunissen, J. Electrochem. Soc. **119**, 351 (1972).

[11] H. J. A. van Dijk and J. de Jonge, J. Electrochem. Soc. **117**, 553 (1970).

[12] M. J. J. Theunissen, J. A. Appels, and W. H. C. G. Verkuylen, J. Electrochem. Soc. **117**, 959 (1970).

[13] A. C. M. Gieles and G. H. J. Somers, Philips Tech. Rev. **33**, 140 (1973).

[14] M. Esashi, H. Komatsu, T. Matsuo, M. Takahashi, T. Takishima, K. Imabayashi, and H. Ozawa, IEEE Trans. Electron Dev. **ED-29**, 57 (1982).

[15] B. Hök and K. Gustafsson, Sensors and Act. **8**, 235 (1985).

[16] R. Memming and G. Schwandt, Surface Sci. **4**, 109 (1966).

[17] M. J. Eddowes, J. Electroanal. Chem. **280**, 297 (1990).

[18] P. C. Searson and X. G. Zhang, J. Electrochem Soc. **137**, 2539 (1990).

[19] H. Gerischer and M. Lübke, Ber. Bunsenges. Phys. Chem. **91**, 394 (1987).

[20] C. J. M. Eijkel, J. Branebjerg, M. Elwenspoek, and F. C. M. van de Pol, IEEE Electron Dev. Letters **11**, 588 (1990).

[21] Z. Shengliang, Z. Zongmin, and L. Enke, Proc. 4th Int. Conf. Solid-State Sensors and Actuators (Transducers '87), Tokyo, Japan, June 2–5, 1987, p. 130.

[22] J. D. Benjamin, in *Silicon Based Sensors, IOP Short Meetings Series, No. 3* (Institute of Physics, London, 1986).

[23] L. Tenerz and B. Hök, Electron. Letters **21**, 1207 (1985).

[24] T. Yoshoda, T. Kudo, and K. Ikeda, Proc. MEMS '92 (Travemünde, Ger., Feb. 4–7, 1992), p. 56.

[25] B. Kloeck, S. D. Collins, N. F. de Rooij, and R. Smith, IEEE Trans. Electron Dev. **36**, 663 (1989).

8

Introduction into dry plasma etching
in microtechnology

In the first part of the book the wet etching of silicon was treated. This second part will highlight a dry technique to etch silicon for its use in microtechnology, the so-called plasma etching. In this chapter, first the title *"Plasma etching in microtechnology"* will be explained. This is not straigthforward because of the misleading names for many well-known plasma and microtechnologies. This results in that sometimes important mechanisms are not recognized. But, especially on an educational level, it is important to have a well-organized structure in the material to be studied. Therefore, a first attempt is made to order all of the various technologies more consistently.

8.1 Microtechnology

The field of microtechnology is a logical follow-up of microelectronics. Microelectronics started its revolution after the invention of the transistor by Shockley in the Bell laboratories in 1947. In years, the size of the transistor decreased and integrated circuits came onto the market. Therefore, the size of sensors and actuators (S&A) created by means of finemechanical techniques became the bottleneck for further miniaturization of devices. Moreover, time is money, thus time-consuming steps were carefully examined in order to reduce them. With the spin-off of the microelectronics it was found that it was possible to create micron-sized mechanical structures, and the micromechanics became an important field within the S&A groups. Almost at the same time, the other disciplines from the S&A groups picked up the same ideas from microelectronics, and microoptics, microchemics, etc. became popular research areas. The integration of all these groups of microspecialism are nowadays known as microtechnology. So, micromachining has become a fundamental tool for the fabrication of microtechnological devices and, in general, miniature sensors and actuators. Micromachining techniques include first the basic processing steps of IC technology, namely, film formation, doping, lithography, and etching. In addition, they incorporate special etching and bonding processes which allow for the sculpturing of three-dimensional microstructures.

Traditionally, microstructuring with resolution in the nanometer range has been performed with silicon. The technologies used originated from the IC technology, and due to the good mechanical characteristics of silicon, these technologies appeared to be useful for the fabrication of a variety of structures. Generally, the resulting structures from the IC process are relatively flat, and the urge to spatial structures for a stiffer construction stimulated the activities with respect to microstructuring by means of different methods. In addition, the brittleness of silicon and the need for materials with specific properties (e.g., optical properties) stimulated the microstructuring of different materials as well. The development of the so-called LIGA technology has given a stimulus with respect to high aspect ratio structuring of different materials. The materials range from metals such as nickel and gold and alloys such as nickel iron up to polymers such as polyoxymethylene, polymethylmethacrylate, polysterene, polycarbonate, and others. However, the use of synchrotron radiation and the resulting complicated fabrication of the mask make the LIGA technology economically less attractive and more time consuming. Therefore, a variety of alternatives has been developed, such as UV lithography, laser ablation, microdrilling, ion milling, and spark erosion. Also, the innovations in dry etching enabled the fabrication of micromoulds in silicon. This micromould can subsequently be electroplated resulting in a metal micromould.

Within the field of microtechnology, a group of systems has been paid special attention in the past decade. This is due to its integratability with respect to the integrated circuit industry; the mechano electrical micro systems or micro electro mechanical systems (MEMS). The name "MEMS" should be used for electro mechanical transducers only, although it is possible for a MEMS device to use a transduction in between the electrical input and the mechanical output, such as the magnetic flux. Nevertheless, the name MEMS is often used for non-electro mechanical devices also, e.g., optical switches, especially in the United States. In Europe, most people insist to give these products the more logical name of microsystems, and the technology to create such devices is then called the micro system technology (MST). So, in fact a MEMS is just a special MST product. In Japan, commonly, MST is called micromachining (MM). This might be caused by their opinion that the MEMS/MST products are just created by fine mechanical techniques, if only a little smaller. This philosophy has resulted in a tendency to downscale fine mechanical equipment such as microdrills, spark erosion, and lasers in Japan, as opposed to the fine tuning of integrated circuit equipment in the USA and Europe. The author will follow the European convention and will call an MST product a MEMS only when the input signal is electrical (say a voltage) and the output is mechanical (say a vibration). But, whatever people call these microtechniques MM, MEMS, or MST, the results are, not rarely, astonishing, and it will influence the future of mankind one way or another.

8.2 Plasmas

During the micromachining of components with the help of microtechnology, materials are continuously deposited (i.e., applied), etched (i.e., removed), modified (e.g., hardened), or bonded (e.g., glued). For the deposition, all kinds of vacuum equipment were invented, resulting in machines which were able to remove materials as well. A special group of these so-called dry etch techniques are the plasmas. In 1992, M. Konuma gave a nice general description with respect to plasmas [1]:

"Ordinarily, we hardly come into contact with plasma in our daily life. Matter, which is normally seen, exists in the solid, liquid, or gas phase. However, the conducting gas in a fluorescent tube or in a neon sign is in the plasma state. For this reason a plasma is sometimes called the fourth state in which a material can exist. The plasma gas is characterised by a high population density of charged particles and therefore a plasma gas is highly conductive. The conduction distinguishes the plasma gas from the normal gas. Lightning and auroras appearing in the polar regions are plasmas in nature. All stars, including the sun, are masses of high-temperature plasmas. The interstellar matter and nebulae are also in the plasma state. So, it has been said that the greater part of the universe is in the plasma state.

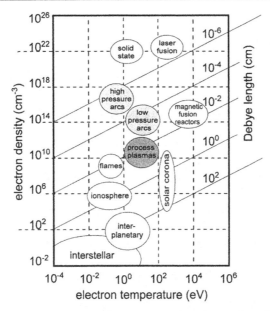

Fig. 8.1. Characteristic electron densities and energies of a number of plasmas. The corresponding Debye lengths are the diagonal lines [13].

It has been demonstrated through studies that various amino acids, nucleic acids, and other organic compounds can be synthesised. This was done under primitive atmospheric conditions generated by a spark and the resultant silent discharge from a mixture of gases containing methane, ammonia, hydrogen, and water vapour. This simulation is considered to be similar to the earth's atmosphere in its early stages, implying that life originated in the plasma state."

In Fig. 8.1 a graph of some typical plasmas with their electron density (i.e., the bulk conductivity) and their electron temperature (i.e., the mean electron velocity) is shown. In the same graph, the Debye length is shown, which is a measure for the distance in which a charged particle is counterbalanced by particles with an opposite charge, for example, the shielding of a positive ion by a cloud of electrons.

So, a plasma is characterized by a high population of free movable electrons. How are these electrons formed? Imaging a particle "P" coming in from the left and colliding with a molecule "M-M" as shown in Fig. 8.2. The incoming particle might be an electron, an ion, a neutral particle, and even a photon. When the collisional energy is too low to excite the molecule, the molecule remains at the same energy level and the collision is called elastic. This means that only kinetic energy can be transferred from the particle onto the molecule as long as the conservation of momentum and energy is fulfilled. However, at a certain level the collisional energy is high enough to cause excitation (electronic, vibrational, or rotational states), dissociation, or ionization. In the electronic excitation process, an electron from an atom is pushed into an orbit at a greater distance from the atom center than before the excitation. At the moment the electron falls back to a lower energy level, it will emit a photon which gives the plasma its characteristic glow. In the dissociation process, a molecule is broken into smaller pieces like atoms. Such atoms or "radicals" are normally quite reactive and responsible for many chemical reactions. In the ionization process, an electron is broken out of the shell which leaves an ion behind and increases the number of

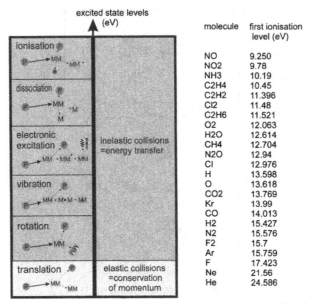

Fig. 8.2. Excited state levels caused by particle collisions with the minimum energy needed for ionization.

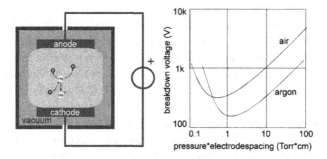

Fig. 8.3. A direct current plasma reactor showing ionization by electrons accelerated by the electric field between two electrodes together with the Paschen curve showing the breakdown voltage of a gas between the two electrodes separated by a distance d at a pressure p.

charged particles. This latter process, ionization, is important for plasmas to be a plasma, i.e., a highly conductive gas. In Fig. 8.2, the energy needed for ionization for a number of gases is given. It is found that this energy normally is between 10 and 20 eV. The ions are, together with the radicals, the most important particles needed for etching produced by plasmas.

Thus, to create a plasma, a flux of highly energetic particles is needed to be able to ionize the gas under this particle bombardment. For this it is possible to use photons, as in the case of laser fusion (Fig. 8.1), the chemical energy from exothermic reactions which produces fast-moving particles (flames), the gravitational energy from heavy objects (solar corona), or electrical fields (neon tube). It is this latter process which is normally used for plasma processes. In Fig. 8.3, a schematic of a direct current (dc) plasma is given. Because the extra electrons formed by ionization will also convert electrical energy into kinetic energy, which will produce even more electrons, the electrical fields are quite effective in creating plasmas.

The moment when the strength of the electrical field is high enough to increase the number of electrons rapidly is called breakdown, and at this moment the gas flashes and becomes a plasma. This moment is depending on the pressure and it is expressed with the help of the so-called Paschen curve, which also is found in Fig. 8.3. Typically, the gas is turned into a plasma at pressures ranging from 1 m Torr up to 1 Torr. At higher pressures, the electrons make too many collisions to be able to gain enough energy necessary for ionization, and only the lower excitation processes take place without an increase in electrons. At the lower pressures, the electrons do not make any collisions with gas particles and will hit the anode with high energy, again without an increase in electrons. Konuma describes the same process in the following way [1]:

"Electrical discharges have commonly been used to generate plasma in the laboratory. This is accomplished by using a glass discharge tube in which two metal electrodes are installed, evacuating the discharge tube to a pressure range between 10^{-4} to 10^{-3} atmosphere and applying a voltage between the two electrodes. As the voltage gradually increases, only a very small electrical current flows. When the voltage has reached a few hundred volts, electric current through the tube shows an abrupt increase, and the tube begins to emit visible orange-red light. This condition in the discharge tube is called plasma. The gas becomes electrically conductive due to its ionization. Generally in such a plasma, positively charged ions and negatively charged electrons move with statistically distributed random velocities. Positive and negative particles drift in opposite direction according to the electrical field. Plasma, such as can be obtained by a direct current glow discharge, can also be generated by a high frequency or a microwave discharge at low pressure. In this plasma, the degree of ionization is typically only 10^{-4}, so the gas consists mostly of neutral but excited species. A characteristic of this plasma is the lack of thermal equilibrium between the electron temperature and the gas temperature. The rf field is able to effectively convert electrical energy into kinetic energy for the electrons. In turn, the electrons convert their kinetic energy into elastic energy for the neutrals only and the mean neutral velocity isn't changed a lot. Hence, this type of plasma is called a nonequilibrium plasma or cold plasma. There are more types of plasmas, such as thermal plasmas, created with the help of an arc discharge. Arc discharges used for fusing metal are generated at near atmospheric pressure or above."

In order to increase the lifetime of the electrons traveling freely in the reactor, an alternating field is used, the so-called radio frequency (rf) plasmas. Therefore, ionization is possible at a lower pressure, as found with a dc glow discharge. Another advantage is the possibility to use electrodes covered with insulating materials because the alternating current (ac) field may pass this dielectricum. For example, many discharges consist of a glass tube covered by the two electrodes on the outside. With this type of plasma it is much easier to etch insulating samples.

8.3 Etching

To etch a material, it is possible to use the energy content of any kind of particles, such as (*i*) *neutrals* (N): sand blasting or neutral beam etching (NBE); (*ii*) *radicals* (R): ozone stripping or plasma/radical etching (RE); (*iii*) *ions* (I): ion beam etching (IBE); (*iv*) *electrons* (E): electron-beam evaporation/etching (EBE); (*v*) *phonons* (T): thermal decomposition/etching (TE); or (*vi*) *photons* (P): laser ablation, i.e., photon beam etching (PBE). These energetic particles are readily produced in a plasma environment, and generally they will interfere with each other synergetically, while the etch rate is increased enormously, as in chemical-assisted ion beam etching (CAIBE) or reactive ion etching (RIE).

In literature, the given techniques are not always labeled consistently. Therefore, the next labeling is introduced. The labeling ends always with the main purpose of the technique; etching (E), deposition (D), modifying (M), or welding (W). If the, e.g., etching is made possible by a collimated stream of particles, the technique will end with "beam etching" (BE). In front of this the type of energetic particle is placed, and if focused, the label is extended with the letter "F." So, focused ion beam etching (FIBE) stands for a focused stream of ions which are etching a substrate. Similarly, the focused electron beam resist technique is labeled with "FEBM." If there is no collimated stream but random flux of etching particles, the labeling only ends with "etching" (E). So, ozone stripping is an example of "radical etching" (RE). When the flux/stream of particles is accompanied by a different source of particles, the labeling is extended with "assisted" (A), and if the flux of particles has more than one energy reservoir, e.g., kinetic and ionic, it is extended with this source. Thus, chemical assisted reactive ion beam etching (CARIBE) stands for a stream of reactive ions assisted by a flux of chemicals.

The plasma etching of polymers is normally fulfilled at elevated temperatures (e.g., 150°C), so a better name would be "thermally assisted radical etching" (TARE). Identically, the term "reactive ion etching" (RIE) is a misleading name. As for the beam-etching techniques, the label should depend on the type of energetic particles and substrate material. In most RIE cases a directed ion flux is accompanied by an isotropic radical flux. So, a more correct name would be "ion beam assisted radical etching" (IBARE).

8.4 Outline

A large number of reviews on plasma etching have been published [1–13]. However, above all, the author is charmed by *"Reactive Ion Etching"* by Oehrlein and this reference is taken as a framework to treat many plasma concepts [2]. This "dry" part is a brief review of plasma etching as applied to pattern transfer, primarily related to silicon technology. Rapid progress is being made in all aspects of plasma etching; therefore, this review focuses on concepts, rather than listing data obtained on all different systems. Although the detailed plasma chemistry of rf discharges used for etching materials of interest in micromechanics is different, many of the basic processes are similar, so they can be cautiously adapted to different plasma-substrate systems. In short, this part is intended to be a guide to the tremendous plasma jungle and it has tried to give an answer to the question:

"How and where to use what plasma chemistry?"

The next chapter, *"Why plasmas?"* will compare the dry etching of silicon with respect to the wet silicon etching. What is the advantage of plasma etching? Chapter 10, *"What is plasma etching?"* will treat the physics behind plasmas. How are they created, what kind of particles exist in the plasma phase, and how can we use them to etch a sample? In Chapter 11, *"Plasma system configurations,"* several commercially available etch machines are summarized. What kind of equipment do we have, what is their characteristics, and when should we use them? The next chapters, *"Contact plasma etching"* and *"Remote plasma etching,"* will treat the main dry etch techniques more in detail. The "contact" chapter especially is treated extensively because it combines the high etch rate of the radical etching with the etch directionality of the ion beam etching. It answers the questions how the etch directionality is obtained, which plasma chemistry can we use to etch silicon, how can we create free-standing structures with plasmas, what parameters will influence our final etch result, what are specific plasma etch problems, how do we solve them, how do we optimize our parameter setting of the equipment, and what are current trends with respect to etch equipment. Chapter 14, *"High aspect ratio trench etching,"* will treat one of the main challenges nowadays: the etching of deep grooves into silicon with profile control. What mechanisms control the profile and etch rate of silicon trenches in plasma etching and what kind of problems are found. The last chapters, *"Moulding of microstructures"* and *"Fabrication of movable microstructures,"* will use the knowledge of the preceding chapters. How can we successfully use plasmas to pattern silicon for, e.g., moulding purposes and to create movable structures and, in particular, MEMS?

References

[1] M. Konuma, *Film Deposition by Plasma Techniques* (Springer-Verlag, Berlin, Heidelberg, Germany, 1992).

[2] G. S. Oehrlein, *Reactive Ion Etching: Handbook of Plasma Processing Technology*, ed. S. M. Rossnagel (Noyes Publications, Park Ridge, New Jersey, 1990), p. 196.

[3] S. M. Irving, U.S. Patent 3,615,956, assigned to Signetics (1971).

[4] A. Reinberg, Proc. Symp. Etching, Electrochem. Soc. 91 (1976).

[5] B. Chapman, *Glow Discharge Processes* (John Wiley & Sons, New York, 1980).

[6] D. L. Flamm and V. M. Donnelly, Plasma Chemistry and Plasma Processing, **1**, 317 (1981).

[7] J. A. Mucha and D. W. Hess, *Plasma Etching* (American Chemical Society, 1983), p. 215.

[8] R. A. Morgan, *Plasma Technology, Vol. 1, Plasma Etching*, ed. L. Holland (Elsevier, Amsterdam, The Netherlands, 1985).

[9] H. W. Lehmann, *Thin Film Processes, II, Plasma-Assisted Etching*, ed. J. L. Vossen and W. Kern (Academic Press, Inc., San Diego, 1991), p. 673.

[10] H. F. Winters and J. W. Coburn, Surface Science Rep. 14, North-Holland (1992), p. 161.

[11] *AEDEPT* (Automatic Encyclopedia of Dry Etch Process Technology), Du Pont Electronics, Wilmington, Delaware (1989).

[12] R. d'Agostino, *Plasma Deposition, Treatment, and Etching of Polymers* (Academic Press, Inc., San Diego, 1990).

[13] J. R. Hollahan and A. T. Bell, *Techniques and Applications of Plasma Chemistry* (Wiley, New York, 1974).

9

Why plasmas?

Dimensional control in etching small geometries, necessary for advanced micromachining, is an important topic in microtechnology. To etch these structures, dry plasma assisted etching is increasingly used. Although the basic investments (buying the equipment) are rather high, it was introduced rapidly due to

(1)   the ability to faithfully transfer lithographically defined photoresist patterns into underlying layers,
(2)   cleanliness and compatibility with vacuum processing technologies, and
(3)   the achievement of etch directionality without using the crystal orientation, as in the case of wet etching of single crystals like silicon, germanium, or gallium-arsenide (although recent developments in anodic wet etching may alter this situation).

The micro system technology (MST) which manufactures sensors and actuators (S&A) has its foundation in microelectronics, and this has resulted in a tendency to use "IC compatible" materials as structural or mask materials. In addition, new methods are adapted in MST a long time after its introduction in microelectronics. However, the MST is not restricted by IC compatibility (and there is no word such as "S&A compatibility") which gives the MST much more freedom with respect to the use of construction materials. When plasma etching is not restricted by any IC compatibility, it is found that it can fulfill some special tasks not easily performed by other techniques, like

(1)   profile control for moulding purposes,
(2)   high-speed etching when using bulk micromachining,
(3)   high-speed isotropic etching in sacrificial layer etching,
(4)   high-speed anisotropic etching for, e.g., high aspect ratio combdrives,
(5)   high-selectivity etch masks for reliable pattern definition,
(6)   smooth surfaces after etching to avoid stress concentration,
(7)   plasma deposition of polymers for various applications, and
(8)   release of movable structures.

Within the field of MST, or more specifically in the S&A branch, the micro-electro-mechanical systems (MEMS) have gained a lot of interest in the past decade. They have become increasingly popular research topics, especially in the car industry (many S&As) and surgery (small S&As). Starting a MEMS process, usually a sandwich of polysilicon on top of silicon dioxide (SiO_2), i.e., glass, is used, and for single crystalline silicon (SCS) structures, silicon on insulator (SOI) wafers are used [1–3]. After etching micromechanical structures in the top silicon layer, they have to be released. This is not straightforward, and many techniques have been proposed. Frequently, in surface micromachining the SiO_2 layer is used as a sacrificial layer which is etched using wet or vapor etchants [4–6]. They are cheap but suffer the so-called sticking problem, due to surface tension of liquids, and many solutions have been proposed to solve this [5–12]. In bulk micromachining some very useful dry plasma release techniques have been proposed [13, 14]. "Sticking during manufacturing" caused by surface tension is not found in dry etching, making this technique more reliable.

After successfully releasing an MEMS, it is still possible that the structures attach and stick to near surfaces. This "sticking during operation" might be caused by water vapor which condenses (adsorbs) onto the structures or by charging of the surface of the structure which is caused by the applied electrical field (voltage source). The adsorption of water is minimized by heating the MEMS, by operating the device in a moisture-free area, or by coating the MEMS with a uniformily hydrophobic material with a low surface free energy like the (downstream) plasma deposition of fluorocarbons (e.g., Teflon). The sticking caused by surface charging is minimized when an overall dc current is prevented from flowing to the MEMS surfaces. This is accomplished by the use of a blocking capacitor in series with the MEMS or balancing the ac signal carefully.

In this chapter, some well-known vapor, wet, and dry etch/release techniques will be treated. In fact, we can distinguish etching into five different groups: solid etching (drilling, sawing), wet etching (acids), vapor etching (VHF), gas etching (erosion), and plasma etching. Although strong motivations are given to use dry or nonwet etch techniques, for the releasing of structures, one specific technique seems to be most suitable: plasmas.

9.1 Vapor etching

In vapor fluoric acid (VHF) etching, the removal of the sacrificial SiO_2 layer is almost directly transformed into the gas phase. Nevertheless, it is believed that condensation of water vapor is needed to uniformly initiate etching, which may cause sticking [5]. Another problem with this technique is that some reaction products may precipitate on the surface reducing the yield. J. Ruzyllo et al. showed that these problems can be suppressed by etching at low pressure in an anhydrous HF/CH_3OH gas mixture at elevated temperature [6].

9.2 Wet etching

Generally, fluoric acid (HF) is used for sacrificial layer etching the SiO_2. However, for long thin beams the yield is low due to

(1) structure damage caused by bubble formation during etching and rinsing steps during cleaning and
(2) structural deformation resulting in stiction during drying.

The first problem is difficult to control, but for the last problem several techniques have been developed, such as freeze drying [7, 8], photoresist-assisted releasing [9–11], and surface modifications [12].

9.2.1 Freeze drying

This technique uses, e.g., cyclohexane as the final rinsing agent. After freezing at 7°C it is sublimated, thus avoiding sticking. This is accomplished by placing the substrate under a forced nitrogen flow on a regulated Peltier element with a temperature just below the freezing point. Nevertheless, for long thin structures a low yield is observed, caused by bubble and rinse damage. And still, in spite of careful "sublimation," stiction may be found.

9.2.2   Photoresist-assisted releasing

This technique uses a resist layer temporally to avoid stiction. There are at least three different approaches found in literature:

(1)   A resist pattern used as "rubber feet" under the microstructures to carry them [9],
(2)   a resist pattern sideways or on top of the structures to hold them [10], and
(3)   a photoresist refill to replace the final rinsing liquid [11].

After sacrificial layer etching and drying, the photoresist layer is removed by an oxygen plasma. A disadvantage of the first two techniques is that an extra mask has to be aligned, the second approach is only possible for special mask designs, and the last technique requires a uniform coating of photoresist beneath the structures for successful releasing. All techniques are less successful for long thin MEMS structures with high aspect ratios (>5) and small gap spaces ($< 2 \,\mu$m).

9.2.3   Surface modification

Important parameters to consider for the sticking effect during drying a microstructure in a wet etchant are the surface free energy and wetting angle. If the wetting angle is lower than $90°$, beams will be attracted to the nearest surfaces. In contrast, at higher angles beams are "pushed away," thus preventing sticking. The magnitude of the force is a function of both the surface free energy and wetting angle. Shortly, to prevent sticking, the etchant or structure surface should be modified to create a wetting angle exceeding $90°$. Although this technique is promising, little activity is found in this area [12].

9.3   Dry etching

Many sensor and actuator principles require high aspect ratios such as combdrive structures [15] and curved electrode actuators [16]. Fluidic devices and chemical analysis systems such as micromachined chromatographs preferably require relatively high aspect ratio structures as well. By using the recent innovations in dry etch processes, structures with variable tapering and high aspect ratios can be fabricated. In the next paragraphs, the following techniques will be briefly discussed: photon beam etching, neutral (beam) etching, plasma etching, which includes focused radical etching, ion beam etching, and ion beam assisted radical etching. There are more techniques which are promising, such as microdrilling (to be exact, like microsawing, this technique should be placed under the heading "solid etching"), but they won't be discussed here.

9.3.1   Photon beam etching

Laser ablation

Material can be sputtered away by means of laser ablation [17]. This can be done by using a mask, just as with UV lithography, or by using optics in order to focus the laser beam.

When using a mask for the image, the improvement with regard to the aspect ratio due to the shorter wavelength as compared to the "normal" optical lithography is not very large (when using excimer laser with, e.g., ArF as laser medium, the resulting wavelength is 193 nm). An advantage of using optics and moving the focus point is the possibility of real three-dimensional processing. When using optics in order to focus the laser beam, there will be a trade-off between the depth of focus and the resolution. The smallest spot size is of the order of a micrometer, and the depth of focus is of the same order of magnitude. Anisotropic structures become more difficult to fabricate at a small depth of focus. In addition, the use of laser beams do not allow batch processing.

9.3.2   Neutral (beam) etching

Sometimes neutrals are useful for etching bulk material to achieve high etch rates. For example, reactive gases are effective to release structures isotropically. For fast anisotrpic etching, sand blasting is a well-known technique, but the particle diameter distribution is probably too broad and the mean particle diameter too big for sand blasting to be useful in micromachining. However, this problem could vanish when using molecular beams such as supersonic nozzles. Other types of selection are also possible. For example, electric fields may be used to deflect polar molecules and to obtain a beam of aligned molecules.

Reactive gas erosion

Sometimes it is possible to etch materials spontaneously with the help of reactive gases, but without the help of a plasma (e.g., the etching of silicon with fluorine gas (F_2) or xenondi-fluoride (XeF_2)). Normally, these processes etch the sample isotropically, and therefore this technique is useful for dry sacrificial layer etching, i.e., the releasing of structures. However, these gases are quite hazardous and special precautions are needed to protect users from inhaling the gas.

Sand blasting

Another example of neutral etching is the use of a collimated stream of small (say 50-nm diameter) particles. Sand blasting is the technique in which particles of glass are directed with high velocity to the sample to be anisotropically etched. The collimation is normally fulfilled with the help of a superatmospheric pressure source. Due to the pressure difference between the source and the sample, the particles are directed and a gas jet is the result.

Supersonic nozzles

There are also more sophisticated devices for generating molecules with a desired velocity. Among the most important are supersonic nozzles, in which the molecules stream out of the source with a very narrow velocity distribution and are then stripped down to a beam by passing through a skimmer shaped like an inverted cone (Fig. 9.1). Although it seems to be more difficult to create a uniformly directed stream of molecules, it has the advantage of simplicity and ignoring electrical fields. The effect of the image force caused by charged species near surfaces is especially not found in neutral (molecular) beams.

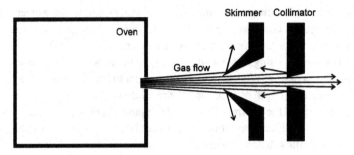

Fig. 9.1. A supersonic nozzle skims off some of the molecules of the beam and leads to a beam with well-defined velocity.

9.3.3 Plasma etching

The use of plasmas is probably the most popular dry etch technique to structure samples in microtechnology. This can be accomplished at near-atmospheric pressures (as in the case of spark erosion), with remote plasmas at extremely low pressures (as in the case of ion milling), or at intermediate pressures (as in the case of ion beam assisted radical etching (IBARE) or radical etching (RE)).

Ion beam etching

An ion beam can be used to ablate material from a substrate and can therefore be used to fabricate anisotropic structures. Besides tilting or rotating the substrate with respect to the ion beam source, there is no freedom in the third dimension. Generally, only positively tapered profiles and low etch rates are obtained. In addition, due to the low selectivity of an ion beam for the mask material with regard to the ablated material, the aspect ratios are directly related to this selectivity. In practice, aspect ratios will therefore not be very high (<6).

Spark erosion

Techniques originating from fine mechanics such as spark erosion are becoming more sophisticated. In the near future it will be possible to fabricate structures of metals with sub-micron resolution and with roughness below 300 nm. Of course, the spark erosion process is not a batch process.

Ion beam assisted radical etching

IBARE has considerable advantages over alternative ways of high aspect ratio processing, such as UV lithography, laser ablation, microdrilling, or spark erosion due to its ability to control the tapering of the etched structures, and is far cheaper than the LIGA process. Recent innovative developments in IBARE using, e.g., the SF_6/O_2 gas mixture at relatively high pressure resulted in the possibility of etching silicon with high selectivity with regard to masking material (up to even more than 100,000), high etch rates (up to several μm/min.) with directional freedom, and, most important, resulted in control over the etch profile, i.e., ranging from isotropic, negative tapering, vertical walls up to positive tapering. Above all, in silicon, positive aspect ratios of over 20 are feasible and negative aspect ratios above 50 are feasible, and in polymer, positive aspect ratios of over 40 are feasible. The aspect ratio is defined as the maximum depth divided by the maximum width and is positive for trenches and negative for mesas. The accuracy and resolution are determined by the standard IC mask-making processes. Roughness lower than 50 nm has been observed. Sidewall bowing, feature size-dependent profiles, and RIE lag, as well as loading and microloading effects, complicate the applicability of IBARE, as will be explained more extensively in Chapter 14. However, proper control of the parameter setting diminishes the observed effects. In addition, design rules for the microstructures may overcome these effects for the aspect ratios mentioned here. It is anticipated that aspect ratio dependent etching will limit the aspect ratios, although at this moment it is unclear where and what this limit will be.

Radical etching

During surface micromaching, the sacrificial layer has to be removed. This is difficult for most dry techniques, and only by using special "complicated" release techniques, like a buried cavity or releasing with RE equipment, can the problem of sacrificial layer etching be avoided.

References

[1] T. Nakamura, 7th Int. Conf. Solid-State Sensors and Actuators, 1993, p. 230.

[2] A. Benitez, J. Esteve, and J. Bausells, Proc. IEEE MEMS, 1995, p. 404.

[3] B. Diem et al., 7th Int. Conf. Solid-State Sensors and Actuators, 1993, p. 233.

[4] R. Legtenberg et al., Sensors and Act. **A43**, 230 (1994).

[5] T. Lober and R. Howe, Proc. IEEE MEMS, 1988, p. 59.

[6] J. Ruzyllo et al., J. Elec. Soc., **140**, L64 (1993).

[7] H. Guckel et al., Proc. IEEE MEMS, 1989, p. 71.

[8] R. Legtenberg and H. Tilmans, Sensors and Act. **A45**, 57 (1994).

[9] C. Mastrangelo and G. Saloka, Proc. IEEE MEMS, 1993, p. 77.

[10] D. Kobayashi et al., 7th Int. Conf. Solid-State Sensors and Actuators, 1993, p. 14.

[11] A. Kovacs and A. Stoffel, Europ. Workshop on Micromach., 1992, p. 114.

[12] R. Alley et al., Proc. IEEE MEMS, 1992, p. 202.

[13] Y. Li et al., Proc. IEEE MEMS, 1995, p. 398.

[14] K. A. Shaw, Z. Zhang, and N. MacDonald, Sensors and Act. **A40**, 63 (1994).

[15] H. V. Jansen, M. de Boer, R. Legtenberg, and M. Elwenspoek, Journal of micromechanics and microengineering nr: (5), (pp. 115–120). ISSN 0960–1317.

[16] R. Legtenberg, E. Berenschot, M. Elwenspoek, and J. H. J. Fluitman, Proc. MEMS '95, Jan. 30–Feb. 2, 1995, Amsterdam, The Netherlands, pp. 37–42.

[17] T. M. Bloomstein and D. J. Ehrlich, Proc. MEMS '91, Jan. 30–Feb. 2, 1995, Nara, Japan, p. 202.

10

What is plasma etching?

Looking at an rf plasma, it is found that it consists of many different particles, such as ions responsible for the bombardment of surfaces, radicals which remove surface atoms due to chemical reactions, and photons emitted from excited atoms giving the plasma its characteristic glow. These particles are generated by way of collisions of highly energetic electrons with neutrals, and the energy needed for such collisions is supplied by an external rf power supply which accelerates electrons to energies up to 10 eV and beyond. Plasma etching proceeds by physical sputtering (ions), chemical reactions (radicals), and synergetic ion-assisted mechanisms (ions + radicals), and therefore it is necessary to estimate the concentration and other characteristics of these particles. For this purpose, three different models are in use: physical, chemical, and electrical.

The physical model, described with the help of the atomic or kinetic gas theory and statistical mechanics, considers collision phenomena on the atomic level, like the collision between two particles. It is the most common language to use when talking about partial pressures, particle fluxes, or plasma oscillations. The chemical model, having its foundation in thermodynamics, is most suited to describe the energy involved in chemical reactions. It uses concepts like enthalpy and entropy. The electrical model is an effective method to describe the ion-enhanced mechanism. For this it uses an electrical equivalent circuit to explain the existence of, e.g., the plasma self-bias voltage. So, the physical model treats the plasma by looking at the individual molecules (atomic scale), whereas the chemical and electrical models are more concerned about how the particles behave as a whole (macro scale). Which model to use strongly depends on the type of (re)action we want to explain. For example, when reactive ions approach a surface, the ion kinetic energy is most easy to calculate with the electrical model, whereas the collimation of the ion beam is best handled with the physical model, and for the reaction probability of the ion with the surface it is more convenient to use the chemical model.

"During the last years of the nineteenth century (1895), a great intellectual battle was waged in the field of scientific theory, between those who believed in atoms as real fundamental particles and those who considered them simply as useful models for mathematical discussions in a world based entirely on transformations of energy. . . . The leader of the forces of atomism was Ludwig Boltzmann of the University of Vienna, whose name is linked with that of Maxwell as cofounder of the kinetic theory of gases, and with that of Gibbs in the development of statistical mechanics. . . . The enemies of traditional atomism, under the leadership of Ernst Mach, called Boltzmann the last pillar of that bold edifice of thought. Boltzmann personally felt each tremor of what was then believed to be a 'tottering edifice.' Boltzmann was subject to sudden changes from happiness to affliction, which others

suggested that they were caused by concern for the atomic theory. In a sudden intensification of depression, he committed suicide by drowning at Duino near Triest during a summer outing in 1906. His memorial is a white marble bust, under which is engraved a short formula: $S = k * \ln(W)$. From our present vantage point, it might seem that the attacks of Mach on atomism were only minor setbacks in a broad advance of the basic theory, which was to achieve some of its greatest successes in the first half of the twentieth century. It may be unduly romantic to call Boltzmann a martyr for the atomic theory," [6]. But what was the battle about?

Suppose we want to calculate equilibrium and nonequilibrium properties of systems, like temperature, entropy, or enthalpy, from the properties of their constituent molecules. This task would result in the coupling of the atomic kinetic theory and the energetic thermodynamic theory. Because the number of molecules in even small volumes, say 1 liter, is overwhelming, any theory that seeks to interpret the behavior of macroscopic systems in terms of molecular properties must rely on statistical methods. The theory that allows us to make this theoretical connection is, therefore, called statistical mechanics. Under reversible conditions, the physical kinetic model and chemical thermodynamic model are not in conflict: The calculated properties of the perfect gas of kinetic theory are the same as the experimental properties of the ideal gas of thermodynamics. In thermodynamics, a distinction is drawn between mechanical work and heat. According to the kinetic theory, the transformation of mechanical work into heat is simply a degradation of a large-scale motion on the molecular scale. An increase in the temperature of a body is equivalent to an increase in the average translational kinetic energy, E_{rms}, of its constituent molecules. This equivalence is expressed mathematically by saying that temperature is a function of, in fact proportional to, E_{rms}. The interpretation of absolute zero by kinetic theory is thus the complete cessation of all molecular motion; the zero point of kinetic energy.

Both theories, however, conflict when considering the Third Law of thermodynamics, which states: "It is impossible by any procedure, no matter how idealized, to reduce temperature of any system to the absolute zero in a finite number of operations." The absolute zero conflict (not really a conflict, of course, because the Third Law is only a postulate, nothing more) between both theories could be solved by improving the kinetic theory with the quantum theory. A curious thing about the statistical quantum mechanics is that we do not know, and indeed cannot know, the values of the mechanical variables for all the molecules in the system. What we can know, however, is the possible values that these mechanical variables may take for any single molecule. Quantum mechanics does not permit a simultaneous fixing of both coordinates and moments to any arbitrary accuracy. They are subject to the famous Heisenberg uncertainty relation: $\Delta q * \Delta p \geq h/2\pi$. Thus, an exact specification of a coordinate ($\Delta q \rightarrow 0$) implies a complete loss of information about its momentum ($\Delta p \rightarrow \infty$) and vice versa. So, we cannot, even in principle, keep track of individual molecules according to the quantum mechanical description, and every system requires a small residual energy. Therefore, the absolute zero is impossible to reach. The uncertainty relation is not only restricted to coordinates and moments. It holds for energy (i.e., a generalized momentum) and time (i.e., a generalized coordinate) too: Mother Nature may lend a restricted amount of energy for awhile, the time given by $\Delta t * \Delta E \geq h/2\pi$. A fundamental difference between classic mechanics and thermodynamics is that thermodynamic systems change with time always toward the equilibrium state. There is nothing in the mechanical properties of an individual molecule to indicate why this should be so. Saying it differently, whereas thermodynamic processes are inherently irreversible, classic mechanical processes are inherently reversible. Would it be straightforward that the uncertainty relation would hold for entropy also? In this case we get $\Delta gq * \Delta S \geq h/2\pi$, where gq [Ks] is the generalized coordinate belonging to the entropy. If Δgq is positive, we would have $\Delta S \geq h/(2\pi \Delta gq) \geq 0$, i.e., the inequality of Clausius. Anyway, it seems to be that the development of the quantum mechanics has restored the foundation of "the tottering edifice" of the atomic kinetic theory. Both atomic as well as quantum physics rely upon statistical mathematics and talk about chances. However, quantum mechanics has added the concept of uncertainty (Heisenberg), which might make it possible to give the irreversible thermodynamics a strong foundation from the uncertainty atomic theory. The same quantum postulate can identify the Third Law (it is impossible to reach $T = 0$) with atomic theory.

The substructure of atomic theory became even stronger after Einstein had "proven" with the help of his scientific milestone, relativistic theory, that energy and mass (atoms) are distinguishable. So, talking about mass or energy is just a matter of taste. Quite ironically (or naturally) the relativistic and quantum theory that helped the atomic theory survive came into conflict with each other. The two most important theories from the twentieth century had their own battle between the great brain in the "experiment of thought" Albert Einstein (relativistic theory) and Niels Bohr (quantum mechanics). The actual experiment of thought was a philosophic version of Young's double split experiment. Imagine there would be a particle in an excited state and falling to its ground state while sending off two identical photons. According to Einstein, it would be possible to measure exactly the coordinate from one of the photons, and at the same time the momentum (or spin) of the other photon could be measured with great precision. In other words, we could deduce both the coordinate and momentum of only one photon with unlimited precision. In the view of quantum physics, the photon acts as a wave function and both photons should still obey Heisenberg's uncertainty relation. "God doesn't play dice" was Einstein's remark against the statistical quantum mechanics. But nowadays, the experiment of thought, as proposed by Bohr and Einstein, is made possible in practice and it strongly indicates that indeed there seems to be a big chance that God indeed does play dice.

So, in case of classic statistical mechanical systems with only a few variables, particles are not subjected to statistical laws: It is not difficult to follow and determine the positions and moments of all the particles. In the perspective of the quantum mechanics, however, even one particle is subjected to the laws of chance and we cannot determine both its position and momentum at the same time. Therefore, although classic mechanical processes are inherently reversible, the quantum processes are irreversible. This property of quantum mechanics makes it possible to connect quantum atomic theory with irreversible thermodynamics. Shortly, we can show the connection between atomic and thermodynamic theories as follows:

classic atomic theory $\Rightarrow$ reversible thermodynamics

quantum atomic theory $\Rightarrow$ irreversible thermodynamics.

For centuries, many different views on the question "What is science?" were eloquently put forth, such as conventionalism, inductivism, and deductivism. In the conventionalist view, the scientist was like a creative artist working with the unorganized sensation of a chaotic world. The inductivist considers science as a procedure to collect and classify sensory input data into a form called observable facts. From these facts, the scientist then drew general conclusions called the laws of nature. Deductivism emphasizes the primary importance of theories. The role of an experiment is to subject a scientific theory to a critical test. According to the deductivist, there is no valid inductive logic, since a scientific theory can never be proved, but it can be disproved. So, the conventionalistic scientist works ad hoc, the inductivistic scientist use a macro-to-micro scale approach, whereas the deductivistic scientist works from micro-to-macro scale. Thermodynamics deals with pressure, volumes, masses, energies, and relations between them, without seeking to elucidate further the nature of these properties. Thermodynamics allows us to derive relationships between large-scale (macroscopic) properties of systems, but in no case can it explain why a given property has a certain numerical value. To understand at all why the macroscopic properties of matter have their actual values, a theory is necessary that explores matter on a finer scale in terms of elementary particles, fields of force, and other principles of structure and interactions, i.e., a microscopic theory. The theory which enables us to do so is the kinetic theory. So, the thermodynamic scientist has a more or less inductivistic point of view, whereas the kinetic scientist uses a typical deductivistic logic.

So, the battle of atomic theory versus thermodynamic theory was in fact a battle between inductivism and deductivism, and at this moment the improved atomic theory holds a stronger card to increase the knowledge of modern science. Anyway, in this chapter we will not take any position in the battle. In fact, it is meaningful to make a combination: Generally, science develops along a path using a combination of all three philosophical views. An experiment, meant to prove a certain theory, gives an unexpected result (conventionalism). Then a theory is postulated to explain the result (inductivism). The consequence of this theory is not only to explain the observed phenomena, but also to predict some new effects not yet observed. By looking at these new effects, the theory can be tested (deductivism). Of course, by Murphy's law, at least one of these testing experiments will again give an unexpected result (conventionalism). This closes the circle and we are able to define science as a never-ending chain of

$$\cdots \rightarrow \text{conventionalism} \rightarrow \text{inductivism} \rightarrow \text{deductivism} \rightarrow \text{conventionalism} \rightarrow \cdots$$

So, Boltzmann needed the thermodynamic theory to test the statistical atomic theory ($PV = NRT$, Third Law entropy, etc.).

In this chapter, the generation of most plasma particles will be explained together with how they react with the sample or substrate material to be etched while using the above-mentioned three models. However, before starting, the plasma etch mechanism is described shortly to give a rough idea on what type of processes do take place, although the denominations will be explained more in detail in the rest of this chapter. For convenience, a short overview of some of the physical constants is listed in Table 10.1(a).

10.1   Principle of plasma etching

The principle of plasma-assisted etching is simple; use a gas glow discharge to dissociate and ionize relatively stable molecules forming chemically reactive and ionic species, and choose the chemistry such that these species react with the solid to be etched to form volatile or gaseous products. A basic reactive ion etch (RIE) system, or better ion beam assisted radical etcher (IBARE), currently the most important plasma configuration, is illustrated in Fig. 10.1 with a list of typical parameters shown in Table 10.1(b). The following processes

Table 10.1a. *List of physical constants.*

Constant	Symbol	SI value	SI units	Constant	Symbol	SI value	SI units
Mass electron	m_e	9.110×10^{-31}	[kg]	Debye	D	3.33×10^{-30}	[Cm]
Mass proton	m_p	1.673×10^{-27}	[kg]	Planck	h	6.626×10^{-34}	[J/s]
Charge electron	q	1.602×10^{-19}	[C]	Stefan-Boltzmann	σ	5.670×10^{-8}	[W/m^{2}K^4]
Permittivity	ε_o	8.854×10^{-12}	[F/m]	Boltzmann	k	1.381×10^{-23}	[J/K]
Permeability	μ_o	$4\pi \times 10^{-7}$	[H/m]	Avogadro	N_A	6.022×10^{23}	[1/mol]
Speed of light	$c = 1/\sqrt{\varepsilon_o \mu_o}$	2.998×10^8	[m/s]	Faraday	$F = q N_A$	9.649×10^4	[C/mol]
Molar volume	V_m	2.2414×10^{-2}	[m^3/mol]	perfect gas	$R = k N_A$	8.314	[J/molK]

Table 10.1b. *Characteristics of a basic RIE system.*

Quantity	Typical value	Unit
rf power density	0.05–1.0	Wcm^{-2}
rf frequency	0.01–27	MHz
Pressure	1–30	Pa
Gas flow	1–200	sccm
Wafer temperature	100–1000	K
Gas temperature	300–600	K
Ion energy		
Bulk	0.05	eV
Sheath	10–1000	eV
Electron energy	3–30	eV
Gas number density	10^{14}–10^{16}	cm^{-3}
Ion density	10^9–10^{10}	cm^{-3}
Electron density	10^9–10^{10}	cm^{-3}
Neutral flux	10^{18}–10^{20}	$cm^{-2}s^{-1}$
Radical flux	10^{16}	$cm^{-2}s^{-1}$
Ion flux		
Bulk	10^{14}–10^{15}	$cm^{-2}s^{-1}$
	~10–100	μAcm^{-2}
Sheath	~100–1000	μAcm^{-2}
Electron flux	10^{17}–10^{18}	$cm^{-2}s^{-1}$
	~10–100	$mAcm^{-2}$

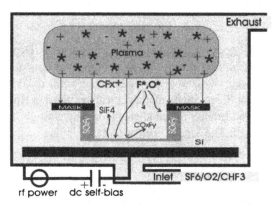

Fig. 10.1. Basic RIE system. 1 eV $\sim$ 8,000 K, 1 Pa $\sim$ 7.5 mTorr, 1 atm $\sim$ 1 bar = 10^5 Pa $\sim$ 760 Torr.

take place in IBARE [4]:

- Generation: An rf glow discharge is used to generate from a suitable feed gas (e.g., SF_6 for Si etching) by electron-impact dissociation/ionization the gas phase etching environment which consists of neutrals, electrons, photons, radicals (F), and positive (SF_5^+) and negative (F^-) ions.
- D.C. bias formation: The sample or wafer is placed on an rf driven capacitatively coupled electrode. Since the electron mobility is much greater than the ion mobility,

after ignition of the plasma the electrode acquires a negative charge resulting in the creation of the so-called dc self-bias voltage.

- Diffusion/forced convection: The transport of reactive intermediates from the bulk of the plasma to the wafer occurs by diffusion. Positive ions from the glow region are forced to the wafer by way of the dc self-bias and will assist the etching.

- Adsorption: Reactive radicals adsorb on the wafer surface. This step can be strongly enhanced by concurrent ion bombardment which serves to produce "active sites" since it aids in the removal of, e.g., an SiO_xF_y layer which otherwise passivates the surface.

- Reaction: A reaction between the adsorbed species and the wafer must take place. In the case of fluorine-based etching of Si, chemical reactions between the F atoms and the surface spontaneously produce either volatile species, SiF_4, or their precursors, SiF_x ($x < 4$). However, in Cl-based etching, Cl atoms are known to adsorb readily on Si surfaces, but the spontaneous etch rate is very slow. Ion bombardment makes it possible for adsorbed Cl atoms to attack the backbones of Si more efficiently and form a volatile $SiCl_4$ molecule. This mechanism is called ion-induced IBARE.

- Desorption: The desorption of the reaction product into the gas phase requires that the reaction product is volatile, thus it should have a high vapor pressure at the substrate temperature. Additionally, there should be no deposited blocking film at the surface. The removal of these films can be greatly accelerated by ion bombardment via sputtering. This mechanism is known as ion-inhibitor IBARE.

- Exhaust: The desorbed species diffuse from the etching surface into the bulk of the plasma and should be pumped out, otherwise plasma-induced dissociation of product molecules will occur and redeposition can take place.

10.2   Physical model

Although the word "atom" goes back into ancient Greek history (Democritus, 420 B.C.), it was Gassendi (1592–1655) who introduced many of the concepts of the present kinetic atomic theory; his atoms were rigid, moved at random in a void, and collided with one another. In 1738, David Bernoulli provided a mathematical treatment of Hooke's elastic theory (1678) and correctly derived Boyle's thermodynamic law by considering the collisions of atoms with a container wall. The kinetic theory of gases is one of the most remarkable theories of matter, for from a very simple model an equation of state of a perfect gas can be deduced and reasonable approximations to its physical properties can be derived. The kinetic theory of *perfect gases* is based on three assumptions:

(1)   The gas consists of molecules of mass m and diameter d in ceaseless random motion.

(2)   The size of the molecules is negligible (in the sense that their diameters are much smaller than the average distance traveled between collisions).

(3)   The molecules do not interact, except that they make perfectly elastic collisions when the separation of their centers is equal to d.

The phrase "perfectly elastic collisions" has to be explained more in detail: Molecules carry with them a certain amount of energy, which is the total energy. This energy can be divided into two different kinds: kinetic and potential. The kinetic energy of particles is due to their motion and is equal to $\frac{1}{2}mv^2$ for translational motion. The potential or internal energy may be in the form of excitation (electronic, vibrational, or rotational states), dissociation, ionization, or even the mass of the particle. The sum of the kinetic and potential energy, the total energy, is always constant. Now, collisions with no change of internal energy are termed elastic collisions, i.e., ones in which there is an interchange of kinetic energy only. Those with an exchange of internal energy between the particles are termed inelastic collisions. According to the tenets of the kinetic theory, both temperature and pressure are manifestations of molecular motion. Temperature is a measure of the average translational kinetic energy of the molecules, and pressure arises from the average force resulting from repeated impacts of molecules with containing walls.

Thus the kinetic model of gases is a "kinetic energy only" model of their properties. Of course a plasma is everything but a perfect gas. In a plasma there are charged and neutral particles which are continuously in motion and frequent collisions occur causing a continual interchange of internal energy. So the third assumption especially is heavily violated in plasmas. In fact, a plasma is a truly nonkinetic gas which involves chemical changes or reactions. Nevertheless, it is useful to use this model to deal with some aspects concerning the rate of physical change, such as expansion or vaporization, that occur continuously with variations in temperature or pressure.

In this section, some important aspects concerning particle collisions will be treated. First, definitions with respect to particle velocity, temperature, particle flux, pressure, and entropy are given. Then, the probability of collision and the mean free path are described followed by theory about elastic and inelastic collisions. After these subsections about collision phenomena, some plasma concepts will be reviewed, such as the plasma density, Debye length, plasma parameter, plasma oscillations, and plasma-to-floating potential. The latter one is especially important because it describes the minimum impact energy of ions to surfaces subjected to plasma. This section ends with some typical examples in plasma environments. The theory is presented as an example of model-building in plasma science, where a model of a state of matter is proposed, expressed quantitatively, and finally compared with experiment. Apart from the equation of state, some properties that relate to the transfer of energy and matter from one location to another are described. Most parts of this section are copied and rearranged from the excellent works of Brian Chapman in *Glow Discharge Processes*, Mitsuharu Konuma in *Film Deposition by Plasma Techniques*, Dennis Manos and Daniel Flamm in *Plasma Etching*, P. Atkins in *Physical Chemistry*, and Walter Moore in *Physical Chemistry* [1–3, 5, 6].

10.2.1 Particle velocity, mean kinetic energy, and temperature

In their constant motion, the molecules of a gas collide many times with one another, and these collisions provide the mechanism through which the (vector) velocities of individual molecules are continuously changing. As a result, there exists a distribution of velocities among the molecules; most have velocities with magnitudes quite close to the average, and relatively few have magnitude far above or far below the average. The distribution of the components of velocities, v_x, of molecules in a perfect gas is known as the Maxwell-Boltzmann (M-B) distribution, representing both Maxwell's contribution (he derived it originally) and Boltzmann's (who proved it rigorously). The complete form is [5]

$$f(v_x)\,dv_x = \left(\frac{m}{2\pi kT}\right)^{1/2} \exp\left(-\frac{mv_x^2}{2kT}\right) dv_x \quad [\text{m/s}], \qquad (10.1a)$$

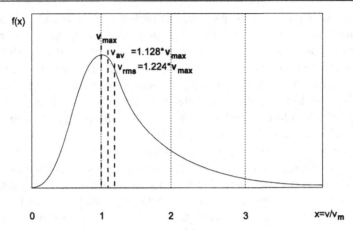

Fig. 10.2. Maxwell velocity distribution (v_{max}: most probable velocity, v_{av}: linearly averaged velocity, v_{rms}: root mean square velocity).

where m is the mass of each particle, $k = 1381 \times 10^{-26}$ [J/K] is Boltzmann's constant, and T[K] is temperature. The probability $f(v)$ that a molecule has a (scalar) speed in the range v to $v + dv$ in the velocity space is the sum of the probabilities that it lies in any of the volume elements $dv_x dv_y dv_z$ in a spherical shell of radius v. The volume of this shell is $4\pi v^2 dv$. Therefore, the distribution of the speeds of molecules irrespective of their motion becomes

$$f(v)\,dv = 4\pi v^2 \left(\frac{m}{2\pi kT}\right)^{3/2} \exp\left(-\frac{mv^2}{2kT}\right) dv \quad \text{[m/s]}. \tag{10.1b}$$

In fact, this distribution is just a special case of problems studied by Boltzmann about the distribution of a large number of particles among the different energy states accessible to them. In the context of quantum mechanics, which came later, these energy states would be discrete quantum stationary states. Several ingenious experiments have been devised to check the M-B distribution; all of them have agreed within experimental error. Figure 10.2 illustrates the M-B distribution function of the particle velocity. The distribution of speeds broadens as the temperature increases. From this distribution function, the following three speeds can be defined:

(1)   The most probable or arithmetic average speed, v_{max}, is the speed corresponding to the maximum of the distribution curve, which is found by differentiating $f(v)$ with respect to v and solving for the speed at which the slope is zero, i.e., $df(v)/dv = 0$:

$$v_{max} = \sqrt{\frac{2kT}{m}} = \sqrt{\frac{2RT}{M}} \quad \text{[m/s]}. \tag{10.2}$$

(2)   The linearly averaged or mean speed v_{av} is derived from

$$v_{av} = \int_0^\infty v \cdot f(v)\,dv = \sqrt{\frac{8kT}{\pi m}} = \sqrt{\frac{8RT}{\pi M}} \quad \text{[m/s]}. \tag{10.3}$$

Table 10.c1. *Average speed, mean free path, and molecular diameter and cross section of gases at STP.*

Gas	AMU	d [nm]	σ [Å^2]	λ [nm]	V_{av} [m/s]	Gas	AMU	d [nm]	σ [Å^2]	λ [nm]	V_{av} [m/s]
H_2	2	0.218	3.73	112.3	1692.0	O_2	32	0.296	6.88	60.0	425.1
He	4	0.200	3.14	179.8	1204.0	Ar	40	0.286	6.42	63.5	380.8
CO	28	0.380	11.34	58.4	454.5	CO_2	44	0.460	16.62	39.7	362.5
N_2	28	0.316	7.84	64.7	454.2	Cl_2	71	0.370	10.75	28.7	283.6

(3) The root mean square speed v_{rms}, defined as the square root of the average value of v^2, is expressed as

$$v_{rms} = \left(\int_0^\infty v^2 \cdot f(v)\, dv \right)^{1/2} = \sqrt{\frac{3kT}{m}} = \sqrt{\frac{3RT}{M}} \quad \text{[m/s]}. \qquad (10.4)$$

Note that $v_{rms}^2 \neq v_{av}^2$, although this approximation is quite accurate. For an SF_6 molecule, the molar mass $M = N_A m = 0.146$ [kg/mol]. At 20°C and with $R = N_A k = 8.314$ ($N_A = 6.023 \times 10^{23}$ [1/mol]), v_{av} has a value of ca. 200 m/s. In Table 10.c1 some typical values for different gases are summarized. The relation between the mean translational kinetic energy E_{rms} and the root mean square speed is

$$E_{rms} = \frac{1}{2} m v_{rms}^2 = \frac{3}{2} kT \quad \text{[J]} = \frac{3}{2} kT/q \quad \text{[eV]}, \qquad (10.5)$$

where m [kg] is the mass of the particle, $q = 1.602 \times 10^{-19}$ [C] is the electron charge, and $k = 1.381 \times 10^{-23}$ [J/K] is Boltzmann's constant. With the help of this expression, we can identify $1\text{eV} \sim 8,000\,\text{K}$. If we would have calculated the mean kinetic energy in case of a distribution along only one direction, i.e., Eq. (10.1a), we would have found that $E_{rms} = kT/2$. This suggests that for each degree of freedom, the kinetic energy is $kT/2$ per particle or $RT/2$ per mole. This result is called the principle of equipartion of energy which is proved severely with the help of statistical quantum mechanics. As a direct consequence of this principle, if the heat capacity of a gas exceeds the value of $C_v = (dU/dT)_v = 3NR/2$, we can conclude that this gas is taking up some form of energy other than translational kinetic energy (such as rotational or vibrational energy states). So, the average energy is independent of the mass of the molecule and is $3RT/2$ per mole of gas. For monoatomic gas, like helium, argon, or mercury vapor, this translational kinetic energy is the total kinetic energy of the gas. For diatomic gases, like nitrogen or chlorine, and polyatomic gases, like methane or water vapor, there is also energy associated with rotational and vibrational motions.

At room temperature the mean kinetic energy becomes $3kT/2q \sim 0.04$ eV. In plasmas, however, electrons are accelerated up to high velocities due to electrical fields. Konuma [2] wrote,

"The mean kinetic energies of electrons, ions, and excited molecules in a plasma are different, in contrast to gas molecules in a system of ordinary mixed gases, which have identical mean kinetic energies regardless of species. Therefore, the temperature of the plasma should be described according to the temperature of the respective particle; the electron

temperature T_e, ion temperature T_i, and gas temperature T_g. In addition, since the gas molecule itself has an internal structure of its own, energies concerning rotation, vibration, and electronic state have to be added. These energies are described in terms of the rotational temperature T_r, vibrational temperature T_v, and electronic temperature T_{elec}. On the other hand, the energy of electrons is described only by T_e since the electron, unlike a molecule, has no internal structure."

10.2.2   Pressure, flux, and rate of effusion

When a collection of particles with random velocities is put into an enclosure, they will exchange energy with the environment, for example, by collisions with the walls. Each time a molecule hits a wall, it will exert a force and the total force per unit wall area due to all the atoms is known as the pressure p [Pa]. The conservation of momentum implies that

$$p = \frac{1}{3}\rho v_{\mathrm{rms}}^2 = \frac{1}{3}nm v_{\mathrm{rms}}^2 = nkT \quad \text{[Pa],} \tag{10.6a}$$

with $\rho = nm$ (kg/m^3) being the density of the gas and $n = N * N_A / V$ [m^{-3}] the number of particles per unit volume, where $N_A = 6.023 \times 10^{23}$ [1/mol] is the number of Avogadro and N [mol] is the number of moles in a volume V [m^3]. This equation is one of the key results of kinetic theory and it connects the "perfect gas" idealization of the kinetic theory to the "ideal gas" approximation from the thermodynamic theory.

Similarly, the flux of molecules passing a unit area per unit of time can be shown as follows:

$$\phi = \frac{1}{4}n v_{av} = \frac{1}{4}n\sqrt{8/3\pi}\, v_{\mathrm{rms}} = p/\sqrt{2\pi mkT} \quad \text{[m}^{-2}\text{s}^{-1}\text{].} \tag{10.7a}$$

With the help of this expression we find that a container filled with nitrogen gas at room temperature and pressure receives about 3×10^{27} collisions per area per unit time. Typical parameter values for an SF$_6$ discharge of electron density 10^{10} cm^{-3} are found with the help of Eq. (10.3): (1) neutrals, $M_n \sim 0.146$ kg/mol and $T_n \sim 300$ K $= 0.04$ eV $\Rightarrow v_{avn} \sim 200$ m/s; (2) ions: $M_i \sim 0.128$ kg/mol and $T_i \sim 500$ K $= 0.07$ eV $\Rightarrow v_{avi} \sim 300$ m/s; and (3) electrons: $m_e \sim 10^{-30}$ kg and $T_e \sim 16{,}000$ K $= 2$ eV $\Rightarrow v_{ave} \sim 2400$ km/s. So, typical current densities are $j_e = q\phi_e = qn v_{ave}/4 = 100$ mA/cm^2 for the electrons and $10\,\mu$A/cm^2 for the ions.

A direct experimental illustration of the different average speeds of molecules of different gases can be obtained from the phenomena called molecular effusion. When a gas at a pressure p and temperature T is separated from a vacuum by a very small hole, the rate of escape of its molecules is equal to the rate at which they strike the area of the hole (Eq. (10.7a)). Therefore, if the area of the hole is A, the number of molecules that escape per unit time is

$$f_{esc} = \phi A = \frac{pA}{\sqrt{2\pi mkT}} \quad \text{[s}^{-1}\text{].} \tag{10.8}$$

The fact that this number is proportional to $1/\sqrt{m}$ is the origin of Graham's law of effusion, that the rate of effusion is inversely proportional to the square root of the molar mass.

Table 10.2a. *Third Law (heat capacity) and statistical (spectroscopic) entropies.*

Substance	S^0_{298} [J/Kmol]	Substance	S^0_{298} [J/Kmol]	Substance	S^0_{298} [J/Kmol]
H_2 (g)	130.59	SO_2 (g)	248.5	C (s, diamond)	2.44
N_2 (g)	191.5	CH_4 (g)	186.2	S (s, monoclinic)	32.6
O_2 (g)	205.1	C_2H_2 (g)	200.8	Na (s)	51.0
Cl_2 (g)	223.0	C_2H_4 (g)	219.5	I_2 (s)	116.1
HCl (g)	186.6	C_2H_6 (g)	229.5	NaCl (s)	72.38
HBr (g)	198.7	NH_3 (g)	192.5	AgCl (s)	96.23
HI (g)	206.7			AgBr (s)	107.1
CO (g)	197.5	Hg (l)	76.02	Ag (s)	42.72
CO_2 (g)	213.7	Br_2 (l)	152	Cu (s)	33.3
H_2O (g)	188.72	H_2O (l)	70.00	Fe (s)	27.2
N_2O (g)	220.0	CH_3OH (l)	127		

10.2.3 Entropy

Imagine the initial state of a system of two reservoirs: One compartment having low kinetic energy (cold) molecules, and the other one having high energetic (hot) molecules. Bringing both compartments into close contact, interdiffusion occurs by virtue of the thermal motions of the molecules, some of the hot molecules entering the cold compartment and vice versa. The final mixture of hot and cold molecules is less ordered than the unmixed state. In fact, a new thermal equilibrium is reached soon with an M-B distribution somewhere between the unmixed states. We say that the mixed state is more probable than the unmixed state, because there are more ways of distributing the hot and cold molecules as to yield mixed states than there are ways to yield the unmixed state. We say that the mixed state has a higher entropy than the unmixed state. Note, that in this case the thermal gradient is said to be causing the entropy to flow from the hot to the cold reservoir, however, that is not all; the total entropy increases as well. The same process occurs when mixing two different gases in thermal equilibrium. The driving force is now the chemical gradient μ as formulated by Gibbs. So in the eyes of classical statistical mechanics, the explanation of the increase in entropy in terms of the properties of the molecules must be related to some property of an entire collection of molecules and cannot be due to any properties of the molecules as individuals. To define a quantitative relationship between entropy S and the number W of distinguishable states of a system, Boltzmann worked out a relation presented in 1896:

$$S = k * \ln(W) \quad [J/K], \tag{10.9}$$

with $k = 1.381 \times 10^{-23}$ J/K. So W is a quantity describing the degrees of freedom, i.e., the number of different ways in which the energy of the system can be achieved by rearranging the molecules among their available states. As Atkins [5] wrote,

"*As an illustration of how this equation may be used, a solid of n HCl molecules at $T = 0$ has its lowest possible energy when all the molecules are perfectly ordered. Then $W = 1$ because there is only one way of achieving a perfectly ordered sample and $S = 0$ (because ln(1)=0). This perfectly ordered system has zero entropy.*"

A number of statistical and Third Law entropies are collected in Table 10.2(a). Using this formulation, entropy is connected to a probability: For a change of state 1 to state

2, $\Delta S = S_2 - S_1 = k * \ln(W_2/W_1)$. The relative probability of observing a decrease in entropy of ΔS below the equilibrium value is, therefore, $W/W_{eq} = \exp(-\Delta S/k)$. For 1 mol of helium at 273 K, $S/k \sim 10^{25}$. The chance of observing an entropy decrease one-millionth of this amount is about $\exp(-10^{19})$. So, the chance of observing a decrease in entropy is much less than observing the opposite. In fact, so small that it is "never" observed. Moore [6] wrote,

"No one, watching a book lying on a desk would expect to see it spontaneously fly up to the ceiling if it experienced a sudden chill. Yet, it is not impossible to imagine a situation in which all the molecules in the book moved in a given direction. Such a situation is only extremely improbable, since there are many molecules in any macroscopic portion of matter. Anyone who sees a book flying spontaneously into the air is dealing with a poltergeist and not an entropy fluctuation (probably). Only when the system is very small, is there an experimental chance of observing an appreciable relative decrease in entropy." Note how closely this phrase is already related to quantum physics.

Of course, quantum mechanics does not allow particles to be traced, even when they are alone, due to the uncertainty relation. Therefore, in the perspective of quantum mechanics the temperature and the entropy can never reach their classical minimum, i.e., $T = 0$ and $S = 1$. So, even for very small systems there is no experimental chance to observe a decrease in entropy, and the individual molecules also show an increase in entropy in time. Note that a stock of only four cards is not a relatively small system. Although it seems to be quite easy to get a certain ordered final result, i.e., a decrease in entropy with respect to the cards, the shuffling process causes friction to occur between the cards. Also, the shuffler puts in lots of energy to make the ordering and, therefore, the total entropy still increases.

Weaver and Shannon were the first to give a quantitative relation between entropy and information. They came to the conclusion that we can interpret entropy as negative information (neginformation) or information as negative entropy (negentropy). The result of mixing can thus be stated in different ways, all meaning the same:

(1)   decrease in order,
(2)   increase in disorder,
(3)   loss of information,
(4)   increase in entropy.

10.2.4   *Probability of collision and mean free path*

The most interesting events in the life of a molecule occur when it makes a collision with another molecule. Chemical reactions between molecules depend upon such collisions. The first theories of gas reactions from Trautz (1916) and Lewis (1918) were based on the kinetic theory of gases. They postulated that during collisions between gas molecules, a rearrangement of chemical bonds would sometimes occur, forming new molecules from the old ones. The rate of reaction was set equal to the number of collisions in unit time multiplied by the fraction of collisions that result in the chemical changes. Important transport processes in gases, through which energy (by heat conduction), mass (by diffusion), and momentum (by viscosity), are transferred from one point to another involve collisions between gas molecules.

The time scale of events in a gas, such as the rates of chemical reactions, is determined by the collision frequency, the number of collisions a molecule makes per unit time. We count a "hit" whenever the centers of two molecules come within some distance d of each other, where d, the collision diameter, is of the order of the actual diameters of the molecules.

In classical kinetic theory, the collision cross section σ is written as $\pi d^2/4 = \pi r^2$ for a collision between hard elastic spherical particles with radius r and is typically 3–20 Å^2 for gases. Suppose there are two kinds of molecules, m_1 and m_2, which interact as rigid spheres with diameters d_1 and d_2. A collision occurs whenever the distance between centers becomes as small as $d_{12} = (d_1+d_2)/2$. Let us also suppose that all the molecules m_2 are at rest and that a molecule m_1 is rushing through the volume with an average speed v_{av1}. We can imagine that this moving gas molecule in unit time sweeps out a cylindrical volume $v_{av1} \cdot \pi (d_{12})^2 = v_{av1} \cdot \sigma_{12}$. If n_2 is the number of stationary molecules per unit volume, the collision frequency for one moving molecule can be estimated as $f_{12} = v_{av1} \cdot \sigma_{12} \cdot n_2$. Actually, m_2 is moving also and it is the velocity of m_1 relative to m_2 that determines the frequency of collisions; the vector difference between the two velocities $v_{av12} = (v_{av1}^2 + v_{av2}^2)^{1/2} = (8kT/\pi\mu)^{1/2}$, with the quantity $\mu = m_1 m_2/(m_1 + m_2)$ called the reduced mass. Thus, the expression for the collision frequency should be

$$f_{12} = v_{av12} \cdot \sigma_{12} \cdot n_2 \quad [s^{-1}]. \qquad (10.10)$$

For only one gas $v_{av11} = (v_{av1}^2 + v_{av1}^2)^{1/2} = v_{av1}\sqrt{2}$ and $\sigma_{11} = 4\sigma_1$. Therefore, $f_{11} = 4\sqrt{2} * v_{av1} \cdot \sigma_1 n_1$. So, knowing the mean speed (450 m/s), the density at STP (3×10^{25} m^{-3}), and the collision cross section ($7\frac{1}{2}$ Å^2) of nitrogen, it is possible to calculate the collision frequency to be ca. 8 kHz at 1 mTorr. If n_1 is the particle density of the gas with molecules m_1, then the number of collisions F_{12} per unit volume and unit time, the collision density, between two kinds of gas particles with Maxwell distribution function characterized by the same temperature is given by

$$F_{12} = f_{12} \cdot n_1 = v_{av12} \cdot \sigma_{12} \cdot n_1 \cdot n_2 \quad [m^{-3}s^{-1}]. \qquad (10.11)$$

For only one gas, $F_{11} = \frac{1}{2} * 4\sqrt{2} * v_{av1}\sigma_1 n_1^2$, where the factor $1/2$ is necessary so as not to count every collision twice. Collision densities may be very large. For example, in nitrogen, with $d = 316$ pm at STP, $F = 9 \times 10^{34}$ [m^{-3}s^{-1}]. For electron collisions with gas molecules m_g, which are considered to be stationary, $f_{eg} = v_e\sigma_g n_g$ and $F_{eg} = v_e\sigma_g n_g n_e$.

Once the collision frequency is found, the mean free path λ, i.e., the average distance a molecule travels between collisions, can be calculated. The collision frequency f_c is the number of collisions per second and, provided that the mean velocity of a particle is v_r, the mean free path can thus be written as

$$\lambda_g = \frac{v_{avg}}{f_{gg}} = \frac{1}{\sigma_g n_g 4\sqrt{2}} \quad [m] \qquad (10.12a)$$

$$\lambda_e = \frac{v_{ave}}{f_{eg}} = \frac{1}{\sigma_g n_g} = \frac{kT}{\sigma_g p_g} \quad [m]. \qquad (10.12b)$$

So, the mean free path of the electron is ca. $4\sqrt{2} \sim 6$ times larger than that of the gas molecule. For nitrogen, the cross section $\sigma_g \sim 7.5$ [Å^2] and $\lambda_g \sim 50$ mm at 1 mTorr and $\lambda_e \sim 300$ mm at 1 mTorr. Some measurements of mean free paths are given in Table 10.c1. Just as is there a distribution of particle velocities and collision frequencies, so is there a distribution of free path lengths between collisions. The probability $P(x)$ of an atom traveling a distance x without colliding is

$$P(x) = \exp(-x/\lambda) \quad [-], \qquad (10.13)$$

where λ [m] is the average distance between the position at which a particle collides with one particle and the position of the next collision, i.e., the mean free path between collision.

Even if the molecules cannot be considered as rigid spheres, the results of various collision processes can be expressed in terms of an effective collision cross section σ_{eff}, which can be calculated from expressions for the forces between molecules. For example, as the temperature increases, the collision cross section becomes smaller. The apparent molecular radius can be considered to obey Sutherland's formula:

$$r = r_\infty \sqrt{1 + \frac{T_S}{T}} \quad [\text{m}], \tag{10.14}$$

where T_S is the Sutherland constant, and r_∞ is the radius of the particle at $T = \infty$.

10.2.5 Elastic collisions

When the potential energies of the two bodies colliding are the same before and after an interaction, the collision is called elastic. Assume that m_2 is initially stationary and that m_1 collides with velocity v_1 at an angle α to the line joining the centers of m_1 and m_2 at the moment of collision. By conservation of linear momentum and energy, the fractional energy E transferred from mass m_1 to mass m_2 is

$$E_2/E_1 = \cos^2(\alpha)\, 4m_1 m_2/(m_1 + m_2)^2 \quad [-]. \tag{10.15}$$

When the collision is frontal and $m_1 = m_2$ this expression has its maximum of 1, i.e., the velocity of mass m_2 is v_1 and mass m_1 has lost all its kinetic energy. However, when an energetic electron strikes an SF_6 molecule, then the transfer function becomes just $4m_e/m_{SF6}$. So, the function has a value of about 10^{-5}, and very little energy can be transferred from the electron to the molecule.

10.2.6 Inelastic collisions

Electron energy, channeled into inelastic electron–neutral particle collisions, maintains the supply of ions and radical species which are continuously lost by reaction and recombination. In turn, these electrons rely on the ionization process for their sustenance. The degree of these processes is typically 10^{-4}, so that essentially we still have a neutral ground state gas that can be described by the gas laws. Despite this, practical glow discharges are rather complex environments, and they are far from being well understood. So, before we launch into the confusing (and sometimes conflicting) detail of specific discharges, let's look at some more collision phenomena, which are common to all of the glow discharge processes covered in this section.

Above, the energy transfer function for elastic binary collisions was established. Now, it is allowed for the collision to be inelastic, so that the molecule struck gains internal energy of ΔU. Then, using the same laws of conservation, we obtain

$$\Delta U/E_1 = \cos^2(\alpha)\, m_2/(m_1 + m_2) \quad [-]. \tag{10.16}$$

So, whereas the maximum elastic energy transfer from an electron to an SF_6 molecule was only 0.001%, by inelastic means this may rise to 99.999%.

It is not only by electron impact that energy is transferred. The process includes thermal and photon activation and ion–neutral and metastable–neutral collisions also. For the relatively "cold" plasma environments, the thermal activation is usually neglected, but photo ionization can be significant. In the next paragraphs, some important activation processes will be treated: excitation, dissociation, and ionization.

Excitation and relaxation

One of the immediately self-evident features of a glow discharge is that it glows. This glow is due to relaxation or de-excitation of excited molecules, i.e., the inverse of the excitation process. In the excitation process, the transfer of energy to the bound electron enables the electron to jump to a higher energy level within the atom with a corresponding quantum absorption of energy. The excited state is conventionally represented by an asterisk superscript, e.g., SF_6^*.

The lifetime of the excited states varies from nanoseconds to seconds. Each transition is accomplished by the emission of a photon whose wavelength corresponds to the energy difference between the excited upper level and the lower level. A useful and reasonable good approximation is $E(eV) * \lambda(\text{Å}) \equiv 12400$, since $E(eV) = h\nu/q = hc/\lambda q$, where $h = 6.626 \times 10^{-34}$ J/s, $q = 1.602 \times 10^{-19}$, and $c = 2.998 \times 10^8$ m/s. So, the visible light with wavelength between 4100 Å (violet) and 7200 Å (red) corresponds to electron transitions of 3.0 eV and 1.7 eV, respectively. However, with suitable detection equipment, photons from deep ultraviolet (atomic transitions) and far infrared (molecular vibrational and rotational transitions) can be detected. Konuma [2] explains,

"In the case of monoatomic gas molecules such as helium, the translational and electronic energy makes up the entire energy. For diatomic molecules such as nitrogen or polyatomic molecules such as SF_6, rotational and vibrational energies must also be considered.... One series of vibrational energy levels belongs to each electronic level, and one series of rotational levels is attached to each vibrational level. The energy differences between vibrational levels are larger than those between rotational levels. When the system is at thermal equilibrium, the distribution of internal energies of the gas particles in an ensemble is given by the Boltzmann distribution law. Thus, the energy of a plasma can be presented using several temperatures/energies and normally each of them is different. This indicates that thermal equilibrium is not applicable among particles in a plasma or for energy states of a desired particle. However, it often happens that the electrons and the ions have separate Maxwell distribution characterized by different temperatures T_e and T_i, and the electrons or the ions themselves are at thermal equilibrium. This is caused by larger collision rates among electrons or among ions than the collision rates between an electron and an ion."

However, we have to keep in mind that charged particles have long-range interactive forces. Therefore, it is still not a perfect gas since the third assumption of the perfect gas is not fulfilled. Only in the case of low-density plasmas are the charged particles far enough away for the forces to be neglected.

Some excited atoms have very long lifetimes (up to a few seconds), and these are known as metastable excited atoms. They arise because the selection rules forbid relaxation to the ground state. All of the noble gases have metastable states. For example, argon has metastable states at 11.5 eV and 11.7 eV. Metastable atoms lose their energy by striking the wall of the reaction chamber or colliding with other atoms or molecules. When a metastable atom collides with a neutral, the neutral can become ionized. This is known as Penning ionization. A collision of two metastable states may result in the production of an ion also. However, the density of metastable states is probably far too low and this event is neglected. The same criterion, although to a much lesser extent, prevents electron–metastable collisions.

Dissociation

In plasma etching a principal requirement is a process to dissociate or break apart relatively stable gas molecules into chemically active species which then can react with the substrate. An oxygen molecule can be dissociated into two oxygen atoms, but an atomic gas such as argon cannot be dissociated at all. A normal result of dissociation is an enhancement of chemical activity, since the products are usually more reactive than the parent molecule. Such enhancement of activity is used to, e.g., oxidize photoresist in the plasma ashing process.

Ionization and recombination

The process in which a primary electron removes an electron from an atom producing a positive ion and an extra electron is called electron impact ionization. The prime feature of discharges is that of ionization, with perhaps as many as 10^{18} electron–ion pairs being produced per second. The degree of ionization is typically 10^{-4}, and the current densities are of the order of 1 mA/cm^2. The extra electron produced by the ionizing collision is gaining energy from the rf field also. It is by this multiplication process that a glow discharge is maintained. There is a minimum energy requirement for this ionization process to occur, equal to the energy to remove the most weakly bound electron from the atom. This is known as the ionization potential found in Table 10.2(b).

In the recombination or deionization process, an electron coalesces with a positive ion to form a neutral atom. However, due to the laws of conservation, this type of recombination is very unlikely. But recombination must occur somehow, because otherwise the ion and electron densities would continuously increase, and this is contrary to experience. Therefore, some more subtle recombination processes must take place, such as three-body collisions, two-stage processes, radiative recombination, or, which is more likely, the recombination of charges at the reactor walls.

Electron attachment

There is a possibility that an electron colliding with an atom may join onto the atom to form a negative ion. This process is known as electron attachment. The noble gases already have filled outer electron shells and so have little or no propensity to form negative ions. However, halogen atoms have an unfilled state in their outer electron shells and high electron affinity and thus readily form negative ions. To complicate matters, the electron attachment is a function of the electron energy. This is another manifestation of the problem of simultaneously satisfying energy and momentum conservation.

So, the physical property that gives some insight into electronic properties of atoms is the electron affinity. The electron affinity of an atom is the energy liberated when an electron is added to a neutral atom. As with ionization potentials, it would be of value to have successive electron affinities, arising from the addition of further electrons to the negative ion. The values of these will always be negative. Electron affinities together with ionization potentials give us a useful way of predicting the attraction one atom will have

Table 10.2b. *Particle excitation, dissociation, ionization, and electron affinities from CRC handbook of chemistry and physics [9].*

Particle	Affinity [eV]	Dissociation [eV] thermal/electrical	Metastable [eV]	Lifetime [seconds]	Ionization [eV]
He	−0.22	—	20.61	0.02	24.586
H	0.754	—	10.20	0.12	13.598
F	3.399	—			17.423
Cl	3.617	—			12.976
Br	3.365	—			
I	3.059	—			
N	0.0	—	2.38	$14 * 10^4$	14.534
O	1.461	—	1.96	110	13.618
S	2.077	—			10.36
Si	1.385	—			8.15
C	1.263	—			11.26
Ar	−0.36	—	11.55	55.9	15.759
O^-	−6.1	—			
S^-	−8.9	—			
H_2	?	4.5/8.8	11.75	0.001	15.427
Fv_2	3.08				15.7
Cl_2	2.38	2.48/3.7			11.48
O_2	0.440	5.08/7	0.98	2700	12.063
N_2		9.8/24.3	6.17	1.3−12	15.576
ON	0.024	6.48/>10			9.250

Particle	Ionization	Particle	Ionization	Particle	Ionization
Ne	21.56	H_2O	12.614	CO_2	13.769
Kr	13.99	CO	14.013	CH_4	12.704

Particle	Affinity	Particle	Affinity	Particle	Affinity
ClO	2.17	Br_2	2.55	I_2	2.55
C_2	3.39	SO	1.126	SiH_3	<1.44
CH	1.238	SiH	1.277	WF_5	1.25
CN	3.821	SF_4	2.35	WF_6	3.36
CS	0.205	SF_6	0.46	CH_2	0.67
OH	1.828	SO_3	>1.70	CH_3	0.08
SH	2.317	SiF_3	<2.95	CO_3	2.69
SiH_2	1.124	HN	0.370	CF_3Br	0.91

for the electron of another atom when forming molecules. For a fluor atom, the electron affinity is 3.399 eV, i.e., the combination of a fluor atom with an electron is "exothermic" by 3.4 eV per bond.

In Table 10.2(b) the atomic electron affinity of some important particles is given [9]. Ions and neutrals can collide with each other (in)elastically to either exchange charges or cause further ionization. Charge transfer might be important in the sheath region, where it has the effect of changing the energy distribution of ions and neutrals on the electrode. The plasma

environment is even more complicated because all the activation processes just described
may or may not be accompanied by another activation process. In summary, some important
activation processes are:

Excitation	$CF_4 + e^- \Rightarrow CF_{4*} + e^-$	[04.0 eV]
Dissociative attachment	$CF_4 + e^- \Rightarrow CF_3 + F^-$	[07.3 eV]
Dissociation	$CF_4 + e^- \Rightarrow CF_3 + F + e^-$	[12.5 eV]
Dissociative ionization	$CF_4 + e^- \Rightarrow CF_{3+} + F + 2e^-$	[14.8 eV]
Direct ionization	$CF_4 + e^- \Rightarrow CF_{4+} + 2e^-$	[15.5 eV]
Cumulative ionization	$CF_{4*} + e^- \Rightarrow CF_{4+} + 2e^-$	[eV]
Penning ionization	$CF_4 + A^* \Rightarrow CF_{4+} + A + e^-$	[eV]

10.2.7   Plasma density

A plasma is a collection of particles consisting of electrons, ions, and excited molecules.
Of these, the charged particles are the electrons and the ions. Ions are normally positively
charged, except for plasmas that contain gases with large electron affinities, such as oxygen
or halogen gases. For a plasma composed of various kinds of positive ions, at densities of
$n_{i1}^+, n_{i2}^+, \ldots$, negative ions at densities of $n_{i1}^-, n_{i2}^-, \ldots$, and an electron density of n_e^-, the relation

$$n_e^- + n_{i1}^- + n_{i2}^- + \cdots = n_{i1}^+ + n_{i2}^+ + \cdots = n \quad [m^{-3}] \tag{10.17}$$

should be satisfied for the plasma as a whole. This shows the electrical neutrality of the
plasma, i.e., the plasma is macroscopically neutral in the equilibrium state, so that n is
known as the plasma density. Such a condition is realized when the characteristic length of
a system is much larger than the Debye length λ_D, which is defined next.

10.2.8   Debye length

Plasmas are conductive and, as a result, can respond to local changes in potential. The distance over which a small potential can perturb a plasma is known as the self-shielding or Debye length [1, 3, and 7] Cohen et al. [3] describes this subject simplified in the following way:
   *"Consider a collection of charged particles arranged in a slab, with equal numbers of
each charge. Try to displace the positive charges to the right and the negative ones to the
left (Fig. 10.3). Energy is required to effect the separation. An electric field develops, which
makes the separation of charges progressively more difficult. Now consider the particles
as they try to separate by simply using own thermal energy, not some externally supplied
source. The distance they can separate depends on their thermal energy. When their thermal
energy is all converted into potential energy, they can separate no further. The potential
energy, $E = q\Delta V$, may be derived from Poisson's equation for the potential (MKSA-units):*

$$\nabla^2 V = \frac{\partial^2 \Delta V}{\partial s^2} = -\rho/\varepsilon_0 = -nq/\varepsilon_0 \quad [V/m^2], \tag{10.18}$$

*where $\rho_e = \rho i = \rho = nq$ is the charge density (in the case that there are no negative ions),
$\varepsilon_0 = 8.85 \times 10^{-12}$ [F/m] the permittivity of vacuum, and $q = 1.6 \times 10^{-19}$ [C] the electron*

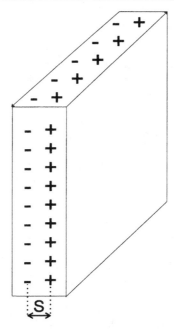

Fig. 10.3. Collection of charged particles in a slab. The positive particles were displaced from the negative. A strong electric field results, which will force the two sets of particles back toward each other.

charge. For a separation S this gives

$$\Delta V = -\frac{1}{2}s^2 nq/\varepsilon_0 \quad [V].$$
(10.19)

This is the potential energy change, $q \Delta V$, as a function of the separation of the charges. By equating the available thermal energy ($kT_e/2$ because of the one-dimensional motion) with the potential energy, we can solve for the maximum distance, λ_D, by which the charges may separate from each other before the restoring force of the electrostatic field pulls them back together again. This distance is the Debye length, λ_D:"

$$\lambda_D = \sqrt{\varepsilon_0 k T_e/nq^2} \quad [m].$$
(10.20a)

Konuma [2] wrote, *"In the case of a glow discharge, typically $n = 10^{10} cm^{-3}$ and $kT_e = 1\,eV$, thus $\lambda_D = 0.1\,mm$ is obtained. So λ_D is significantly smaller than the dimensions of the discharge tube and therefore a glow discharge is considered neutral as a whole and is regarded as a plasma. This criterion is one of the important conditions for an ionized gas to be a plasma. The Debye length is an important physical parameter for a plasma. If a piece of charged matter is inserted into a plasma, a cloud of ions will surround the object when the object is negatively charged, and a cloud of electrons will surround a positively charged object. No electric field will be present in the bulk of plasma outside of these clouds. This phenomenon is called the Debye shielding and the clouds are called sheaths. The Debye length gives the thickness of the sheath in which the shielding is almost complete. Thus only over the outside of the sheath can macroscopic electrical neutrality hold."*

Another way to calculate the Debye length is by asking how charges rearrange themselves around a test particle in response to its electric field. After finding the distribution of charges, we again use Poisson's equation to see how much the Coulomb field is reduced [6]:

$$V = V_0 \exp(-r/\lambda_D) = (Q/4\pi\varepsilon_0 r)\exp(-r/\lambda_D) \quad [V]. \tag{10.20b}$$

So, the answer we obtain is that the electric field is reduced to $1/e$ of its Coulomb value in a distance equal to the Debye length. Thus, the electric field does penetrate into a plasma further than one Debye length.

10.2.9   Plasma parameter

We are now able to find out how many particles are still affected by the shielded field of the test particle. This can be estimated by calculating the number of particles within a sphere of radius λ_D:

$$N_D = \frac{4}{3}\pi\lambda_D^3 n = \frac{4}{3}\pi q^{-3}\sqrt{(\varepsilon_0 kT_e)^3/n} \quad [-]. \tag{10.21}$$

This number, commonly called the Debye sphere or plasma parameter, must be larger than 100 for collective effects to be important. In the case of a glow discharge, typically $n = 10^{10}$ cm^{-3} and $kT_e = 1$ eV, thus $\lambda_D = 0.1$ mm and $N_D = 40,000$ is obtained.

10.2.10   Plasma oscillations

Generally, the density distribution of gas which is in the equilibrium state is macroscopically uniform as a whole. However, by means of fluctuations the uniformity can be broken at a given time and position. Also, in a plasma the particle density distribution has fluctuations. This is evidenced through oscillations built up by charge of the particle.

"*As just stated, when two slabs of charge are separated (Fig. 10.3), a restoring force is developed. This pulls the charges back toward each other. They accelerate and then pass through their equilibrium positions and separate in the opposite sense. Now the positive charges are to the left and the negative to the right. It is easy to estimate the frequency of this motion. It is simply the inverse of the time it takes particles to move the Debye length at their thermal velocity,*" (Cohen et al. [3]). (The thermal energy for one-dimensional motion is $kT_e/2 = mv^2/2 \Leftrightarrow v^2 = kT_e/m$):

$$\omega_{pe} = V_e/\lambda_D = \sqrt{nq^2/m\varepsilon_0} \quad [\text{rad/s}]. \tag{10.22}$$

Plasma oscillations are classified into plasma–electron oscillations and plasma–ion oscillations. The frequency of plasma–electron oscillation is usually very high since the mass of an electron is small. For a plasma of density $n = 10^{10}$ cm^{-3}, $f_{pe} = 898$ MHz, which is in the microwave range. The extent of the fluctuation of the density distribution is about the range of the Debye length.

10.2.11   Plasma-to-floating potential

Consider the influence of an electrically isolated object suspended in a plasma. Initially, electrons and ions will flow into the object with current densities predicted by Eq. (10.7a): $j_e \sim 100$ mA/cm^2 and $j_i \sim 10\,\mu$A/cm^2. Since $j_e \gg j_i$, the object immediately starts to build a negative charge and hence negative potential with respect to the plasma. Soon a steady state is reached where the greatly reduced flow of repelled electrons is balanced by the undisturbed flow of ions. Now the plasma body is virtually electric-field-free due to Debye shielding and so is equipotential: the so-called plasma (space) potential V_p. Similarly, we can associate a floating potential V_f with the isolated object, which is always negative with respect to the plasma potential. Since electrons are repelled by the potential difference ~ $V_f - V_p$, it follows that the object will attain a net positive charge around it (the space charge or sheath) and the electron density decreases in the sheath. Therefore, it doesn't glow as much; the object is surrounded by a relatively dark space.

Let us now try to estimate the magnitude of $V_f - V_p$. To surmount this barrier, an electron needs $q(V_f - V_p)$ of potential energy. Hence, only electrons that enter the sheath from the plasma with kinetic energies in excess of $q(V_f - V_p)$ will reach the object. The M-B distribution function gives us the fraction that can do this:

$$\frac{n'_e}{n_e} = \exp\left(\frac{q(V_f - V_p)}{kT_e}\right) \quad [-]. \tag{10.23}$$

If the density n'_e just achieves balance with the ion density at the surface of the object, then

$$\phi_i = \phi_e \Leftrightarrow n'_e v_{eav} = n_i v_{iav} \quad [\text{m}^{-2}\text{s}^{-1}]. \tag{10.24}$$

Substituting for n'_e (Eq. (10.23)) together with $n_e = n_i$ and $v_{av} = \sqrt{(8kT/\pi m)}$ (Eq. (10.3)) gives

$$V_f - V_p = \frac{kT_e}{2q}\ln\left(\frac{m_e T_i}{m_i T_e}\right) \quad [\text{V}]. \tag{10.25a}$$

The calculation of the floating potential is slightly controversial. In the foregoing discussion the ion flux was not disturbed by the barrier because it is only limited by the random arrival of ions at the plasma–sheath interface. However, this model is oversimplified. It was assumed that the sheath terminated at the plane where the ion and electron density became equal, to become undisturbed again. In fact, between these two regions there is a quasi-neutral transition region or presheath of low electric field. This field directs ions toward the object which increases the ion velocity and thus the current density entering the sheath. This effect has come to be known as the Bohm sheath criterion. An odd result is that the ion velocity on entering the sheath–presheath boundary must be greater than $v_{bi} = \sqrt{(kT_e/m_i)}$, known as the sound speed, which means that the electron and ion motions are coupled. Therefore, the potential drop across the presheath, $V_b - V_p$, is $kT_e/2q$. Putting this into Eq. (10.23), $n_{eb}/n_e = \exp(-1/2) \sim 0.6$. Then the ion flux becomes, together with $n_{eb} = n_{ib}$,

$$n_{ib}v_{ib} = 0.6n_e\sqrt{kT_e/m_i} \quad [\text{m}^{-2}\text{s}^{-1}]. \tag{10.26}$$

Typical parameter values for a discharge of density 10^{10} cm^{-3} are $kT_e = 2$ eV and $m_i \sim 10^{-25}$ kg. Thus, an ion current density of 200 μA/cm^2 (instead of 10 μA/cm^2 found earlier) is obtained, which is more realistic as found experimentally. The effect of the Bohm criterion is to increase the ion flux to a floating substrate. Chapman [1] wrote that: "*Using a similar derivation to before, the criterion for net zero current becomes*"

$$V_f - V_p = \frac{kT_e}{2q}\left(\ln\left(2\pi\frac{m_e}{m_i}\right) - 1\right) \sim \frac{kT_e}{2q}\ln\left(2.3\frac{m_e}{m_i}\right) \quad [\text{V}]. \tag{10.25b}$$

Cohen, instead, uses $v_b = \sqrt{[k(T_e + T_i)/m_i]}$ as the sound speed at the sheath–presheath boundary. Therefore, the potential drop across the presheath becomes $(V_b - V_p) = k(T_e +$

$T_i)/2q$ and a different final result is obtained:

$$V_f - V_p = \frac{kT_e}{2q}\left(\ln\left[2\pi\frac{m_e}{m_i}\left(1 + \frac{T_i}{T_e}\right)\right] - 1 - \frac{T_i}{T_e}\right) \quad [\mathrm{V}]. \tag{10.25c}$$

However, as normally is the case, $T_i/T_e \ll 1$ and Eq. (10.25b) is a reasonable approximation. In case of an argon ion, the ion mass M_i is 40×1836 times the electron mass M_e and we get $V_f - V_p \sim -4.7 * kT_e/q$. For SF$_5$ it is $V_f - V_p \sim -5.2 * kT_e/q$. So, typically the floating potential is 5 times the electron temperature. Thus, the ions that reach the surface will have an energy equal to what they have in the plasma plus what they have gained in passing through the presheath and sheath, $(4.7 + 0.5)kT_e. \sim 5kT_e$. This latter statement means that ions will hit surfaces in a plasma environment severely, even when they are only floating due to their acceleration in the plasma presheath and sheath. In case of a typical electron temperature of $kT_e \sim 2\,\mathrm{eV}$, the ion energy is already 10 eV, which is enough to break most chemical bonds.

10.2.12   Examples

Now that we have treated a lot of theory, let us try to give some indications of how we can use this theory. First, the fluxes of all the different plasma particles will be estimated. After that the respective densities are roughly calculated. This section will end with some approximations concerning the particle energies and a comprehensive overview is found in Table 10.1(c).

Particle fluxes

- Neutral flux: To find the neutral flux in equilibrium, Eq. (10.7a) is used. In equilibrium, the flux of particles leaving a volume equals the amount of particles arriving in the volume at a rate

$$\phi_{b_n} = p_n/\sqrt{2\pi M_n RT} = p_n/\sqrt{2\pi m_n N_A RT} \quad [\mathrm{mol/m^2s}], \tag{10.7b}$$

  with p_n [Pa] being the partial pressure of the neutrals, $N_A = 6 \times 10^{23}\,\mathrm{mol^{-1}}$ the number of Avogadro, $m_n = 146 * 1.6 \times 10^{-27}\,\mathrm{kg}$ the mass of an SF$_6$ molecule (i.e., 1 mole SF$_6$ gas $\Rightarrow M_n = 0.146\,\mathrm{kg/mol}$), and $R = kN_A = 8.3\,\mathrm{J/molK}$ ($k = 138 \times 10^{-24}\,\mathrm{J/K}$ the constant of Boltzmann). At $T = 300\,\mathrm{K}$ and $p_n = 1\,\mathrm{Pa}$, we calculate $\phi b_n = 2 \times 10^{-2}\,\mathrm{mol/m^2s}$.

- Radical flux: To estimate the net radical flux ϕb_r, it is important to know the etch rate, because the Si is consuming radicals and therefore radicals will move to the Si surface. Assume an Si wafer, covering the complete cathode area, is etching with 10 Å/s at 20 W input power. The Si interatomic distance is ca. $2\frac{1}{2}$ Å, i.e., 0.064 [atom/Å^3], 6.4×10^{18} atom/Å m^2, or 10^{-5} mol/Å m^2; therefore, 10^{-4} mol/m^{2}s Si is removed. The reaction end product of Si etching is SiF$_4$, so four radicals are needed to remove one Si atom and $\phi b_r = 4 \times 10^{-4}$ mol/m^{2}s. Generally, the etch rate will be much higher when a smaller piece of Si is brought into the reactor. This is an indication that the radical concentration in the plasma is depleted, the ion density is changed, or radicals are supplied via surface migration.

- Ion flux: For the ion flux ϕb_i the same technique is used as for the radical flux. Assume there is one ion needed to remove one Si atom (i.e., the ion is able to break all the

Table 10.1c. *Basic RIE variables of the plasmafab 340 for Si HART's with SF6/O2, 1 eV ~ 8,000 K, 1 Pa ~ 8,000 K, 1 Pa ~ 7.5 mTorr.*

Quantity	Symbol	Typical value	Unit
RIE setting			
Power	P	20–200	W = J/s
Pressure	p_n	1–30	Pa
Gas flow	ϕ	1–200	sccm
Wafer temperature	T	300–1000	K
dc self-bias	U_{dc}	10–1000	V
Equipment parameters			
rf frequency	f	13.56	MHz
Anode area	A_a	0.05–0.075	m^2
Cathode area	A_c	0.025	m^2
Anode/cathode ratio	R_{ac}	2–3	
Reactor volume	V	0.002	m^3
Glow characteristics			
Power density		10^4–10^5	W/m^3
Neutral density	ρg_n	$4*10^{-4}$–10^{-2}	mol/m^3
Radical density	ρg_r	10^{-6}–10^{-5}	mol/m^3
Ion density	ρg_i	10^{-9}–10^{-8}	mol/m^3
Electron density	ρg_e	10^{-9}–10^{-8}	mol/m^3
Neutral energy	Eg_n	0.05 + 0.03	eV
Radical energy	Eg_r	0.05 + 2	eV
Ion energy	Eg_i	0.05 +15	eV
Electron energy	Eg_e	3–30	eV
Boundary characteristics at ER = 1nm/s			
Energy flux		10^3–10^4	$J/m^2 s$
Neutral flux	ϕb_n	$2*10^{-2}$–0.6	$mol/m^2 s$
Radical flux	ϕb_r	10^{-4}–10^{-3}	$mol/m^2 s$
Ion flux	ϕb_i	10^{-6}–10^{-5}	$mol/m^2 s$
Electron flux	ϕb_e	10^{-6}–10^{-5}	$mol/m^2 s$
Sheath characteristics			
Ion energy (sheath)	Es_i	10–1000	eV
IADF			
IEDF			

bonds and enable fluor atoms to remove it); then $\phi b_i = 10^{-4}$ mol/m²s. However, from experiments it is found that the ion flux approximates 10^{-6}–10^{-5} mol/m²s. So, after removing the inhibitor a great number of F atoms are etching the bare silicon before oxygen is blocking the surface again.

- Electron flux: This flux, equals the ion flux, because otherwise the plasma glow would be charged more, and more which is not true in equilibrium.

Particle densities

To estimate the particle densities in a plasma environment, the perfect gas together with the kinetic gas theory are functional. For the neutrals it is enough to consider the perfect gas theory only. In a perfect gas, only elastic collisions occur. This means that only momentum (i.e., kinetic energy) is transferred and no energy is incorporated into excitation levels like molecular vibrations. Generally, only the monoatomic gases are near this ideal, because they suffer for vibrational and rotational excitation levels and only electronic excitation or ionization is possible. These latter energy levels are normally quite high and therefore only possible at extremely high temperatures (like the plasma of the sun) or pressures.

- Neutral density: The neutral density ρg_n [mol/m^3] can be expressed in the pressure p_n [Pa] with the help of Eq. (10.6a), the perfect gas theory:

$$p_n V = NRT \Rightarrow \rho g_n = N/V = p_n/RT \quad \text{[mol/m}^3], \tag{10.6b}$$

  with V [m^3] being a volume containing N [mol] particles. At $T = 300$ K and $p_n = 1$ Pa, we calculate $\rho g_n = 4 \times 10^{-4}$ mol/m^3.

- Radical density: To find the radical density, again the formula of kinetic gas theory is used but now backward. At first, it could be concluded that the calculated vapor pressure is a lower limit because the radical flux is a net flux whereas the neutral flux is an equilibrium flux. In other words, the equilibrium flux is higher than the net flux (when there is no driving force other than the thermal vibration, of course). However, when radicals are transported from the surrounding surfaces and the net flux equals the equilibrium flux, this calculation becomes an upper limit. So, the best estimation is found when the complete target is made of silicon and the surface migration is minimal (i.e., the etch rate is measured in the center of the silicon wafer). Under these circumstances the radical density can be estimated as

$$\rho g_r = p_r/RT = \phi b_r (2\pi M_r/RT)^{1/2} = \phi b_r (2\pi m_r/kT)^{1/2} \quad \text{[mol/m}^3]. \tag{10.7c}$$

  With $m_r = 19 * 1.6 \times 10^{-27}$ kg, the atomic mass of the F atom, and $\phi b_r = 10^{-4} - 10^{-3}$, we calculate $\rho g_r = 10^{-6} - 10^{-5}$ mol/m^3.

- Ion density: The ion density is even more difficult to estimate, because the plasma has a net positive charge. Due to the coulomb force, these ions are located on the plasma boundary. When ions are depleted due to etching, they are supplemented by the internal glow. However, this supply is forced by the electrical fields and kinetic gas theory can't be used. Nevertheless, we are able to give an upper limit. The amount of ions in the plasma glow is always smaller than the amount of radicals because it takes more energy from a colliding electron to create an ion, i.e., a radical needs ca. 2 eV (SF$_6$ $\Rightarrow$ SF$_5$ + F), whereas the ion needs 15 eV (SF$_6$ $\Rightarrow$ SF$_5^+$ + F$^-$). Using this limit, we estimate the ion density to be (much) smaller than 10^{-6} mol/m^{2}s, let's say $10^{-9} - 10^{-8}$ mol/m^{2}s. When a better estimation is necessary, the electron energy distribution function (EEDF) should be used.

- Electron density: For a plasma composed of electrons and various kinds of ions the electrical neutrality as a whole should be satisfied in the equilibrium state and the electron density matches the ion density. (Such a condition is realized when the characteristic length of a system L is much larger than the Debye length λ_D, which was derived with the help of Eq. (10.20a).)

- Photon density: The photon density is even more difficult to estimate than the other densities. This is caused by the enormous broad spectrum in which this particle may exist. Collisions between low energetic electrons and polyatomic molecules can easily excite the molecule into a vibrational or rotational energy level. These low energetic excitation levels will relax in a short time, emitting a photon in the far infrared. The lifetime of the excited levels and the energy of the photons are in practice almost impossible to calculate. Only for the very simple molecules, like hydrogen, are these figures in stock.

- Energy density: This figure is calculated with the help of the particle densities and their energy content. For example, the neutral density is 10^{-2} mol/m^3 at 20 Pa with a thermal energy of $3k\Delta T/2$ J per neutral (ΔT is the temperature difference between the plasma and the surroundings). So, the neutral energy density is

$$\rho g_n * N_A * 3k\Delta T/2 = \rho g_n * R * 3\Delta T/2 = 25 \quad [\text{J/m}^3] \qquad (10.7\text{d})$$

for $\Delta T = 200$ K and $p_n = 20$ Pa. In total, the particle energy density is

$$\rho g_E = \Sigma(\rho g_x * N_A * e * Eg_x) \sim 10^{5*}\Sigma(\rho g_x * Eg_x) \quad [\text{J/m}^3], \qquad (10.7\text{e})$$

and we are able to estimate $\rho g_E < 50$ J/m^3. For a reactor volume of 2dm^3, this is 0.1 J. At 100 W input power this means that the reactor chemicals are refreshed every millisecond.

- Power density: This can be calculated when the volume of the glow is known. The volume of the glow is almost the reactor volume of the plasmafab 340, i.e., a commercially available IBARE system, which is 2×10^{-3} m^3, and therefore $P_V = 50,000$ W/m^3 when the power is 100 W.

Particle energies

The thermal/kinetic energy of the plasma particles is normally expressed in eV or K, where 1 eV = 8,000 K. The only particles in the glow region having a different kinetic energy are the electrons because they consume energy from the rf power supply. This situation is expressed in the EEDF. In most conventional RIE systems, the electrons have an energy up to 30 eV, i.e., 240,000 K.

In order to initiate a plasma, a typical peak-to-peak voltage is several hundreds up to several thousands volts over a distance of 10 cm. Thus, the electrical field is typically 10^3–10^4 V/m. The mean free path of an electron is typically $\lambda_e = 30/(p[\text{mTorr}])$ [cm], i.e., 6 mm at 50 mTorr, and the electrons will transfer 6–60 eV electrical energy into kinetic energy. After ignition, the bulk of the plasma is highly conducting and, therefore, the electrical fields will concentrate along the boundary regions where the electrons are continuously consumed by the reactor walls. So, during processing most plasma particles are generated quite near the outer regions of the plasma glow.

The energy of the other particles can be estimated with the help of the balance of energy. The plasma is fed with the rf power supply: P_{supply} [W]. This energy is leaving the plasma glow as five major fluxes: (i) the energy necessary to etch the Si with radicals, (ii) ions, (iii) the energy connected with electrons balancing the ion flux, (iv) the thermal energy transported with the neutrals due to collisions with the cold reactor walls, and (v) the energy leaving the plasma region as photon and black body radiation, i.e.,

$$P_{\text{supply}} = P_{\text{radicals}} + P_{\text{ions}} + P_{\text{electrons}} + P_{\text{thermal}} + P_{\text{radiation}} \quad [\text{W}]. \qquad (10.7\text{f})$$

- Power supply: The power unit is the main supplier of energy. When the power is 100 W = 100 J/s = 6×10^{20} eV/s, we know the upper limit of most processes. Not more than 6×10^{20} eV/s can be used for any process in the reactor.

- Neutral/phonon energy: The energy transported as thermal energy with the neutrals could be high because their population is so high in spite of their low-energy content. Every particle colliding with the cold reactor walls, with a total area of the anode and cathode $A_a + A_c = 0.1 m^2$, will lose an energy of $3k\Delta T/2$ [J], with ΔT the difference in temperature between the plasma and the walls. The loss of power due to this mechanism is $P_n = 3\phi b_n (A_a + A_c) R \Delta T/2 = 0.75 \Delta T$ at $p = 30$ Pa. So, when the temperature of the plasma is 100 K higher than the surroundings, the loss of power is 75 W.

- Radical energy: A part of the energy flux is moving with the radicals to the Si surface. The energy needed to break an SF_6 molecule into $SF_5 + F$ is approximately 2 eV. With $\phi b_r = 10^{-3}$ mol/m^{2}s and $A = 0.025$ m^2 the area of the consuming Si surface, we calculate a power loss of only 5 W.

- Ion energy: The next flux of energy is transported with the ions. The energy needed to create an ion from SF_6 is approximately 15 eV. The ion is gaining extra energy during its travel trough the plasma sheath. When the voltage between the plasma and the cathode is 85 V and the voltage between the plasma and the anode is 10 V (i.e., the dc self-bias is 75 V), the total energy transported with an ion going to the cathode is 100 eV and to the anode is 25 eV. With the cathode area of 0.025 m^2 and the anode area of 0.075 m^2, the total power lost is 5 W. This statement also implies another effect. When the energy transported with the help of ions is 5 W, then it is possible to create a high flux of low-energy ions or a low flux of high-energy ions. In other words, when the dc self-bias is high, the ion flux is correspondingly lower, as well as the etch rate, to fit the input power.

- Electron energy: There is a net flow of electrons traveling to the reactor walls to compensate for the electrical charge carried with the ions (in fact, the mechanism is just the opposite, i.e., ions are forced to the walls to compensate for the electrons). However, the electrons will be forced backward in the dark space and will arrive with minor kinetic energy. So, the energy lost due to electrons is practically zero.

- Photon energy: A part of the energy is leaving the glow as photons and black body radiation: $Q_{rad} = \sigma A(T_g^4 - T_s^4)$, with $\sigma = 5.7 \times 10^{-8}$ W/m^{2}K^4 being the Stefan-Boltzmann constant, $A \sim 0.1$ m^2, T_g the temperature of the plasma glow region, and T_s the temperature of the surroundings. When $T_g = 400$ K and $T_s = 300$ K, we calculate a power loss of $P_p = 100$ W. In other words, the temperature of the plasma is always lower than 400 K, because there is less than 100 W available for radiation (e.g., at 350 K the power loss is 40 W).

10.3   Chemical model

In plasma chemistry, reactions are encountered that can be controlled to supply heat and work, reactions whose liberated energy might be wasted but which form the plasma processes and give products we need. Thermochemistry or plasma chemistry, a branch of the thermodynamics, is the study of the heat effects that accompany chemical reactions, the formation of solutions, and changes in state of aggregation, like melting. Thermodynamics is the study of the transformation of energies, which enables us to discuss all these matters quantitatively.

One reason for studying the rates of reactions is the practical importance of being able to predict if a certain reaction is favorable with respect to another. This might depend on variables under our control, such as the pressure, the rf power, and the presence of a catalyst, and we may be able to optimize it by the appropriate choice of conditions. The first stage in the kinetic analysis of reactions is to establish the stoichiometry of the reaction and identify any side reactions. The basic data of chemical kinetics are then the concentration of the reactants and products at different times after a reaction has been initiated. Since the rates of chemical reactions are generally sensitive to the temperature, the temperature of the plasma must be held constant throughout the course of the reaction, for otherwise the observed rate would be a meaningless average of rates at different temperatures. This requirement puts severe demands on the design of a plasma experiment.

This section is split into two basic parts. One deals with the usual equilibrium, reversible, or ordinary thermodynamics, i.e., chemical statics or thermostatics. The other is concerned with the nonequilibrium, irreversible, stationary, or steady state thermodynamics, i.e., chemical dynamics or kinetics. With the help of this base we will try to estimate the products of a few examples of plasma reactions. Most parts from this section are transferred from the excellent textbooks *Physical Chemistry* by P. W. Atkins (1992 edition) and *Physical Chemistry* by W. J. Moore (1972 edition) [5, 6]. However, before we start we will give some definitions to be able to write down most postulates and formulations in a more compact way.

10.3.1 Thermodynamic language

A *system* is a part of the world separated from the rest of the world by definite boundaries. The world outside of the system is called its *surroundings*. The experiments that we perform on a system are said to measure its *properties*, these being the attributes that enable us to describe it with all requisite completeness. This complete description is said to define the *state of the system*. When a system shows no further tendency to change its properties with time, it is said to have reached a *state of equilibrium*. The condition of a system in equilibrium is reproducible and can be defined by a set of properties, which are *functions of the state*, i.e., which do not depend on the history of the system before it reaches equilibrium. Temperature T, entropy S, volume V, and pressure P are all state functions like the energy U.

A simple mechanical illustration will clarify the concept of equilibrium. Figure 10.4 shows four different equilibrium positions of a person on a racing bike in the mountains. In both positions A and D, the biker is lower in altitude than in any slightly displaced position, and if the racer is trying to get uphill (s)he will tend to return spontaneously to

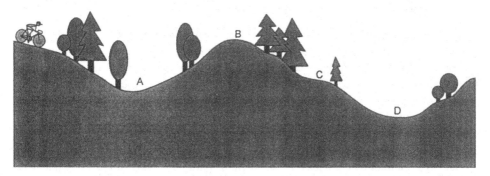

Fig. 10.4. The concept of equilibrium. Position D is stable, position A is metastable, and positions B and C are unstable equilibrium states.

the original equilibrium position. Therefore, it is said that the gravitational potential energy U of the racer in position A or D is at a minimum, and both positions represent *stable equilibrium* states. Yet, it is apparent that position D is more stable than position A, and it is easier for the biker to reach D from A than backward. In position A, therefore, the biker is said to be in local stable or *metastable equilibrium*. Position B is also an equilibrium position, but it is a state of *unstable equilibrium*, as anyone who has tried to balance a pencil on the top a finger will agree. The biker in position B is higher in altitude than in any slightly displaced position, and the tiniest displacement will send the biker downhill. The potential energy at a position of unstable equilibrium is a maximum, and such a position would be realized only in the absence of any disturbing force. Position C is also an unstable position, although it gains potential energy in the upward direction. However, driving a little uphill, the racer will be forced back and (s)he will overshoot the equilibrium position C and arrive in position D soon. As can be found in the figure, for an equilibrium position, the slope dU/dr of the curve for U vs. displacement is equal to zero, and one may write the equilibrium conditions as $(dU/dr)_{r0} = 0$, with $r_0 = $ A, B, C, or D. Examination of the second derivative will indicate whether the equilibrium is stable, $(d^2U/dr^2) > 0$, or unstable, $(d^2U/dr^2) < 0$. Although these considerations have been presented in terms of a simple situation, similar principles will be found to apply in the more complex physical–chemical systems. In addition to purely mechanical changes, such systems may undergo temperature changes, changes of state of aggregation, and chemical reactions. The problem of thermodynamics is to discover or invent new functions that will play the role in these more general systems which the potential energy plays in mechanics.

Mathematically, we distinguish two classes of differential expressions. Those such as dU, dV, dP, dS, or dT are called *exact differentials*, since they are obtained by differentiation of some state function. Those such as dq or dw are *inexact differentials*, since they cannot be calculated by differentiation of a function of the state of the system alone. Conversely, dq or dw cannot be integrated to yield the heat q or work w. When a system is carried out along a path of different states, but is returned at the end to the same state that it initially occupied, the entire process has constituted a complete *cycle*. The following statements are identical:

(1)   The function U is a function of the state of a system.
(2)   The differential dU is an exact differential.
(3)   The integral of dU about a closed path, i.e., a cycle, is equal to zero.
(4)   The change in U between any pair of initial and final states is independent of the path between them.

To specify precisely the state at equilibrium of a substance, we must measure its properties. Since there are equations of a substance that give *relations* between properties, it is not necessary to specify the values of each and every property to define exactly the state of a substance. For example, to specify the state of a pure gas, we may focus attention on three thermodynamic variables: pressure P, volume V, and temperature T. If any two of these are fixed, it is found experimentally that the third will also be fixed ($PV = NRT$). The properties of a system can be classified as *extensive* or *intensive*. Extensive properties or capacity factors, such as the volume and mass, are additive; their value for the whole system is equal to the sum of their values for the individual parts. Intensive properties or intensity factors, such as the temperature and pressure, are not additive; the temperature of any small part of a system in equilibrium is the same as the temperature of the whole. The concept of temperature was first introduced in connection with the study of thermal equilibrium: When two bodies at different temperatures are placed in contact, energy in the form of heat flows from one to the other until a state of thermal equilibrium is reached. The two bodies are then at the same temperature. Identically, when two bodies are placed in contact with different pressures, energy in the form of work will flow until pressure equilibrium is reached. So, the concepts of thermal and pressure equilibrium are identical with respect to the fact that they are the intensity factors forced to equilibrium. Note that two bodies can be placed in close contact with different volume and entropy without energy flowing between them. The bodies at different volume and entropy can stay in equilibrium; no energy will flow as long as they have the same temperature and pressure. So, the intensity factors are controlling the state of equilibrium, whereas the extensive factors determine the amount of energy that may flow when the system is not at equilibrium.

If the boundaries of a system do not permit any change to occur in the system as a consequence of a change in the surroundings, the system is said to be *isolated*. Any completely isolated system is restricted to *adiabatic* processes, i.e., no heat can either enter or leave the system. So, an adiabatic wall is defined as a wall that separates two systems so that they are prevented from coming to thermal equilibrium with each other. A *closed* system is one for which there is no transfer of mass across the boundaries, although the transfer of heat is allowed. Processes that release energy as heat are classified as *exothermic*, and those that absorb energy as heat are called *endothermic*. An *isothermal* process is a process for any fixed value of T; *isobaric* processes are fulfilled at constant pressure, and *isochoric* or *isometric* processes are considered for constant volume.

With the help of the foregoing discussion we are able to define energy, work, and heat more precisely. The *energy U* transferred between two systems is formulated as the total sum of all the products of the intensity factors with their coupled extensive factors. The intensity factor can be treated as a generalized force (gF) and the extensive factor as a generalized displacement (gD). If a pressure difference (gF) exists between two systems, the volume (gD) of both systems will change. We say that energy has been transferred from one system to the other in the form of *work* w; $dw = -PdV$. If a temperature difference (gF) exists between two systems in thermal contact, entropy (gD) will flow between them. Now energy has been transferred in the form of *heat* q; $dq = TdS$. Or, more accurately, the heat q transferred in a given process is defined as the difference between the work w_{ad} done on the system along an adiabatic path ($dq = 0$) from A to B and the work w done along the given path from A to B: $q = w_{ad} - w$. Note that work and heat are strictly prohibited for the transfer of energy. We speak of the work done by the system and the heat absorbed, but not of the heat or work of the system. This is a direct consequence of dw and dq not being exact differentials. However, we can and do speak of the energy of the system. This subtle difference is a direct consequence of the First Law of thermodynamics, as will be pointed out in this chapter. So, work, heat, and energy are the basic concepts of thermodynamics. Energy is the ability or capacity of a system to do work or to produce heat. When work is done on an otherwise thermodynamically isolated system, its energy is increased. When the energy of a system changes as a result of a temperature difference between it and its surroundings, we say that energy has been transferred as heat. In molecular terms, work is the transfer of energy that makes use of organized motion (e.g., the electrons in an electric current move in an orderly direction) and the process of heating is the transfer of energy that makes use of the difference in thermal motion (i.e., the chaotic and random motion of the molecules) between the system and its surroundings.

The work done by a system in going from one state to another is a function of the path between the states and $\oint dw$ is not in general equal to zero. The reason is readily apparent when a reversible gas process is considered. Then $\int dw = -\int P\,dV$. The differential expression $P\,dV$ cannot be integrated when only the initial and final states are known, since P is a function not only of the volume V but also of the temperature T ($PV = NRT$), and this temperature may change along the path of integration. On the other hand, as will be pointed out in this section, we can use the principle of the conservation of energy to define a function U called the internal energy.

We already formulated the definition for a perfect gas in the kinetic (gas) theory. The definition of its counterpart in thermodynamics, the ideal gas, is as follows:

(1)   The internal pressure $(\delta U/\delta V)_T = 0$.
(2)   The gas follows the equation of state $PV = NRT$.

The first assumption, the internal energy is not changing with volume at constant temperature, is identical to the kinetic language that the size of the molecules is negligible and that the molecules do not interact inelastically. The second assumption is a consequence of the ceaseless random motion together with the perfectly elastic collisions.

A *reversible path* is one connecting intermediate states all of which are equilibrium states. A process carried out along such an equilibrium path is called a *reversible process*. If a process is not reversible, it must be an *irreversible process*. "*For example, to expand a gas reversibly, the pressure on the piston must be released so slowly, in the limit infinitely slowly, that at every instant the pressure everywhere within the gas volume is exactly the same and is just equal to the opposing pressure on the piston. Only in this case can the state of the gas be represented by the variables P and V. ... In contradiction ... consider the situation if the piston were drawn back suddenly. Gas would rush in to fill the space, pressure differences would be set up throughout the gas volume, and even a state of turbulence would ensue. The state of the gas under such conditions could no longer be represented by the two variables, P and V. Indeed an enormous number of variables would be required, corresponding to the many different pressures at different points throughout the gas volume. Such a rapid expansion is a typical irreversible process; the intermediate states are no longer equilibrium states,*" (Moore [6]). Reversible processes are never realizable in actuality because they must be carried out infinitely slowly. All processes that occur naturally are therefore irreversible. The conditions of reversibility can, however, be closely approximated in certain experiments.

10.3.2 Reversible thermodynamics

We shall begin our study of physical chemistry in plasmas with reversible thermodynamics, which is based on concepts common to the everyday world. The word itself is evidently derived from dynamics, which is a branch of mechanics dealing with matter in motion. Mechanics is founded on the work of Isaac Newton (1642–1727), and usually begins with the equation $F = ma$. The equation states the proportionality between a vector quantity F, called the force applied to a particle of matter, and the acceleration a of the particle. The proportionality factor m is called the mass. In mechanics, if the point of application of a force F moves, the force is said to do work. The amount of work done by a force F whose point of application moves a distance dr along the direction of the force is $dw = F\,dr$. In 1669, Huygens discovered that if each mass m is multiplied by the square of its velocity v^2, the sum of these products was conserved in all collisions between elastic bodies, i.e., collisions in which the potential energy of all colliding particles is separately conserved. This statement can be given a mathematical formulation. Consider a particle at position r_0 and apply to it a force $F(r)$ that depends only on its position. In the absence of any other forces, the work done on the body in a finite displacement from r_0 to r_1 is

$$w = \int_{r0}^{r1} F(r)\,dr = \int_{t0}^{t1} F(r)(dr/dt)\,dt = \int_{t0}^{t1} m(dv/dt)v\,dt = m \int_{v0}^{v1} v\,dv$$

$$= 1/2mv_1^2 - 1/2mv_2^2 = E_{k1} - E_{k0}, \tag{10.27a}$$

where the kinetic energy is defined by $E_k = 1/2mv^2$. The work done on the body equals the difference between its kinetic energies in the final state and the initial state. If the force is a function of r alone, the integral defines another function of r, the potential energy $F(r)\,dr = -dE_p(r)$:

$$w = \int_{r0}^{r1} F(r)\,dr = E_{p0} - E_{p1}. \tag{10.27b}$$

Combining both energy functions, we find $w = E_{k1} - E_{k0} = E_{p0} - E_{p1} \Leftrightarrow E_{k1} + E_{p1} = E_{p0} + E_{k0}$. The sum of the potential and the kinetic energies, $E_p + E_k$, is the total mechanical energy of the body, which remains constant during the motion: the mechanical principle of the conservation of energy. So, both E_{pot} and E_{kin} have many possible values, depending on exactly how the system passes from one state into another, but their sum is invariable and independent of the path. So, in the special case of Huygens's elastic collisions, the internal energy for every particle is separately conserved and, therefore, the total sum is also. But the kinetic energy is allowed to exchange freely between all the particles as long as the summation over all the particles is constant.

   If a force depends on velocity as well as position, the situation is more complex. Take, for example, a body which speeds through water. The higher the velocity, the greater the frictional resistance and the mechanical energy is no longer conserved. However, from the dawn of history it has been known that the frictional dissipation of energy is attended by the evolution of something called heat, and it is possible to include heat among the ways of transforming energy. So, through frictional phenomena, even primitive peoples knew the connection between heat and motion (= work). By about 1840, the law of conservation of energy was accepted in purely mechanical systems, the interconversion of heat and work was well established, and it was understood that heat was simply a form of motion of the smallest particles composing a substance. Yet, a more inclusive principle of the conservation of energy to include heat changes had not yet been clearly obtained. But, by the beginning of 1842, Mayer could equate heat to kinetic energy and potential energy: "*that the fall of a weight from a height of about 365 meters corresponds to the warming of an equal weight of water from 0 to 1°C.*" Mayer was able to state the principle of the conservation of energy including heat, and the First Law of thermodynamics was born. So, the First Law is an extension of the principle of the conservation of mechanical energy (Jean Bernoulli, 1735). Such an extension became reasonable after it was shown that expenditure of work could cause the production of heat. Thus, both work and heat were seen to be entities that described the transfer of energy from one system to another. The two characteristics of the internal energy (U), work (w) and heat (q), are summarized in the so-called *First Law of thermodynamics*: The internal energy of an isolated system is constant unless it is changed by doing work or by heating:

$$dU = dw + dq \quad \text{[J]}. \tag{10.27c}$$

Hence, for any cyclic process the integral of dU vanishes: $\oint dU = 0$. If any part of an isolated system increases its energy, the remaining part must decrease by an exactly equal amount. The equation indicates that when the inexact differential dq is added to the inexact differential dw, it becomes an exact differential. The integrand $\int_c^h dq_{rev}$ is dependent on the path, whereas $\int_c^h (dq_{rev} + dw)$ is independent of the path. So, the First Law states that although dq and dw are not exact differentials, their sum $dU = dq + dw$ is an exact differential. Note that the interconversion of mass and energy can readily be measured in nuclear reactions. The First Law should, therefore, become a law of the conservation of mass-energy.

In this section, some important concepts of plasma chemistry are briefly reviewed, like the temperature, pressure, entropy, and the four thermodynamic functions, i.e., the internal energy, enthalpy, free energy (Helmholtz), free enthalpy (Gibbs), and their underlying relations (Maxwell, Legendre). After this, the influence of the temperature will be discussed with the help of the activation energy.

10.3.2.1   Temperature

The relative temperature can be measured by noting the volume change of fluids with temperature at constant pressure. Although the first primitive gas thermometer was invented by Galilei Galileo, in 1631 Jean Rey used a liquid thermometer, a glass bulb having a stem partly filled with water, to follow the progress of fevers in his patients. A calibration based on two fixed points, the boiling point of water ($100°C$) and the melting of ice ($0°C$) at constant pressure (1 atmosphere), was introduced by Elvius in 1710 and they define the centigrade scale or Celsius scale. So, the temperature θ defined and measured in this way is defined entirely in terms of the mechanical properties of pressure and volume, $\theta = \theta(P, V)$, which suffice to define the state of the pure fluids. The first detailed experiments on the variation with temperature of the volumes of gases at constant pressure were published by Joseph Gay-Lussac from 1802 to 1808. He found that an ideal gas expands by $1/273.15$ of its volume at $0°C$ for each degree rise in temperature at constant pressure: $V = V_o(1 + \theta/273.15)$. Moreover, it states that the volume decreases to zero at $\theta = -273.15°C$ for an ideal gas. (Of course, as we all know, at lower temperature an ordinary or real gas will become first a liquid, and at even a lower temperature it turns into a solid: The dimensions of the molecules are not zero and there are interactions.) Kelvin defined a new temperature scale with the temperature denoted by $T = \theta - T_o$ and $T_o = 273.15$ K and called it the absolute temperature in Kelvin. So, he arrived at $V/V_o = T/T_o$.

10.3.2.2   Pressure and molar volume

A differential element of work was defined as $dw = F dr$. Now consider a fluid confined in a cylinder with a movable, frictionless piston. The external pressure on the piston of area A is $P_{ex} = F/A$. If a piston is displaced a distance dr in the direction of the force, the element of work is $dw = (F/A) * A dr = P_{ex} dV$. This is the work done by the pressure. In mechanics the work is always associated with the force; however, in thermodynamics we focus attention upon the system and its surroundings. We speak of work done on the system and work done by the system on its surroundings. By convention, work done on the system is positive and work done by the system is negative. Therefore, we write for the work done on the system: $dw = -P_{ex}\,dV$. Since dV is negative for a compression, the work done is positive.

The pressure can be measured in the following way. Consider a fluid contained in a cylinder with a frictionless piston. The piston is attached to the other fixed side of the cylinder by way of a spring with a certain spring constant k. We can calculate the absolute pressure in the fluid by dividing the force on the spring by the area of the piston facing the fluid: $P = F/A$. However, a much more practical method to measure the pressure was invented by Evangelista Torricelli in 1643, the well-known mercury barometer. In 1662, Boyle found that at constant temperature, the volume of a given sample of gas varies inversely as the pressure, $PV = $ constant.

Any two of the three variables, P, V, and T, suffice to specify the state of a given amount of an ideal gas and to fix the value of the third variable. $PV = $ constant, Boyle's Law, is an expression for the variation of P with V at constant T, and $V/T = $ constant, the law of Gay-Lussac, is an expression for the variation of V with T at constant P. We can readily combine these two relations with Avogadro's Law: According to Avogadro's principle, the molar volume of all ideal gases should be the same. At 0°C and 1atm, the molar volume would be 22414 cm^3. The number of molecules in 1 mol is now called Avogadro's number, $N_A = 6.02 \times 10^{23}$. The combination results in the equation of state of an ideal gas, which is one of the most useful relations in physical chemistry:

$$PV = NRT \quad \text{[J]}, \tag{10.28}$$

with $R = 8.31431$ J/Kmol being the gas constant per mole and N [mol] the molar number of particles.

10.3.2.3 Entropy

Although Mayer was the philosophic father of the First Law, Joule's experiments firmly established the law on an experimental or inductive foundation. The experiment of Joule in 1840 on the heating effect of an electrical current showed, as in the case of frictional energy transfer, that heat was not conserved in physical processes, since it could be generated by mechanical work, the Joule heat. The reserve transformation, the conversion of heat into work, had been of interest to the practical engineer ever since the development of the steam engine by James Watt in 1769. Only in 1824 was the theory of this English machine taken up by Sadi Carnot. Carnot devised a cycle to represent the operation of an idealized engine, in which heat q_h is transferred from a hot reservoir at temperature θ_h, partly converted into work w, and partly discarded as heat q_c to a colder reservoir at temperature θ_c. The gas through which these operations are carried out is returned at the end to the same state that it initially occupied, so that the entire process constitutes a complete cycle. With the help of the First Law he could calculate the efficiency of the engine: $\varepsilon = -w/q_h = (q_h + q_c)/q_h = 1 + q_c/q_h$. William Thomson (Kelvin) proved that the efficiency could be written also as $\varepsilon = 1 + F(\theta_c)/F(\theta_h)$ and took the functions $F(\theta_c)$ and $F(\theta_h)$ to define the thermodynamic temperature function T. Thus, a temperature ratio on the Kelvin scale was defined as equal to the ratio of the heat absorbed to the heat rejected in the working of a reversible Carnot cycle: $q_c/q_h = -T_c/T_h$, in the limit as $T_c \to 0$, then

$\varepsilon \to 1$. This equation may be written as $q_h/T_h + q_c/T_c = 0$ and can be extended to any reversible cycle to obtain

$$\oint \frac{dq_{rev}}{T} = 0 \quad [\text{J/K}].$$

(10.29)

This statement holds true for any reversible cyclic process whatsoever. The vanishing of the cycle means that the integrand is a perfect differential of some function of the state of the system. Thus, we can thus define a new state function S, the *Second Law of thermodynamics* for reversible processes, by

$$dS = \frac{dq_{rev}}{T} \quad [\text{J/K}].$$

(10.30a)

The function S was first introduced by Clausius in 1850 and he called it entropy. Hence, for any cyclic process the integral of dS vanishes: $\oint dS = 0$. If any part of an isolated reversible system increases its entropy, the remaining part must decrease by an exactly equal amount. The equation indicates that when the inexact differential expression dq is multiplied by the integration factor $1/T$, it becomes an exact differential. The integrand $\int_c^h dq_{rev}$ is dependent on the path, whereas $\int_c^h dq_{rev}/T$ is independent of the path.

Now, after we have treated both the First Law and the Second Law, we will look at them from a different, mathematical, perspective. It was found that dw as well as dq were inexact differentials. By adding dw and dq, the First Law was obtained: the exact differential $dU = dw + dq$. By dividing dw with the pressure, another exact differential is obtained: $dV = dw/P$. Similarly, dividing dq by the temperature gives the Second Law: $dS = dq_{rev}/T$.

Let us recollect: The statistical definition of entropy, which was treated in the section of the physical model, concentrates on the absolute value of the entropy and makes it possible for us to calculate the degree of disorder:

$$S = k * \ln(W) \quad [\text{J/K}].$$

(10.9)

In contradiction, the thermodynamic definition of entropy concentrates on the relative value of the entropy, i.e., the change in entropy dS during a process:

$$dS = \frac{dq_{rev}}{T} = [\text{Eq. (10.9)}] = k\frac{dW}{W} \quad [\text{J/K}].$$

(10.30b)

To be able to compare the absolute value of S from both equations, we should know the integration constant S_o when integrating Eq. (10.30b).

"The entropy has no physically fixed zero level, but to set $S_o = 0$ would be equivalent to adopting such a definite zero level for the entropy in all cases. As can be shown, the constant is indeed independent of the parameters of the system, so that the difference in entropy ΔS between any two states of the system differing in values of the defining parameters (V, P, etc.) will approach zero at $T = 0$. Furthermore, this statement that $\Delta S \to 0$ at $T = 0$ applies also to any possible chemical changes in the system," (Moore [6]).

Thus at $T = 0$, the random orientations are effectively frozen. The best values from the Third Law (measurements of heat capacities upward from very low temperatures) are almost identical with statistical entropies (from spectroscopic data on the properties of molecules), as they should be.

10.3.2.4 Internal energy (constant volume and entropy)

From the First Law ($dU = dw + dq$) and Second Law ($dq = T dS$), we obtain an important relation, sometimes called the *Combined First and Second Laws*:

$$dU(V, S) = -PdV + TdS \quad \text{[J]}. \tag{10.31a}$$

An isolated system is not allowed to interact with its surroundings. Because, from the First Law, $dU = dq - PdV = 0$ and $dq = 0$ for the isolated system, we must have $dV = 0$. Thus, the necessary constraints on an isolated system are that U and V cannot change. If this restriction does not apply, the system may either do work on its surroundings or have work done on itself. Thus, only a part of the heat added to a substance causes its temperature to rise, the remainder being used in the work of expanding the substance. No mechanical work is done during a process carried out at constant volume, since $dV = 0$ means $dw = 0$. Identically, no heat is exchanged when the entropy is constant, since $dS = 0$ means $dq = 0$. This relation applies to any system of constant composition in which only PdV work is considered. Of course, the change of work can also be found as a surface expansion ($\gamma d\sigma$), extension (Fdl), diffusion (μdN), electrical charging (Udq), etc. But, historically, the volume expansion has been paid extra attention for its use in heat engines and, of course, explosives. This equation gives a new equation for the temperature and the pressure:

$$\left(\frac{\partial U}{\partial V} \right)_S = -P \quad \text{and} \quad \left(\frac{\partial U}{\partial S} \right)_V = T \quad \text{[J]}. \tag{10.32}$$

By means of these equations, the intensive variables P and T are given in terms of extensive variables of the system, U, V, and S. So, the infinitesimal change in heat (dq) is written as TdS, and the change in work (dw) is written as a volume expansion ($-PdV$). Impose a reversible process under the condition that the heating occurs at constant volume ($PdV = 0$), then

$$dU_V = dq_V = TdS \quad \text{[J]}. \tag{10.31b}$$

Thus, the heat of reaction measured at constant volume, e.g., in a bomb calorimeter, is exactly equal to the change in internal energy, $\Delta U_V = q_V$.

We shall now begin to unfold the consequences of dU being a state function. Suppose U is given as a function of volume and temperature, i.e., $U = U(V, T)$, then

$$dU(V, T) = dU = \left(\frac{\partial U}{\partial V} \right)_T dV + \left(\frac{\partial U}{\partial T} \right)_V dT = \pi_T dV + C_V dT \quad \text{[J]}. \tag{10.33a}$$

The interpretation of this equation is that in a closed system of constant composition, any infinitesimal change in the internal energy is proportional to the infinitesimal changes of volume and temperature, the coefficients of proportionality being the partial derivatives. Most of the partial derivatives we meet have an easily identifiable physical meaning. For example, if there are no interactions between the molecules of a substance, the internal energy should be independent of their separation and hence independent of the volume the sample occupies. This suggests that for an ideal gas, the so-called internal pressure $\pi_T = 0$, i.e., the energy of an ideal gas is a function of its temperature alone. Suppose the system under consideration is an ideal gas or is constrained to have constant volume ($\pi_T dV = 0$). The heat required to bring about a change in temperature dT is then

$$dU_V = dq_V = C_V dT \quad \text{[J]}, \tag{10.33b}$$

where C_V is the heat capacity at constant volume. Identifying Eq. (10.31b) with Eq. (10.33b), it is found that $T dS = C_V dT$, that is, $\int dS = S_2 - S_1 = \Delta S = C_V \int [dT/T] = C_V \ln(T_2/T_1)$ at constant volume.

10.3.2.5 Enthalpy (constant pressure and entropy)

When a system is free to change its volume against a constant external pressure, the change in internal energy is no longer equal to the energy supplied as heat. In effect, some of that energy is converted into the work required for volume expansion or gained by volume compression, so $dU_P \neq q_P$. However, at constant pressure the heat supplied is equal to the change in another thermodynamic property of the system, the enthalpy H defined as

$$H = U + PV \quad [J]. \tag{10.34}$$

H is a function of the state alone, since U, P, and V are all state functions. Using the Combined Laws,

$$dH(P, S) = dU + d(PV) = -P dV + T dS + P dV + V dP = V dP + T dS \quad [J]. \tag{10.35a}$$

Now we impose the condition that the heating occurs at constant pressure, then

$$dH_P = dq_P = T dS \quad [J]. \tag{10.35b}$$

So, when a system is heated at constant pressure, and only work due to volume expansion can occur (e.g., in a calorimeter), the change in enthalpy is equal to the energy supplied as heat, $\Delta H_P = q_P$. We conclude that the heat of a chemical reaction depends upon the conditions that hold during the process and that there are two particular conditions that are important because they lead to heats of reaction equal to changes in thermodynamic functions: $\Delta_V = q_V$ and $\Delta H_P = q_P$.

Note that we have to bear in mind that the enthalpy of a substance increases as it is heated. Hence, a reaction enthalpy changes with temperature because the enthalpy of each substance in a reaction varies in a characteristic way. Suppose the enthalpy is a function only of the pressure and temperature, $H = H(P, T)$, then

$$dH(P, T) = \left(\frac{\partial H}{\partial P}\right)_T dP + \left(\frac{\partial H}{\partial T}\right)_P dT = \left(\frac{\partial H}{\partial P}\right)_T dP + C_P dT \quad [J]. \tag{10.36a}$$

So, the constant-pressure heat capacity C_P expresses how the enthalpy varies with temperature at constant pressure. Suppose the system is constrained to have constant pressure ($dP = 0$). The heat required to bring about a change in temperature dT is then

$$dH_P = dq_P = C_P dT \quad [J]. \tag{10.36b}$$

Identifying Eq. (10.35b) with (10.36b), it is found that $T dS = C_P dT$, that is, $\Delta S = C_P \ln(T_2/T_1)$ at constant pressure. The heat capacity at constant pressure C_P is usually larger than that at constant volume C_V, because at constant pressure, part of the heat added to a substance may be used in the work of expanding it, whereas at constant volume, all the added heat produces a rise in temperature. An exception is water between 0°C and 4°C, which expands its volume with a decrease in temperature at constant pressure.

10.3.2.6   Helmholtz function and free (internal) energy
(constant volume and temperature)

Chemical reactions are rarely studied under conditions of constant entropy or constant energy. Usually, the chemist places his system in thermostats and investigates them under conditions of approximately constant temperature and pressure. Sometimes changes at constant temperature and volume are followed. It is most desirable, therefore, to obtain criteria for thermodynamic equilibrium that are applicable under these practical conditions. A system under these conditions is called a closed system, since no mass can be transferred across the boundary of the system, although transfer of energy is allowed.

Let us first consider a closed system under a condition of constant volume and temperature. Such a system would have perfectly rigid walls so that no PdV work could be done on it. The rigid container would be surrounded by a heat bath of virtually infinite heat capacity at a constant temperature T, so that heat transfers could take place between the system and the heat bath without changing the temperature of the latter. In equilibrium, the temperature of the system would be held constant at T. Helmholtz introduced a new state function especially suitable for discussions of the system at constant T and V, or, indeed, whenever we wish to specify its state in terms of the independent variables T, V, and appropriate composition variables. This function is called the Helmholtz free energy A, sometimes found as F, and is defined as

$$A = U - TS \quad \text{[J]}. \tag{10.37}$$

Therefore,

$$dA(V, T) = dU - d(TS) = -PdV + TdS - TdS - SdT = -PdV - SdT \quad \text{[J]}. \tag{10.38}$$

Generally, if the value of dA for a change is known, then the maximum amount of work the system can do can also be stated. This is why A is sometimes called the (maximum) work function (A = Arbeit, the German word for work). This is the origin of the name free energy for A, since dA is that part of the change in internal energy that is free to use to do work.

10.3.2.7   Gibbs function and free enthalpy (constant pressure and temperature)

Probably the most frequently encountered condition for a closed system is that of constant temperature and pressure. To a first approximation, this would be the common operating condition in a thermostat at atmospheric pressure. The special function most suitable for such conditions, in which the state of the system is to be specified by T, P, and the necessary composition variables, was invented by J. Willard Gibbs. It has been called the Gibbs free energy, the free enthalpy, or simply the Gibbs function, and is denoted G. The definition of G is

$$G = U + PV - TS = H - TS = A + PV \quad \text{[J]}. \tag{10.39}$$

Therefore,

$$dG(P, T) = dU + d(PV) - d(TS) = VdP - SdT \quad \text{[J]}. \tag{10.40}$$

Changes such as the melting of ice, the solution of sugar in tee, or the transformation of graphite to diamond are called changes in state of aggregation or phase changes. They are characterized by discontinuous changes in certain properties of the system at definite temperature and pressure. If a system contains more than one component in a given phase, we cannot specify its state without some specification of the composition of that phase. This is because the equation given above reveals that $dG(P, T) = 0$. In addition to P, V, S, and T, we need to introduce new variables to measure the amounts of the different chemical constituents in the system. As usual, the mole will be chosen as the chemical measure, with N_i representing the number of moles of component 1, 2, ..., i in the particular phase we are considering. It then follows that each thermodynamic function depends on these N_is as well as on P, V, S, T, and $G = G(P, T, N_i)$, but also $U = U(V, S, N_i)$, etc. Consequently, a complete differential becomes

$$dG = \left(\frac{\partial G}{\partial T}\right)_{P,N_i} dT + \left(\frac{\partial G}{\partial P}\right)_{T,N_i} dP + \sum_i \left(\frac{\partial G}{\partial N_i}\right)_{T,P,N_k} dN_i \quad k \neq i \quad [\text{J}]. \quad (10.41)$$

$$dG = V dP - S dT + \sum \mu_i dN_i \quad [\text{J}]. \quad (10.42a)$$

The coefficient $\mu_i = (\delta G/\delta N_i)_{T,P,Nk}$ was introduced by Gibbs, who called it the chemical potential. At constant T and P, this equation becomes

$$dG_{P,T} = \sum \mu_i dN_i \quad [\text{J}]. \quad (10.42b)$$

It is the change in the Gibbs free energy of the phase with a change in the number of moles of component i, the temperature, pressure, and number of all other components being kept constant. The chemical potentials, therefore, measure how the Gibbs free energy of a phase depends on any changes in its composition. If we consider the phase to be closed, so that no transfer of mass across its boundaries is allowed, it becomes $dG = 0$. We might, however, consider the entire system of, say, two phases (water and ice) to be closed. We then have $\mu_1 dN_1 + \mu_2 dN_2 = 0$. So, we can still have transfer of mass between two phases, although the system as a whole is closed for the transfer of mass.

10.3.2.8   Maxwell's relations, Legendre transformations, and thermodynamic potentials

Let us summarize the four important relations for differentials of thermodynamic functions without $\sum \mu_i dN_i$ work:

$$dU(V, S) = \left(\frac{\partial U}{\partial V}\right)_S dV + \left(\frac{\partial U}{\partial S}\right)_V dS = -PdV + TdS \quad [\text{J}] \quad (10.31a)$$

$$dH(P, S) = \left(\frac{\partial H}{\partial P}\right)_S dP + \left(\frac{\partial H}{\partial S}\right)_P dS = +VdP + TdS \quad [\text{J}] \quad (10.35a)$$

$$dA(V, T) = \left(\frac{\partial A}{\partial V}\right)_T dV + \left(\frac{\partial A}{\partial T}\right)_V dT = -PdV - SdT \quad [\text{J}] \quad (10.38)$$

$$dG(P, T) = \left(\frac{\partial G}{\partial P}\right)_T dP + \left(\frac{\partial G}{\partial T}\right)_P dT = +VdP - SdT \quad [\text{J}]. \quad (10.40)$$

If we apply Euler's relation to the differentials, we obtain Maxwell's relations:

$$\left(\frac{\partial T}{\partial V}\right)_S = -\left(\frac{\partial P}{\partial S}\right)_V \qquad \left(\frac{\partial S}{\partial V}\right)_T = +\left(\frac{\partial P}{\partial T}\right)_V$$

$$\left(\frac{\partial T}{\partial P}\right)_S = +\left(\frac{\partial V}{\partial S}\right)_P \qquad \left(\frac{\partial S}{\partial P}\right)_T = -\left(\frac{\partial V}{\partial T}\right)_P. \tag{10.43}$$

Finally, we derive two equations called "thermodynamic equations of state," because they give U and H in terms of P, V, and T:

$$\left(\frac{\partial U}{\partial V}\right)_T = \left(\frac{\partial (A + TS)}{\partial V}\right)_T = \left(\frac{\partial A}{\partial V}\right)_T + T\left(\frac{\partial S}{\partial V}\right)_T = -P + T\left(\frac{\partial P}{\partial T}\right)_V \tag{10.44}$$

$$\left(\frac{\partial H}{\partial P}\right)_T = \left(\frac{\partial (G + TS)}{\partial P}\right)_T = \left(\frac{\partial G}{\partial P}\right)_T + T\left(\frac{\partial S}{\partial P}\right)_T = +V - T\left(\frac{\partial V}{\partial T}\right)_P.$$

Another useful relation between partial differential coefficients can be derived as follows. Since the volume is a function of both T and P, a differential change in volume or pressure can be written as

$$dV = \left(\frac{\partial V}{\partial T}\right)_P dT + \left(\frac{\partial V}{\partial P}\right)_T dP = \alpha V dT - \beta V dP \quad \text{and} \quad dP = \left(\frac{\partial P}{\partial V}\right)_T dV + \left(\frac{\partial P}{\partial T}\right)_V dT, \tag{10.45}$$

with α the fractional change in V with T, the thermal expansivity, and β the isothermal compressibility of a substance. Combining both gives

$$\left(\frac{\partial P}{\partial T}\right)_V = \alpha/\beta \quad [\text{N/m}^2 \text{ K}]. \tag{10.46}$$

Let's also give an important equation for the difference $C_P - C_V$ of the heat capacities. We defined the heat capacities at constant pressure and volume as

$$C_V = dq_V/dT = dU_V/dT \tag{10.33b}$$

and

$$C_P = dq_P/dT = dH_P/dT \quad [\text{J/K}]. \tag{10.36b}$$

Since $H = U + PV$ and

$$dU = \left(\frac{\partial U}{\partial V}\right)_T dV + \left(\frac{\partial U}{\partial T}\right)_V dT = \pi_T dV + C_V dT \quad \text{and} \quad dV = \left(\frac{\partial V}{\partial T}\right)_P dT + \left(\frac{\partial V}{\partial P}\right)_T dP, \tag{10.47}$$

by substituting the expression for dV into the expression for dU and comparing coefficients and substituting the result into $C_V - C_P$, we finally may arrive at

$$C_P - C_V = [P + \pi_T]\left(\frac{\partial V}{\partial T}\right)_P \quad [\text{J/K}]. \tag{10.48}$$

The term P is the contribution to the heat capacity caused by the change in volume of the system against the external pressure P. The term π_T, the internal pressure, is the contribution for the energy required for the change in volume against the internal cohesive or repulsive forces of the substance, represented by a change of the energy with volume at constant temperature. In the case of liquids or solids, which have strong cohesive forces, π_T is large. In the case of gases, π_T is usually very small with respect to P.

The total differentials of the thermodynamic functions can be related by means of Legendre transformations. Let us apply a Legendre transformation to the basic equations:

$$U = U(S, V)$$

$$dU = -PdV + TdS \quad [\text{J}] \tag{10.31a}$$

so that

$$H = U - \left(\frac{\partial U}{\partial V}\right)_S V = U + PV \quad [\text{J}] \tag{10.34}$$

$$dH = dU + PdV + VdP = VdP + TdS \quad [\text{J}] \tag{10.35a}$$

$$A = U - \left(\frac{\partial U}{\partial S}\right)_V S = U - TS \quad [\text{J}] \tag{10.37}$$

$$dA = dU - TdS - SdT = -PdV - SdT \quad [\text{J}] \tag{10.38}$$

$$G = H - \left(\frac{\partial H}{\partial S}\right)_P S = H - TS \quad [\text{J}] \tag{10.39}$$

$$dG = dH - TdS - SdT = VdP - SdT \quad [\text{J}]. \tag{10.40}$$

Mathematically, therefore, the introduction of the new thermodynamic functions, H, A, and G, is achieved by performing Legendre transformations on the basic $U(V, S)$.

In mechanics, the potential energy U serves as the potential function in terms of which equilibrium can be specified and from which forces acting on the system can be derived. If the potential U is given as a function of variables $r_1, r_2, \ldots$, the so-called generalized coordinates, defining the state of the system, a generalized force acting on the system is $F_j = -(\delta U/\delta r_j)_{r \neq r_j}$. The force is the gradient of the potential. So, $U(V, S)$, $H(P, S)$, $A(V, T)$, and $G(P, T)$ can be used as thermodynamic potentials. We can consider the gradient of any of these thermodynamic potentials as generalized forces. For example, in systems at constant T and P, it is convenient to think of gradients of G as driving forces for chemical and physical processes.

10.3.2.9   Activation energy

At this point we shall leave the subject of classical chemical kinetics and turn to what has been called the theory of absolute reaction rates, by which is meant the theory of the rate constants of chemical reactions. The effect of varying the temperature has been the most important key to the theory of rate processes. In 1889, Arrhenius pointed out that a reasonable equation for the variation of rate constant k_r with temperature might be $k_r = A * \exp(-E_a/RT)$. Here, A is called the frequency factor and the quantity E_a the activation energy of the reaction. This is the famous Arrhenius equation for the rate constant. It follows that a plot of logarithm of rate constant against reciprocal of absolute temperature should be a straight line. The validity of the Arrhenius equation has been excellently confirmed in this way for a large number of experimental rate constants. According to Arrhenius, molecules must acquire a certain critical energy E_a before they can react, the Boltzmann factor, $\exp(-E_a/RT)$, being the fraction of molecules that managed to obtain the necessary energy. The activation energy is the potential-energy hill that must be climbed to reach the activated state. It also is evident that the heat of reaction at constant volume, ΔU_V, is the difference between the activation energy of forward and backward reactions, $\Delta U_V = E_{af} - E_{ab}$. Although our ultimate aim would be to calculate the rate constant of any elementary reaction from the structure of the reactive molecules and the properties of the medium in which they are reacting, this task is still beyond the reach of present-day theory.

10.3.2.10 Data banks and data acquisition

To specify the heat of reaction, it is necessary to write the exact chemical reactions and to specify the states of all the reactants and products, noting particularly the constant temperature at which the measurement is made. Since most reactions are studied in calorimeters, ΔH is usually the heat of reaction stated. An example of an exothermic reaction is the burning of hydrogen:

$$H_2(g) + \frac{1}{2}O_2(g) \rightarrow H_2O(g) \qquad \Delta H_{291} = -241.750 \text{ kJ/mol} = -57.741 \text{ kcal/mol}$$

The heat is given from the system and therefore is written with a negative sign. Note 1 cal = 4.1868 J.

Changes in enthalpy when a system undergoes a physical or chemical change are normally reported for the process taking place under a set of standard conditions. Normally a temperature of 298.15 K and a pressure of 1 bar = 10^5 Pa are taken as STP. Otherwise the absolute temperature is written as a subscript or in parentheses. In most discussions the standard enthalpy change ΔH° is considered, i.e., the change in enthalpy for a process at STP. The superscript "$\circ$" indicates we are writing a standard enthalpy of formation. So the pressure for reactants and products is 1 bar. Note that 1 atm = 101,325 Pa = 760 Torr = 760 mmHg.

Thermochemical data are often reported in terms of a compound relative to its elements under standard conditions. Thus, the standard enthalpy of formation, ΔH°_f, of a substance is the reaction enthalpy for its formation from its elements in their reference states; the reactants and products all being in the standard state. The reference state of an element is its most stable state (solid, liquid, or gas) at the specified temperatures and 1 bar pressure. Consequently, the enthalpies of formation of elements in their reference states are zero at all temperatures. For example,

$$2Al + 1\frac{1}{2}O_2 \rightarrow Al_2O_3 \qquad \Delta H^\circ_{(298.15)} = -1669.8 \text{ kJ/mol}.$$

Thermochemical data are conveniently tabulated as standard enthalpies of formation, and a few examples are given in Table 10.2c, 10.2d and 10.2e. The standard enthalpy of any reaction at STP is then readily found as the difference between the tabulated enthalpies of formation of the products and of the reactants.

As an immediate consequence of the First Law, ΔH or ΔU for any reaction is independent of the path, that is, independent of any intermediate reaction that may occur. So, we can combine the standard enthalpies of individual reactions to obtain the enthalpy of another reaction. This application of the First Law is widely called the law of constant heat

Table 10.2c. *References: J = Janaf, B = Barin & Knacke, L = Landolt-Bornstein, K = Kubaschenski, H = Handbook of Chemistry. 1 eV ~ 23 kcal/mol.*

Specie	dH kcal/mol	dS cal/molK	dH-TdS kcal/mol	dH-TdS eV	Ref	Specie	dH kcal/mol	dS cal/mol K	dH-TdS kcal/mol	dH-TdS eV	Ref
H	52.1	27.4	43.9	1.91	J	Si	108.9	40.1	96.87	4.21	H
H_2	0	31.2	−9.3	−0.41	J	SiO	−23.8	50.6	−38.98	−1.69	H
F	18.9	37.9	7.5	0.33	J, B	SiO_2	−77.0	(50)	(−92)	(−4.0)	H
FH	−65.1	41.5	−77.6	−3.37	J	SiH	86.3	(40)	(74.3)	(3.23)	H
F_2	0	48.5	−14.6	−0.63	J	SiH_4	8.2	48.9	−6.47	−0.28	H
F_2H_2	−136.9	57.1	−154.0	−6.70	K	Si_2H_6	19.2	65.1	−0.33	−0.01	H
Br	26.7	41.8	14.2	0.62	J	SiF	1.7	53.9	−14.47	−0.63	H
Br_2	7.4	58.6	−10.2	−0.44	J	SiF_2	−148.0	60.4	−166.12	−7.22	H
Cl	28.9	39.5	17.1	0.74	J	SiF_4	−386.0	67.5	−406.25	−17.66	H
Cl_2	0	53.3	−16.0	−0.70	J	SiCl	45.4	(60)	(27.4)	(1.19)	H
I	25.5	43.2	12.5	0.55	J	$SiCl_2$	−39.6	67.0	−59.7	−2.60	H
I_2	14.9	62.3	−3.8	−0.16	J	$SiCl_4$	−157.0	79.0	−180.7	−7.86	H
O	59.6	38.5	48.1	2.09	J	SiBr	50.0	(70)	(29)	(1.26)	H
OH	9.5	43.8	−3.6	−0.16	J	$SiBr_4$	−99.3	90.3	−126.39	−5.50	H
OH_2	−57.8	45.1	−71.3	−3.10	J, H	SiS	26.9	53.4	10.88	0.47	H
OF	26.0	51.8	10.5	0.45	J, B	H^-	33.4	(30)	(24.4)	(1.06)	H
OFH	−31.0	54.0	−47.2	−2.05	J	H_2^+	357.2	(30)	(348.2)	(15.14)	H
OF_2	5.9	59.1	−11.8	−0.51	B	F^-	−64.7	(40)	(−76.7)	(−3.33)	H
O_2	0	49.0	−14.7	−0.64	J	Si_3N_4					
O_2F	3.00	61.9	−15.6	−0.68	B, K	SiL_4					

Species					Notes
O2F	-3.00				J
OF2	5.9				B, K
OF2	-4.4				J
N	113.0	36.6	102.0	4.43	J, H
NO2	7.9	57.3	-9.29	-0.40	J, H
N2	0	45.8	-13.7	-0.60	J, H
C	107.9	37.8	96.6	4.20	J, B
CH	142.0	43.7	128.9	5.60	J, B
CH2	92.1	43.3	79.1	3.44	J
CH3	34.8	46.4	20.9	0.91	J
CH4	-17.9	44.5	-31.3	-1.36	J
CF	39.0	50.9	23.7	1.03	L
CFH	30.0	53.4	14.0	0.61	J
CFH2	-7.8	(50)	(-22.8)	(-0.99)	H
CFH3	-56.0	53.3	-72.0	-3.13	J
CF2	-41.0	57.5	-58.3	-2.53	J, B
CF2H	-59.2	(60)	-77.2	-3.36	H
CF2H2	-107.7	58.9	-125.4	-5.45	J
CF3	-115.7	62.4	-134.4	-5.84	J, B
CF3H	-166.6	62.0	-185.2	-8.05	J
CF4	-223.0	62.5	-241.8	-10.51	J
SO2F4					
SiOF2	SiCF2				

Species					Notes
N2O	19.6	52.6	3.82	0.17	J, H
NO	21.6	50.3	6.51	0.28	J, H
NO3	17.0	60.4	-1.12	-0.05	J
CO	-26.4	47.2	-40.6	-1.76	J
COH	10.4	53.7	-5.71	-0.25	J
COF	-41.0	59.0	-58.7	-2.55	J, B
COFH	-90.0	59.0	-107.7	-4.68	J
COF2	-151.7	61.9	-170.3	-7.40	J, B
COCl2	-94.1	51.1	-109.4	-4.76	J
CO2	12.5	52.4	-3.2	-0.14	J
C2H4	-20.2	54.9	-36.7	-1.59	L
C2H6	-157.4	71.7	-178.9	-7.78	J
C2F4	-321.1	79.4	-344.9	-15.00	J
C2F6	-289.0	69.7	-309.9	-13.47	J
SF6	-185.2	69.8	-206.1	-8.96	H
SF4	66.6	40.1	54.57	2.37	H, J
S	1.5	53.0	-14.4	-0.63	H
SO	6.9	66.6			J
SO					
SO2	-70.9	59.3	-88.69	-3.86	H, J
SO3	-94.6	61.3			J
S2O	-13.5	63.8			J
SOF2	-135	67.0			J
SO2F2	-205	68.9			J
S2	30.8	54.5			J
SOF4					
SH2	-4.9	49.1	-19.63	-0.85	H
SOCl2	-50.8	74.0	-73.0	-3.17	H

Table 10.2d. *Standard enthalpies of formation at 298.15 K. Selected from a compilation of the National Bureau of Standards (NBS).*

Compound	State	ΔH_f^o [kJ/mol]	Compound	State	ΔH_f^o [kJ/mol]
H_2O	g	−241.826	H_2S	g	−20.63
H_2O	l	−285.830	H_2SO_4	l	−814.00
H_2O_2	g	−133.2	SO_2	g	−296.8
HF	g	−271.1	SO_3	g	−395.7
HCl	g	−92.312	CO	g	−110.523
HBr	g	−36.40	CO_2	g	−393.513
HI	g	+26.48	$COCl_2$	l	−205.9
HIO_3	c	−238.6	S_2Cl_2	g	−23.85
NO	g	+90.25	NH_3	g	−46.11
N_2O	g	+82.05	HN_3	g	+294.1

Table 10.2e. *Enthalpies of formation of gaseous hydrocarbons derived from heats of combustion (NBS). The standard state of carbon has been taken to be graphite.*

Substance	Formula	ΔH_f^o [kJ/mol]	Substance	Formula	ΔH_f^o [kJ/mol]
methane	CH_4	−74.75	1-butene	C_4H_8	+1.60
ethane	C_2H_6	−84.48	cis-2-butene	C_4H_8	−5.81
propane	C_3H_8	−103.6	trans-2-butene	C_4H_8	−9.78
n-butane	C_4H_{10}	−124.3	2-methylpropene	C_4H_8	−13.41
isobutane	C_4H_{10}	−131.2	acetylene	C_2H_2	+226.9
ethylene	C_2H_4	+52.58	methylacetylene	C_3H_4	+185.4
propylene	C_3H_6	+20.74			

summation, or simply Hess's law. G. H. Hess was the first to establish this principle in 1840. From this law we may arrive at reaction enthalpies from reactions not measured directly in practice, just by adding and/or subtracting other reactions which were measured. Similarly, reactions can be divided into subreactions in different ways; then we might obtain a cycle, a closed path, known as a Born-Haber cycle. The sum of enthalpy changes around a cycle is zero because the enthalpy is a state function. It follows that if all but one of the enthalpy changes in a cycle are known, the unknown may be deduced from the others using the Born-Haber cycle.

In the next sections, some important enthalpy changes in plasmas will be treated, such as those accompanying a physical change, chemical change, dissociation, and ionization.

Enthalpy of physical transition

The enthalpy change that accompanies a change of physical state is called the enthalpy of transition and denoted ΔH_{trs}, such as the enthalpy of vaporization (liquid → gas), ΔH_{vap}, fusion (solid → liquid), ΔH_{fus}, sublimation (solid → gas), ΔH_{sub}, and solution (solid dissolving in liquid), ΔH_{sol}. Since the change of enthalpy is independent of the path between two states, we can conclude that $\Delta H_{sub}(T) = \Delta H_{fus}(T) + \Delta H_{vap}(T)$. As an example, the standard enthalpy of vaporization is the enthalpy change per mole when a pure liquid at 1 bar pressure vaporizes to a gas at 1 bar pressure, as in

$$H_2O(l) \Rightarrow H_2O(g) \qquad \Delta H^\circ_{vap}(373\,K) = 40.66\ kJ/mol.$$

A similar example of a change in state of aggregation is the melting of a solid. At a fixed pressure, the melting point is a definite temperature T_m at which solid and liquid are in equilibrium. To change some of the solid to liquid, heat must be added to the system. As long as both solid and liquid are present, this added heat does not change the temperature of the system, but it is absorbed by the system as the latent heat of fusion or melting, $\Delta H_{fus} \equiv \Delta H_m$, of the solid. Since the change occurs at constant pressure, the latent heat, by $\Delta H = q_P$, equals the difference in enthalpy between liquid and solid. Per mole of substance, $\Delta H_m = H(\text{liquid}) - H(\text{solid})$. At the melting point, the addition of a little heat would melt some of the solid, but the equilibrium between solid and liquid would be maintained. The latent heat of the melting point is necessarily a reversible heat, because the process of melting follows a path of equilibrium states. We can therefore evaluate the entropy of melting ΔS_m at the melting point by a direct application of the relation $\Delta S_m = \Delta q_{rev}/T = S(\text{liquid}) - S(\text{solid}) = \Delta H_m/T_m$. For example, for ice, $\Delta H_m = 5980$ J/mol, so that $\Delta S_m = 21.90$ J/molK. By a similar argument, the entropy of vaporization ΔS_v, the latent heat of vaporization ΔH_v, and the boiling point T_b are related by $\Delta S_v = \Delta H_v/T_b$.

Enthalpy of chemical transition

Many of our thermochemical data have been obtained from measurements of heat of combustion. If the heat of formation of all its combustion products are known, the heat of formation of a compound can be calculated from its heat of combustion. The enthalpy of combustion, ΔH_c, is the reaction enthalpy for the oxidation of an organic substance to CO_2, H_2O, and N_2. Similarly, the standard reaction enthalpy is the change in enthalpy when the reactants in their standard states change to products in their standard states, as in

$$CH_4(g) + 2O_2(g) \Rightarrow CO_2(g) + H_2O(l) \qquad \Delta H^\circ_c(298.15\,K) = -890\ kJ/mol.$$

The enthalpy of hydrogenation is the reaction enthalpy for the hydrogenation of an unsaturated organic compound such as ethene and benzene. The lattice enthalpy, ΔH°_L, is the enthalpy change accompanying the formation of a gas of ions from the crystalline solid and is always positive: $MX(s) \Rightarrow M^+(g) + X^-(g)$. The lattice energy is the lattice enthalpy at $T = 0$. The enthalpy of formation of a solid may be analyzed into several contributions.

Enthalpy of ionization

The enthalpy of ionization, ΔH_i, is the enthalpy change for the removal of an electron $E(g) \Rightarrow E^+(g) + e^-(g)$. Since 1 mol of gaseous reactants gives 2 mol of gaseous products, the internal energy and enthalpy of ionization differ by about RT. The ionization energy, E_i, is the change in internal energy for the same process at $T = 0$. Since the ionization energy at ordinary temperatures is quite similar to that at $T = 0$, it is normally good enough to use the relation $\Delta H^\circ_i = E_i + RT$. Since the ionization energies are typically more than 100 times larger than RT, $\Delta H^\circ_i \sim E_i \sim \Delta U^\circ_i$.

The enthalpy change accompanying electron attachment to an atom, ion, or molecule in the gas phase is the electron gain enthalpy $\Delta H_{ea} : E(g) + e^-(g) \Rightarrow E^-(g)$. The negative of the corresponding internal energy change at $T = 0$ is widely called the electron affinity E_{ea}. With this change of sign, a positive electron affinity corresponds to exothermic electron gain.

For example, the attachment of one electron to an O atom is exothermic, so O has a positive electron affinity. As in the case of cation formation, the internal energy change and the enthalpy change differ by RT: $\Delta H_{ea}^{\circ} = -E_{ea} - RT$. The electron affinity is a state function and, therefore, measuring the ionization energy of a negative ion can be used to determine the electron affinity of the neutral species.

Enthalpy of dissociation

Ever since the time of van't Hoff, chemists have sought to express the structure and properties of molecules in terms of bonds between the atoms. In many cases, to a good approximation, it is possible to express the heat of formation of a molecule as an additive property of the bonds forming the molecule. This formulation has led to the concepts of bond energy and bond enthalpy. The bond dissociation enthalpy, $\Delta H_{(A-B)}$, is the reaction enthalpy for the process in which the A—B bond is broken: $A - B(g) \Rightarrow A(g) + B(g)$. It is important to realize that for a given type of bond, the enthalpy depends on the particular molecule in which the bond occurs and on its particular situation in that molecule. For instance,

$$\Delta H(HO—H) = 499 \text{ kJ/mol} = 119 \text{ kcal/mol} = 5.17 \text{ eV/bond}$$

and

$$\Delta H(O—H) = 428 \text{ kJ/mol},$$

where

$$1.000 \text{ eV/bond} \sim 23.05 \text{ kcal/mol} \sim 96.49 \text{ kJ/mol}.$$

The difference is caused by the different electronic structure of the molecule after the first atom is removed. As in the case of the ionization energy, the dissociation energy is only slightly affected by temperature. The enthalpy change that accompanies the separation of all the atoms in a substance is called the enthalpy of atomization, ΔH_a. For example, for water we calculate

$$\Delta H_a^{\circ} = \Delta H(HO—H) + \Delta H(O—H) = 927 \text{ kJ/mol}.$$

For an elemental solid that evaporates to a monoatomic gas, $\Delta H_a^{\circ} = \Delta H_{sub}^{\circ}$. The mean bond enthalpy B(A—B) is the value of the bond dissociation enthalpy of the A—B bond averaged over a series of related compounds. Mean bond enthalpies are useful because they let us make estimates of enthalpy changes in reactions where data might not be available. As another example, consider the molecule CH_4, and imagine that the H atoms are removed from it one at a time:

$$
\begin{array}{ll}
CH_4 \rightarrow CH_3 + H & \Delta H = 422 \text{ kJ/mol} \\
CH_3 \rightarrow CH_2 + H & \Delta H = 364 \text{ kJ/mol} \\
CH_2 \rightarrow CH + H & \Delta H = 385 \text{ kJ/mol} \\
CH \rightarrow C + H & \Delta H = 335 \text{ kJ/mol}.
\end{array}
$$

If we could imagine the carbon atom reacting with four hydrogen atoms to form methane, we could set one quarter of this overall enthalpy of reaction equal to the average bond enthalpy. In most cases, it is not difficult to obtain the ΔH for converting the elements to monoatomic gases. In the case of metals, this ΔH is simply the heat of sublimation to the monoatomic form. For example,

$$Mg(c) \rightarrow Mg(g) \qquad \Delta H_{298} = 150.2 \text{ kJ/mol}.$$

Table 10.3. *Standard enthalpies of atomization of elements.*

Element	ΔH_{298} [kJ/mol]	Element	ΔH_{298} [kJ/mol]
H	217.97	N	472.70
O	249.17	P	314.6
F	78.99	C	716.68
Cl	121.68	Si	455.6
Br	111.88	Hg	60.84
I	106.84	Ni	425.14
S	278.81	Fe	404.5

In other cases, the heats of atomization can be obtained from the dissociation energies of diatomic gases. For example,

$$\frac{1}{2}Br_2(g) \rightarrow Br(g) \qquad \Delta H_{298} = 111.9 \text{ kJ/mol}$$

$$\frac{1}{2}O_2(g) \rightarrow O(g) \qquad \Delta H_{298} = 249.2 \text{ kJ/mol}.$$

Remember that, for example, the oxygen molecule has a double bond between its two constituent atoms. Some standard enthalpies for conversion of elements from their standard states to the form of monoatomic gases (ΔH of atomization) are given in Table 10.3 (data from NBS Circular 500 and NBS Technical Notes 270-1 and 270-2). With these data, it is possible to calculate average bond enthalpies from standard enthalpies of formation. Determine, for example, the average bond enthalpy of the two O—H bonds in water:

$$H_2 \rightarrow 2H \qquad \Delta H = 436.0 \text{ kJ/mol}$$
$$O_2 \rightarrow 2O \qquad \Delta H = 498.3 \text{ kJ/mol}$$
$$H_2 + \frac{1}{2}O_2 \rightarrow H_2O \qquad \Delta H = -241.8 \text{ kJ/mol}.$$

Therefore,

$$2H + O \rightarrow H_2O \qquad \Delta H = (-241.8) - (436.0) - \frac{1}{2}(498.3) = -927.0 \text{ kJ/mol}.$$

This is the standard enthalpy for the formation of two O—H bonds, so that the standard average bond enthalpy can be taken as 463.5 kJ/mol. This value is somewhat different from the dissociation enthalpy for HOH $\rightarrow$ H + OH, which is 498 kJ/mol. We already found that $\Delta H(\text{HO—H}) = 499$ kJ/mol and $\Delta H(\text{O—H}) = 428$ kJ/mol, so, indeed $\frac{1}{2}(499 + 428) = 463.5$ kJ/mol.

The standard bond enthalpy of a bond A—B will be approximately constant in a series of similar compounds. This fact makes it possible to compile tables of average bond enthalpies, which can be used for estimating the enthalpy values for chemical reactions. The enthalpies so obtained are often close enough to the experimental values to provide rapid estimates of reaction enthalpies and enthalpies of formation. Of course, different bond types must be distinguished, e.g., single, double, and triple bonds in the case of C—C compounds. Table 10.4 is a summary of average single-bond enthalpies as given by Pauling in *Nature of the Chemical Bond* (Cornell University Press, 1960) [8]. Table 10.5 gives a few individual values for specified molecules.

Table 10.4. *Average single-bond enthalpies [kJ/mol].*
100 kJ/mol ~ 1 eV/bond ~ 23 kcal/mol.

	S	Si	I	Br	Cl	F	O	N	C	H
H	339	339	299	366	432	563	463	391	413	436
C	259	290	240	276	328	441	351	292	348	
N					200	270		161		
O		369			203	185	139			
F		541	258	237	254	153				
Cl	250	359	210	219	243					
Br		289	178	193						
I		213	151							
Si	227	177								
S	213									

Table 10.5. *Single- and multiple-bond enthalpies [kJ/mol].*

Triple	ΔH^o	Double	ΔH^o	Single	ΔH^o	Single	ΔH^o
N≡N	946	$H_2C{=}CH_2$	682	H_3C—CH_3	368	H_3C—H	435
HC≡CH	962	$H_2C{=}O$	732	H_2N—NH_2	243	H_2N—H	431
HC≡N	937	$O{=}O$	498	HO—OH	213	OH—H	498
C≡O	1075	$HN{=}O$	481	F—F	159	F—H	569
		$HN{=}NH$	456	H_3C—Cl	349	H_3C—NH_2	331
		$H_2C{=}NH$	644	H_2N—Cl	251	H_3C—OH	381
				HO—Cl	251	H_3C—F	452
				F—Cl	255	H_3C—I	234
						F—I	243

10.3.2.11   Data utilization and data manipulation

It is often necessary to use data obtained with a bomb calorimeter, which give ΔU, in order to calculate ΔH, ΔA, or ΔG. From the definition, $\Delta H = \Delta U + \Delta(PV)$. By $\Delta(PV)$ we mean the change in PV for the entire system, or, in particular, the PV of the products minus the PV of the reactants for the indicated chemical reaction. If all the reactants and products are liquids or solids, these PV values change only slightly during the reaction at low pressure (say 1 atm). In such cases, $\Delta(PV)$ may be neglected and $\Delta H \sim \Delta U \Leftrightarrow q_V \sim q_P$. For reactions in which gases occur, the values of $\Delta(PV)$ depend on the change in the number of moles of gas as a result of reaction. From the ideal gas equation, we can write $\Delta(PV) = RT\Delta n$ and therefore $\Delta H = \Delta U + RT\Delta n$. By Δn we mean the number of moles of gaseous products minus reactants.

Consider as an example the reaction

$$SO_2(g) + \frac{1}{2}O_2(g) \rightarrow SO_3(g) \qquad \Delta U_{298} = -97.030 \text{ kJ/mol},$$

as measured in a bomb calorimeter. $\Delta n = 1 - 1 - \frac{1}{2} = -\frac{1}{2}$, and therefore

$$\Delta H = \Delta U - \frac{1}{2}RT = -97030 - \frac{1}{2} * 8.3 * 298 = -98.270 \text{ kJ/mol}.$$

So, even in reactions accompanied by gas products, the difference between ΔU and ΔH is normally (i.e., at the lower temperatures) still rather small. Now, consider the same system that changes at constant temperature. From $A = U - TS$, for an isothermal change, $\Delta A = \Delta U - T\Delta S$. The ΔS can be found from Table 10.2c: $\Delta S_{298} = 61.3 - 59.3 - \frac{1}{2} * 49$ cal/molK $= -0.0942$ kJ/molK. Thus,

$$\Delta A = \Delta U - T\Delta S = -97030 + 94.2 * 298 = -69.0 \text{ kJ/mol.}$$

This is ΔA for the reaction at 298 K and 1 atm, irrespective of how it is carried out. The change in the Gibbs free energy of a system in an isothermal process leading from state 1 to state 2 is given as $\Delta G = \Delta H - T\Delta S$. If we consider a process at constant pressure, $\Delta G = \Delta A + PV$. Therefore,

$$\Delta G = \Delta A + P\Delta V = -69.0 - 1.24 = -70.2 \text{ kJ/mol.}$$

So, $\Delta H > \Delta U > \Delta G > \Delta A$ and ΔH is the maximum energy to be subtracted from the oxidation of sulfur dioxide.

Let us consider another example, that the change in question is the oxidation of iso-octane at 298 K and 1 atm:

$$C_8H_{18}(g) + 12\frac{1}{2}O_2(g) \rightarrow 8CO_2(g) + 9H_2O(g).$$

We can find ΔU for this reaction by measuring the heat of combustion of C_8H_{18} in a bomb calorimeter. It is

$$\Delta U_{298} = -5109 \text{ kJ/mol.}$$

The ΔS can be found by calorimetric methods: $\Delta S_{298} = 422$ J/molK. Thus,

$$\Delta A = \Delta U - T\Delta S = -5109 - 298 * 0.422 = -5235 \text{ kJ/mol.}$$

The change in the Gibbs free energy, with $P\Delta V = RT\Delta n = (17 - 13\frac{1}{2})RT = 3\frac{1}{2}RT = 3\frac{1}{2} * 8.314 * 298 = 8680$ J, is

$$\Delta G = \Delta H - T\Delta S = \Delta A + P\Delta V = -5235 = -5226 \text{ kJ/mol.}$$

The enthalpy can be found with the help of the relation $\Delta H + \Delta A = \Delta U + \Delta G$:

$$\Delta H = \Delta G + \Delta U - \Delta A = -5100 \text{ kJ/mol.}$$

So, $\Delta A > \Delta G > \Delta U > \Delta H$, which tells us that 5235 kJ is the maximum work that can possibly be obtained from the oxidation of 1 mol iso-octane at 298 K and 1 atm. Note that this work is actually greater than $-\Delta U$ for the change, simply because ΔS is positive. This result does not contradict the First Law, because the system is not isolated. There is no practical way to obtain $-w_{rev} = 5235$ kJ, however, since to achieve the reversible process it would be necessary to eliminate all frictional losses and to carry out the process in the cell infinitely slowly. Nevertheless, it is useful to know that $-\Delta A$ for this chemical reaction gives the upper limit for the work that can be obtained.

As found in the previous sections, figures of major concern in thermodynamics are the standard enthalpy $\Delta H°$ [J/mol] and entropy $\Delta S°$ [J/molK]. In Table 10.2a–e some values of interest for plasma etching are given. A plasma reactor is a system more or less working at constant pressure and temperature. An appropriate choice for the driving force behind the plasma reactions would thus be the Gibbs free energy or free enthalpy. When the standard enthalpy and entropy of the molecules which are in equilibrium in a reaction are known, the standard free enthalpy exchange at constant pressure and temperature can be expressed as $\Delta G° = \Delta H° - T\Delta S°$.

With the help of this expression we are able to estimate how easy it is to create a certain molecule. Saying it thermodynamically, with the use of the standard free enthalpy it is possible to calculate if a reaction is spontaneous ($dG_{T,P} \leq 0$). Of course, we have to consider the activation energy also, but this is theoretically difficult to calculate. Moreover, the activation energy might be drastically lowered by catalytic actions of certain species. For example, the reaction of hydrogen gas with oxygen gas into water is exothermic and the free enthalpy decreases: $dG \sim -237$ kJ/mol. However, the reaction is proceeding very slowly because the activation energy is much too high. Another example, the dissolving of sugar into water, is endothermic (your coffee gets colder), but the reaction is spontaneous because the entropy increases enormously. In this part we will not incorporate the activation energy. As a direct consequence, it might be possible that energetically favorable processes will not occur due to a much too high activation energy. But, endothermic reactions are generally always difficult, not depending on the activation energy. So, the results of this chapter are only indicative. When two processes might occur in a (binary) particle collision, one should notice that normally only one of the transitions will take place due to a lower activation energy. The chance that a reaction molecule occurs is expressed with the help of the constant of dissociation. In the case the dissociation of a gas is in equilibrium, we are able to define a constant of dissociation. Take for example the reaction

$$CHF_3 \Leftrightarrow CF_2 + HF.$$

Then k is derived from the partial pressures as

$$k = [CF_2] * [HF]/[CHF_3],$$

with [x] the pressure or density of the gas "x". The constant k is related to the heat as

$$\ln(k) = -\Delta G°/RT = \Delta S°/R - \Delta H°/RT.$$

In other words, when the enthalpy and entropy of a gas reaction is known, it is possible to calculate the partial pressure of the reaction gases.

Now, after the most important binary particle interactions have been found with their energy exchange, it is possible to give a rough idea of what type of particles may exist in a particular plasma. This section will give just a few examples of the reaction products of a particular collision, and the results are found in Table 12.1 contact plasma etching. The work is quite tedious because all kinds of activation levels must be taken into account, such as ionization, dissociation, and excitation which produce ions, radicals, and metastable states. The energy involved in the reaction together with the conservation of momentum will finally give information about which particles are favorable. In order to examine the particles which may exist in a plasma, the next collisions are considered.

(1) First-order collisions: Collisions of the primary gas with the cathode surface, for example, the collision of an SF_6 molecule with the silicon substrate. These collisions are proportional to the pressure of the molecule in the gas phase (SF_6). The first-order collisions are overwhelming in events with respect to the higher-order collisions and give us the pressure readout from the pressure sensor connected to the reactor.

Table 10.6. *Inelastic energy exchange in electron-neutral collisions.*

Prim. gas	Products	Electron energy needed [eV]
$CF_4 + e^- \Rightarrow$	CF_3+F+e^-	$-(-10.51) - 5.84 + 0.33 = 5.00$
	$CF_2+F_2+e^-$	$-(-10.51) - 2.53 - 0.63 = 7.35$
	$CF_2+F+F+e^-$	$-(-10.51) - 2.53 + 0.33 + 0.33 = 8.64$
	$CF+F_2+F+e^-$	$-(-10.51) + 1.03 - 0.63 + 0.33 = 11.24$
	$C+F_2+F_2+e^-$	$-(-10.51) + 4.20 - 0.63 - 0.63 = 13.45$
	CF_3+F^-	$-(-10.51) - 5.84 + 0.33 - 3.40 = 1.60$
	CF_3^-+F	$-(-10.51)+ <-5.84 + 0.33 =<5.00$
$F_2 + e^- \Rightarrow$	$F+F$	$-(-0.63) + 0.33 + 0.33 = 1.29$
	$F+F^-$	$-(-0.63) + 0.33 + 0.33 - 3.40 = -2.11$
	F_2^-	$-(-0.63) - 0.63 - 3.40 = -3.40$
$Br_2 + e^- \Rightarrow$	$Br+Br^-$	$-(-0.44) + 0.62 + 0.62 - 3.37 = -1.69$
$O_2 + e^- \Rightarrow$	$O+O^-$	$-(-0.64) + 2.09 + 2.09 - 1.46 = 3.36$
	$O_2^*+e^-$	$-(-0.64) + 0.98 = 0.98$
$N_2 + e^- \Rightarrow$	$N+N^-$	$-(-0.60) + 4.43 + 4.43 - 0.0 = 9.46$
	$N_2^*+e^-$	$-(-0.60) + 6.17 = 6.17$
$CHF_3 + e^- \Rightarrow$	CF_3+H^-	$-(-8.05) - 5.84 + 1.91 - 0.75 = 3.37$
	CHF_2+F^-	$-(-8.05) - 3.36 + 0.33 - 3.40 = 1.62$
	CF_2+HF^-	$-(-8.05) - 2.53 - 3.37-? =<2.15$
$SF_6 + e^- \Rightarrow$	SF_5+F^-	$-(-13.47)-? + 0.33 - 3.40 =?$
	$SF_4+F_2^-$	$-(-13.47) - 8.96 - 0.63 - 3.08 = 0.8$
	$SF_4^-+F_2$	$-(-13.47) - 8.96 - 2.35 - 0.63 = 1.53$

(2) Second-order collisions: Collisions between the primary gases themselves, for example, the collision of an SF_6 molecule with another SF_6 molecule or an O_2 molecule. These collisions are proportional to the square of the pressure of the primary gas molecules. Due to collisions between the molecules, electrons exist in the gas phase. These electrons are able to efficiently convert electrical energy from the rf power source into kinetic energy. The highly energetic electrons are able to excite/dissociate/ionize the primary molecules. The ionization process supplements extra electrons (chain reaction), and when the electron density is enough, a plasma is created. The electron-particle collisions are the only collisions in the plasma phase which can be endothermic. So, the reaction is forced into one direction whereby the electron loses its momentum. All the other collisions (reactions) should be exothermic or not too endothermic in order be possible.

(3) Third-order collisions: Collisions between the molecules of the primary gas and the activated particles. The particles generated from the second-order collisions may collide with the primary gas molecules. These collisions are third-order with respect to the pressure and will result in extraordinary particles.

(4) Fourth-order collisions: Collisions between the activated particles are seldom and are therefore assumed to be unimportant.

(5) Electron-neutral collisions: To calculate the electron energy needed to cause an inelastic collision the electron affinity (Table 10.2(b)) and the heat exchange (Table 10.2(c)) are used. Clearly, the reactions incorporating electron attachment are most likely. Table 10.6 gives typical results of some ordinary gases in plasma etching. In this

table, the lower or the more negative the value for the electron energy needed, the more likely the reaction will occur when an electron hits the molecule of the primary gas. Note that the attachment of an electron with an F_2 molecule is forbidden due to law of the conservation of momentum. Also, nitrogen is much easier excited than dissociated, just like oxygen.

(6)   Neutral-neutral collisions: Sometimes, when two gases are mixed in a reactor, the gases react spontaneously. An example of such a reaction is the CF_4/H_2 mixture which reacts into $CHF_3 + HF$:

$$CF_4 + H_2 \Leftrightarrow CHF_3 + HF \qquad \Delta G^o_{298} = -(-10.51) - (-0.41) - 8.05 - 3.37 = -0.5.$$

So, the reaction is exothermic by 0.5 eV and is therefore thermodynamically spontaneous.

(7)   Neutral-active collisions: Many sorts of collisions are thinkable between activated particles and the primary gas, and the reaction products are miscellaneous. As an example, the CF_4/O_2 mixture is taken:

$$CF_4 + O \Leftrightarrow COF_2 + F_2 \quad \Delta G^o_{298} = -(-10.51) - 2.09 - 7.40 - 0.63 = 0.39$$
$$CF_4 + O_2^* \Leftrightarrow COF_2 + OF_2 \quad \Delta G^o_{298} = -(-10.51) - (-0.64)$$
$$- 0.98 - 7.4 - 0.51 = 2.26$$
$$CF_3 + O_2 \Leftrightarrow COF_2 + OF \quad \Delta G^o_{298} = -(-5.84) - (-0.64) - 7.4 + 0.45 = -0.47$$
$$CF_2 + O_2 \Leftrightarrow CO_2 + F_2 \quad \Delta G^o_{298} = -(-2.53) - (-0.64) - 4.76 - 0.63 = -2.22$$
$$\Leftrightarrow COF_2 + O \quad \Delta G^o_{298} = -(-2.53) - (-0.64) - 7.40 + 2.09 = -2.14$$
$$CF_4 + H \Leftrightarrow CF_3 + HF \quad \Delta G^o_{298} = -(-10.51) - 1.91 - 5.84 - 3.37 = -0.61$$
$$H_2 + F \Leftrightarrow \Leftrightarrow HF + H \quad \Delta G^o_{298} = -(-0.41) - 0.33 - 3.37 + 1.91 = -1.38.$$

(8)   Active-silicon collisions: These collisions are of prime importance because the active particle may adhere (physisorb) and react (chemisorb) with the surface atoms. The products will be volatile and desorb or will be involatile and act as an inhibitor for following collisions. Important collisions from this type are the fluor atoms which react with silicon into the volatile silicon tetrafluoride and the oxygen radicals which react with the silicon into the silicon oxide inhibitor layer.

10.3.3   Steady state thermodynamics

A question to be answered is whether plasmas fulfill the rigorous requirements of thermodynamics. The problem is quite tricky, since the plasma system is not at thermodynamic equilibrium to begin with, nor is it at equilibrium when the plasma is used for etching a material. Therefore, as will be explained in this section, although we can define the energy U and volume V of the isolated system, we cannot define its entropy S in the nonequilibrium state. Thus, we have no method for calculating the entropy change. The same difficulty is illustrated by the problem arising if we finish etching and we ask ΔS for the change, etching substrate $\rightarrow$ etched substrate. Although the etched substrate is in a state of equilibrium, the etching substrate certainly is not. At best, it is in a steady state with respect to input and output of power, oxygen, reaction products, etc. A solution to the basic problem might be found if we could define also the entropy for nonequilibrium states. To a certain extent, this may be possible for steady state processes with the help of statistical mechanics. Reversible thermodynamics can therefore tell us nothing directly about plasma processes, although, of course, it can provide much useful information about the properties of various chemicals composing such systems. If we wish to apply thermodynamics to plasma under steady state conditions, we shall need to formulate a science of irreversible thermodynamics.

In all the discussions of thermodynamics up to this point, it has been formulated only for reversible systems. In case of isolated systems, reversible thermodynamics has its foundation on a few definitions of state functions (P, V, T, and S) and upon three basic postulates:

- First Law – The total internal energy remains always constant: $dU = dq + dw = 0$.
- Second Law – The production of entropy equals the amount of heat supplied divided by temperature: $dS = dq/T$.
- Third Law – It is possible to come near $T = 0$ [K], but it is impossible to reach it exactly.

The laws of thermodynamics are postulates that are stated to be universally valid; they are certainly not restricted only to reversible processes or systems in equilibrium. In this section, steady state processes will be treated. First, the inequality of Clausius will be explained and after that its consequences for the four thermodynamic potentials: U, H, A, and G. Second, a short discussion on the analysis of steady state processes is given. Because this theory is far from being completed, it is of limited use for plasma systems and, therefore, we will not go too far into detail.

10.3.3.1   Inequality of Clausius

In the section about reversible processes we found that in case of any reversible Carnot cycle, $\oint dq_{rev}/T = 0$. However, the efficiency of an irreversible cycle is always less than that of a reversible cycle: $\varepsilon < \varepsilon_{max} \Leftrightarrow (q_h + q_c)/q_h < (T_h - T_c)/T_h$. So we have $q_h/T_h + q_c/T_c < 0$. This relation is extended to the general irreversible cycle, giving the *inequality of Clausius*:

$$\oint \frac{dq_{irrev}}{T} < 0 \quad [\text{J/K}]. \tag{10.47}$$

Moore [6] wrote that *"The proof is based on the fact that the efficiency of an irreversible Carnot cycle is always less than that of a reversible cycle operating between the same two temperatures. In the reversible cycle, the isothermal expansion yields the maximum work and the isothermal compression requires the minimum work, so that the efficiency is highest for the reversible case."* Although the writer is not able to understand why this would be a "proof," he will accept its form and use it in the rest of this section. Now consider a perfectly general irreversible process in an isolated system, leading from state A to state B, which is returned to its initial state A by a reversible path. Since the entire cycle is in part irreversible, Eq. (10.47) applies, and writing the cycle in terms of two sections, we therefore obtain

$$\int_A^B \frac{dq_{irrev}}{T} + \int_B^A \frac{dq_{rev}}{T} < 0 \quad [\text{J/K}]. \tag{10.48}$$

The change in entropy in going from an equilibrium state A to an equilibrium state B is always the same, irrespective of the path between A and B, since the entropy is a function of the state of the system alone. It makes no difference whether the path is reversible or irreversible. Only if the path is reversible, however, is the entropy change given by $\int dq/T$:

$$\Delta S = S_B - S_A = -\int_B^A \frac{dq_{rev}}{T} \quad [\text{J/K}]. \tag{10.49}$$

We can combine Eqs. (10.48) and (10.49) into one equation,

$$\Delta S \geq \int \frac{dq}{T} \quad [\text{J/K}], \tag{10.50}$$

in which the equality refers to reversible processes and the inequality to irreversible ones. Applied to an isolated system, for which $dq = 0$, it yields

$$(\Delta S)_{V,U} \geq 0 \quad [\text{J/K}]. \tag{10.51a}$$

If the heat transfer had been carried out irreversibly (e.g., by placing two reservoirs in direct thermal contact and allowing heat q to flow along the finite temperature gradient thus established), the entropy of the isolated system would have increased by the amount $\Delta S = q/T_c - q/T_h > 0$ (Why is $q_h = q_c = q$?). The equation "proves" that the entropy of an isolated system always increases during an irreversible process (unless energy has been transferred into some unknown energy reservoir). Since the entropy is defined only for equilibrium states, the fact that it can increase in an isolated system implies that we still can produce certain changes in such a system, even though it remains isolated. Moreover, all naturally occurring processes are irreversible, any change that actually occurs spontaneously in nature is accompanied by a net increase in entropy. This conclusion led Clausius to his famous concise statement of the laws of thermodynamics:

> *"The energy of the universe is a constant;*
> *the entropy of the universe tends always toward a maximum."*

So, the Second Law combined with the inequality of Clausius gives us the key to predict the direction of a spontaneous change. The entropy of an isolated system, that is, constant T and U, increases in the course of a spontaneous change: $(\Delta S_{\text{tot}})_{V,U} > 0$, where S_{tot} is the total entropy of all parts of the isolated system. Cooling to the temperature of the surroundings and the free expansion of gases are irreversible, that is spontaneous, processes. Hence, they go together with an increase in entropy. Spontaneous processes always produce entropy. Reversible processes are practically impossible because they may transfer entropy from one part to another but may never produce it after performing a cycle, i.e., $\oint S_{\text{tot}} = 0$. Now let us examine the state in equilibrium: There is no possible variation that would allow the isolated system to move to an equilibrium state of lower entropy. Therefore, at equilibrium,

$$(\delta S)_{V,U} \leq 0 \quad [\text{J/K}], \tag{10.51b}$$

that is, in a system at constant energy and volume ($=$ isolated), the entropy is a maximum.

For a reversible process, no entropy is produced and $dq = TdS$. However, in case of irreversible processes, from Eq. (10.50), $dq < TdS$: Clausius inequality. So the entropy, a state function, is the basic concept for discussing the direction of natural change, but in order to use it we have to investigate changes in both the system and its surroundings. Generally, it is simple to calculate the entropy change in the latter and it is possible to devise a simple method for taking that contribution into account for the system. In this way we reach the already familiar thermodynamic functions.

10.3.3.2   *Clausius and the thermodynamic potentials*

Combining the First Law ($dU = -PdV + dq$), the expressions for the other thermodynamic potentials ($dH = VdP + dq$, $dA = -PdV + dq - TdS - SdT$, and $dG = VdP +$

$dq - TdS - SdT$), and the inequality of Clausius ($dq \leq TdS$), the reversible equations turn into irreversible inequalities:

$$dU \leq -PdV + TdS \qquad dU_V \leq TdS \qquad dU_{V,S} = 0 \quad \text{[J]} \qquad (10.31')$$

$$dH \leq VdP + TdS \qquad dH_P \leq TdS \qquad dH_{P,S} = 0 \quad \text{[J]} \qquad (10.35')$$

$$dA \leq -PdV - SdT \qquad dA_V \leq -SdT \qquad dA_{V,T} \leq 0 \quad \text{[J]} \qquad (10.38')$$

$$dG \leq VdP - SdT \qquad dG_P \leq -SdT \qquad dG_{P,T} \leq 0 \quad \text{[J]}. \qquad (10.40')$$

Because $dU \leq -PdV + TdS \Leftrightarrow TdS \geq dU + PdV$, and similar arguments for the other functions, we find

$$dS_{V,U} \geq 0 \quad \text{[J/K]} \qquad\qquad\qquad (10.52)$$

$$dS_{P,H} \geq 0 \quad \text{[J/K]} \qquad\qquad\qquad (10.53)$$

$$dT_{V,A} \leq 0 \quad \text{[J/K]} \qquad\qquad\qquad (10.54)$$

$$dT_{P,G} \leq 0 \quad \text{[J/K]}. \qquad\qquad\qquad (10.55)$$

These are the criteria for natural changes in terms of properties relating to the system. For example, at constant volume and internal energy the entropy of the system must increase in a spontaneous change, for there can then be no change in entropy of the surroundings. Also, if a system is at constant pressure and has a constant Gibbs free energy, then the temperature decreases in a spontaneous change. As a last example, the Gibbs function G is a minimum at equilibrium. That is, a change under the conditions of constant pressure and temperature is spontaneous if it corresponds to a decrease in free enthalpy. Such systems move spontaneously toward states of lower G if a path is available and the criterion of equilibrium is $dG_{P,T} = 0$. It must be understood that the tendency of a system to move to lower A and G is due to its tendency to move toward states of higher entropy and not because they tend to lower internal energy. After all, the inequality is caused by dq and not dw. For example, $dU_S = -PdV$ and not $dU_S \leq -PdV$.

The necessary constraints for an isolated system are that U and V cannot change. Therefore, the criteria for thermodynamic equilibrium in an isolated irreversible system can be stated as follows: In a system at constant energy and volume, the entropy is a maximum at equilibrium, $(\delta S)_{V,U} \leq 0$. The equilibrium condition in terms of a constant displacement from equilibrium δA becomes $\delta A_{V,T} \geq 0$. Thus, in a closed system at constant T and V with no work done on the system, the Helmholtz function is a minimum at equilibrium. In a closed system at constant T and P with only PV work allowed, $\delta G_{P,T} \geq 0$. These are just the conditions applicable in ordinary mechanics, in which thermal effects are excluded.

Under condition of constant T and P, any spontaneous change in a system proceeds from a state of higher Gibbs free energy to a state of lower Gibbs free energy. For this reason, it became natural to think of the Gibbs function G as a thermodynamic potential and to think of any change in a system as a passage from a state of higher to a state of lower potential. Of course, the choice of G as the potential function is due to the choice of the condition of constant T and P (phase changes). At constant T and V the suitable potential function would be A; at constant P and S (calorimeter) it would be H; and at constant V and S (bomb calorimeter) it would be U.

The First and (irreversible) Second Laws have often been stated in terms of the universal human experience that it is impossible to construct a perpetual motion machine. The Second Law combined with the inequality of Clausius is expressed precisely in various equivalent forms. The principle of Thomson: It is impossible to devise an engine which, working in a cycle, shall produce no effect other than the extraction of heat from a reservoir and the performance of an equal amount of heat. The principle of Clausius: It is impossible to devise an engine which, working in a cycle, shall produce no effect other than the transfer of heat from the colder to a hotter body. The phrase working in a cycle means that the system returns to exactly its initial state, so that the process can be carried out repeatedly. It is easy enough to convert heat into work isothermally if a cyclic process is not required: Simply expand a gas in contact with a heat reservoir. "*An isothermal conversion of heat into work without any concomitant change in the system has never been observed. Think what it would imply: A ship would not need to carry any fuel; it could propel itself by taking heat from the ocean. Such an continuous extraction of work from the heat of the environment has been called perpetual motion of the second kind, whereas the production of work by a cyclic process with no change at all in the surroundings was called perpetual motion of the first kind. The impossibility of the latter is postulated by the First Law; the impossibility of the former is postulated by the Second Law,*" (Moore [6]).

10.3.3.3   Steady state theory

Although the extended Second Law gives an indication of the direction of a spontaneous process, it is everything but a theory: It is identical with the statement that the weight of a cow increases when eating some grass. A theory should give an answer to the question of how much a certain property is changing. This brings us to the theory of irreversible steady state processes, or the subject called nonequilibrium thermodynamics, or the thermodynamics of irreversible processes. There are many measurable properties of nonequilibrium systems (thermal conductivity, diffusion coefficient, viscosity, to name a few) which are similar to thermodynamic properties, such as temperature, density, or entropy, in that their definitions are not based upon any model for the structure of matter. Such properties are often called phenomenological. It is natural to ask, therefore, whether any theory exists that can provide relations between such nonequilibrium phenomenological properties.

If the variables of state are examined, it is found that they fall into two classes. Some of them can be used to describe nonequilibrium states with no difficulty of any kind. Examples of this class are volume V, mass m, concentration c, amount of substance n, and energy U. These functions are perfectly well defined for any system or any part of a system, whether or not it is in equilibrium. On the other hand, serious difficulties arise when we attempt to use such functions as P, T, and S in the description of nonequilibrium states. Consider, for example, two heat reservoirs at T_h and $T_h > T_c$ connected by a metal bar along which heat is being conducted from the hotter to the colder reservoir, a typical irreversible process. We do not have any way to define the temperature at any point along the bar, since T has been defined only for a system in equilibrium. To overcome this roadblock so as to proceed with thermodynamic calculations on nonequilibrium systems, we must introduce a new postulate that will allow us to define P and T at any point in a system in which an irreversible process is occurring. We therefore introduce a forth postulate: the postulate of microscopic or local equilibrium.

Fourth Law – A system can be divided (conceptually) into cells so small that each cell effectively corresponds to a given point in the system, but so large that each contains thousands of molecules. At a time t, the matter in a given cell is isolated from its surroundings and allowed to come to equilibrium, so that at t + δt, the P and T of the cell can be specified.

The postulate of local equilibrium is that we can take P and T at any point in the original nonequilibrium system at time t to be equal to the P and T in the corresponding cell when equilibrium is reached at $t + δt$. All the relations between state functions that were derived for equilibrium states also hold good for the functions we have now defined for nonequilibrium states. We do not contend, however, that the postulate can be applied to all irreversible systems, but it should be a range of validity, comprising systems in which the properties are not varying too rapidly with time. Thus, we would feel confident in applying the postulates to steady state processes, such as the heat conduction across a thermocouple junction, but would feel uneasy about their application to a nuclear explosion, that is, not reversible nor steady state processes.

The definition of entropy S might seem to present a special problem in steady state thermodynamics, since the function is introduced into ordinary thermodynamics by means of a definition $dS = dq_{rev}/T$ explicitly in terms of a reversible transfer of heat. In practice, however, the entropy of various substances is calculated by means of other equations, which were derived from the initial definition. We can therefore use our fourth postulate to calculate the entropy of each cell of a system undergoing a steady state process, and hence the entropy per unit mass $s = S/m$ at each point in the system. For example, a heat-conducting bar can be divided into thin slices normal to the temperature gradient, and in each slice T and P are defined and constant. Then we can calculate S for any slice by using the equation that relates the entropy of the metal to its heat capacity C_P: $S = \int_0^T C_P d \ln T$.

"As an example of a non-equilibrium system, consider a mixture of H_2 and O_2 gases in a container at T and P in the presence of a suitable catalyst for the reaction $H_2 + \frac{1}{2}O_2 \rightarrow H_2O$. At any time t, the container would contain definite amounts of H_2, O_2, and H_2O. Since we know the molar entropies of each of the substances, we could calculate the entropy of the system by adding up the entropies for the gases present at any instant and including the calculated entropy of mixing on the assumption that the composition was uniform throughout the container. Since the reaction does not proceed in the absence of catalyst, we could even experimentally stop the reaction at any instant and make any measurement necessary to determine deviation from ideal gas behavior in the partially reacted mixture of gases. Since the example chosen does not differ in any fundamental way from the generality of chemical reactions, there seems to be no reason why we could not determine S as a function of time t, and hence dS/dt, the rate of production of entropy during the reaction process," (Moore [6]). To maintain the temperature of the system constant, it might be necessary to have heat flows into or out of the container during the reaction, so that we could write for the differential change in entropy $dS = dS_{int} + dS_{ext}$, where dS_{int} is due to changes within the system, and dS_{ext} is due to flow of entropy into the system from exterior.

An adequate formulation of thermodynamics for steady state processes was first provided by Lars Onsager in 1931. The formulation can be summarized in the following statements.

(1)   Under equilibrium conditions any process and the reverse of that process will be taking place on the average at the same rate: the principle of microscopic reversibility.

(2) One can write thermodynamic equations of motion for various transport processes, in which the rates of flow or fluxes (heat, matter, etc.) are equal to a sum of terms, each of which is proportional to a thermodynamic force (temperature gradient, chemical potential gradient, etc.). The proportionality factors are called the phenomenological coefficients (direct- and drag-coefficients).

The Onsager formulation has been confirmed experimentally in a wide variety of situations departing considerably from equilibrium.

We have already discussed that in an equilibrium state the rate of entropy production is zero, $dS/dt = 0$. In a stationary state, on the other hand, the rate of entropy production has the minimum value consistent with the external restraints imposed upon the system. This theorem appears to have been first derived by Prigogine in 1947. Its validity is subject to the conditions that the phenomenological equations are linear, their coefficients are constant, and the so-called Onsager reciprocal relation holds. Such nonequilibrium stationary states are stable with respect to small perturbations of variables defining the system. So, although S is increasing for steady state processes, its increase in time is constant, $dS/dt = $ constant, and we can speak of a steady state equilibrium condition.

10.4   Electrical equivalent circuit

An effective method to describe the ion-enhanced mechanism in a plasma is possible with the help of an electrical equivalent circuit. It consists of four main units (Fig. 10.5): the reactor which contains the plasma glow discharge, the rf power supply to support the electrons their kinetic energy, the matching unit which matches the electrical impedance of the power unit

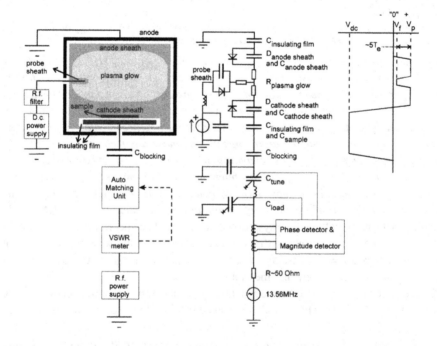

Fig. 10.5. (left) Simplified schematic of a parallel plate reactor with its (middle) electrical equivalent circuit and (right) the forming of the dc self-bias.

to that of the plasma, and a voltage standing wave ratio (VSWR) measuring unit to make automatic matching possible. With the help of this equivalency it is possible to arrive at the way the self-bias is developed. This is an extremely important parameter in plasma etching with IBARE systems.

10.4.1   Discharge unit

Konuma [2] writes: "*The capacitively-coupled internal electrode type has been widely used at production levels in plasma reaction systems employing a high-frequency discharge. The reaction chamber is schematically shown in Figure 10.5 together with its electrical equivalent circuit. The conducting plasma glow can be represented by resistors R and the so-called dark spaces or sheaths can be represented by a parallel combination of a rectifier D, which indicates the asymmetrical characteristic for negative electron current and positive ion current, and a capacitor C, owing to the* (lack of electron) *space charge in the sheaths. If films, deposited on the chamber walls and on the substrate holder, are not conducting, they will act as capacitors.*" For example, during the etching of glass with fluorocarbon gases like CHF_3 and CF_4, an insulating fluorocarbon film is deposited on surfaces with a low ion energy $*$ flux bombardment.

10.4.2   Rf power supply and matching unit

"*The frequency most often chosen is 13.56 MHz. This frequency is allowed for industrial, scientific and medical uses by international communications authorities. At this frequency the plasma impedance is relatively low (partly capacitive), on the order of 100–1000 Ohms. On the other hand, commercial high-frequency generators are designed to have a purely resistive output impedance of 50 Ohms to be able to match properly with the characteristic impedance of rf cables (50–75 Ohms). Therefore, it is necessary to match the impedance between the generator and the load, to minimize the reflected power versus the forward power by using a matching network. For high frequencies about ~1 MHz, a π-type configuration is often employed as found in Figure 10.5,*" (Konuma [2]).

10.4.3   VSWR measuring unit

"*A matching network with auto-tuning capability helps to obtain power-level stability despite changing discharge conditions. For example, the impedance of the plasma bulk and dark spaces vary with the frequency used and the discharge condition, mostly depend on the type of gas and the pressure. The power value which is obtained after subtracting the reflected power from the forward power is usually quoted to describe the plasma condition. This is because it is very difficult to measure the exact power that is actually dissipated in the plasma. The efficiency of power delivery to the plasma is strongly influenced by the design of the matching network and also by the selection of electric circuit components. Large stray capacitances, which exist especially in the matching network or in the space between electrodes and grounded shields, can cause large power losses. Both the forward power which is supplied from the generator to the load and the reflected power which is reflected from the load can be measured by a through-line Watt meter,*" (Konuma [2]). The meter is connected in series between the generator and the matching network by way of coils which contain the rf cable. The output of the phase and magnitude detection electronics is used to tune the load and tune capacitors from the matching unit.

10.4.4   Self-bias potential

"Spatial distribution of the potential between the powered electrode (cathode) and the grounded electrode (anode) in a conventional parallel plate reactor are shown schematically in Figure 10.5 right. This potential is given as the time averaged potential of high-frequency voltages compared to ground. The cathode is floating because no overall charge can be transferred through the blocking capacitor. In this case a potential V_{dc}, which is negative compared to ground level, develops and this voltage is called the d.c. self-bias," (Konuma [2]). This self-bias is developed as follows.

Before ignition, the gas in the reactor is nonconducting, and the total rf voltage drop will be across this gas and nothing special will happen. At a certain voltage, the breakdown voltage, free electrons in the gas phase pick up sufficient rf energy between two collisions to be able to ionize the gas molecules and a rapid increase in the electron density occurs, i.e., ignition of the plasma. This highly conductive plasma glow region can be represented by a resistor. At the same time, the electrons are able to excite the molecules giving the plasma its characteristic glow. The free-moving charges in the plasma phase which are close enough to the electrodes are collected by them during one cycle of the electrical rf field. Therefore, a small region near the electrodes will be depleted from charges and becomes nonconducting. Therefore, this region can be represented by a capacitor in which one of the capacitor plates is the electrode (anode, cathode, or probe) and the other is the conducting plasma glow. Due to the lack of electrons in these regions, no excitation processes take place and the regions are not glowing. These regions are therefore known as the dark space regions or sheath regions.

The mass of an ion is considerably higher than the electron mass (e.g., the mass of an $SF_5^+ \sim 127 * 1836 * m_e \sim 2 * 10^5 * m_e$). Using Eq. (10.22), $\omega_p = \sqrt{[nq^2/m\varepsilon_0]}$, the plasma-ion oscillation frequency is much lower, about 2 MHz, and electrons will, but ions won't, respond to the frequency of 13.56 MHz from the rf power supply. Therefore, this effect may be represented by a diode in parallel with the sheath capacitor: Electrons can easily move from the boundary of the plasma glow region into the anode/cathode electrode, but there they are trapped and won't travel back. So, there is continuous current of electrons to the electrodes giving it a diode characteristic. This unique feature of the difference in plasma resonance between electrons and ions is responsible for the formation of the plasma potential V_p and the self-bias voltage V_{dc} in IBARE systems. *"The I-V characteristics of the plasma resemble that of a leaky rectifying diode, due to the great difference in mobility between electrons and ions. During the half of the cycle in which positive voltage is applied to the electrode, a very large current of light electrons flows into the electrode. During the other half of the cycle, in which negative voltage is applied, a small current of heavy ions flows into the electrode. After a while, at steady state, the electrode will be negatively self-biased which repels electrons and zero net current is attained (equal numbers of electrons and ions reaching the electrode),"* (Konuma [2]).

Due to the high electron density, the plasma glow has a small electrical resistance and the major part of the rf voltage will drop across the sheath capacitors. Moreover, generally the anode surface facing the plasma glow region has a much bigger area than that of the cathode. Therefore, the anode sheath capacitor is much bigger and only a small part of the rf voltage will drop across the anode sheath. This voltage is rectified due to difference in inertia (mobility) between the electrons and ions, and this dc voltage is called the plasma voltage V_p, which is now labeled with respect to the grounded level. The voltage rectified by the cathode sheath diode is big due to the small area of this electrode with respect to the anode surface. Since the blocking capacitor in the rf cable obstructs a dc charge from flowing out, the cathode acquires a strong negative voltage, and this dc voltage is known as the dc self-bias V_{dc}.

Let us look once again to Fig. 10.5, where a (Langmuir) probe is inserted into the reactor in direct contact with the plasma glow. In all cases, a floating potential will develop across the probe sheath of ca. five times the electron temperature, which will give the ions an energy of approximately 10 eV. Now the probe can be driven by a dc or ac voltage source. When a negative potential with respect to the grounded electrodes is applied, ions will gain an extra energy, conform the applied voltage, and, due to the probe sheath diode, the plasma potential is not altered. However, when a positive voltage is applied, the probe diode, is biased in the forward direction and the applied voltage passes the sheath and increases the plasma potential. This voltage will not affect the ion velocity toward the probe, instead it will increase the energy of ions colliding with the other nonfloating surfaces, i.e., the cathode and anode. Of course, the preceding situations will only occur when the probe is not covered by an insulating film which would block the dc current. When the probe is driven by an ac source, the current passes the film and the plasma potential is lifted. This configuration is known as the triode arrangement. With respect to the applied dc source, the triode has the advantage of being unaffected by insulation films, and substrate charging due to dc currents is less of a problem. However, the mean ion energy is much better defined in the case of dc coupling.

References

[1] B. Chapman, *Glow Discharge Processes* (John Wiley & Sons, New York, 1980).
[2] M. Konuma, *Film Deposition by Plasma Techniques* (Springer-Verlag, Berlin Heidelberg, Germany, 1992).
[3] D. M. Manos and D. L. Flamm, *Plasma Etching: An Introduction* (Academic Press, Inc., San Diego, 1989).
[4] H. F. Winters, J. Appl. Phys. **49**, 5165 (1978).
[5] P. W. Atkins, *Physical Chemistry* (Oxford University Press, Oxford, Great Britain, 1990).
[6] W. J. Moore, *Physical Chemistry* (Longman Group Limited, London, 1972).
[7] J. L. Cecchi, *Introduction to plasma concepts and discharge configurations: Handbook of Plasma Processing Technology*, ed. S. M. Rossnagel (Noyes Publications, Park Ridge, New Jersey, 1990), p. 21.
[8] Pauling, L., *The Nature of the Chemical Bond* (Cornell University Press, 1960).
[9] D. R. Lide, *Handbook of Chemistry and Physics* (CRC Press, Boca Raton, Florida, 1994).

11

Plasma system configurations

There are many ways to categorize plasma equipment according to chemistry, generator frequency, electrode arrangement, load capacity and technique, plasma-sample distance, operating pressure, or type of etching species. Hardware choices range from manual, large batch, high-pressure oxygen, chemical barrel reactors to computer-controlled, single-wafer, low-pressure, ion-intensive, complex chemistry etchers. The best choice will depend on many factors, including material to be patterned, whether it is to be used in development or production, required geometry control, and, not insignificantly, the available budget.

The hardware choice depends on the chemistry choice. Equipment for handling relatively inert gases may be completely inadequate for corrosives. The generator frequency directly influences the lifetime of charged particles in the plasma phase and the efficiency of energy transfer. The electrode arrangement determines how the electrical energy is coupled into the reactor to form a plasma. By load capacity, we refer to the amount of wafers which are simultaneously etched in the same chamber, and by load technique, we refer to the way the wafers are facing the plasma. It is also useful to distinguish between different distances between the plasma glow region and the sample to be processed. This determines the type of particles that will reach the sample. The pressure in the vessel during etching determines many plasma (etch) characteristics such as the (relative) particle densities, particle energies, and their collimation. As a last differentiation, we consider groups depending on what types of species are used from the plasma glow region for etching.

Most plasma systems have sufficient automation to remove wafers from a cassette or line, feed them into the processing unit, and return them to a receiver. Usually the size of the reactor batch, commonly 25 wafers, and the cassette identity are matched to that of other process equipment. There are many variations in construction, with major differences in the wafer-handling mechanism. Air tracks, where wafers are moved on an air cushion, "rubber bands," which move wafers on stretched rubber-like bands driven by pulleys, and mechanical pick and place are all used. Most systems come complete with gas handling hardware, gauging, rf generators, and sufficient automation to operate without attendance once wafers have been placed in the input. It is also desirable to heat other portions of the reactor and its exhaust system to prevent condensation of low vapor pressure etch products. Proper selection of electrode material and other structural parts in contact with plasma is important to prevent sputtering of contaminants onto wafers. Trace amounts of ferrous metals can substantially degrade the performance of most ICs. For this reason, electrode materials such as silicon, graphite, glass, silicon carbide, or various polymers are preferred over metals. Most commercial systems make use of special coatings, which, though they sputter, are usually less harmful than metals to device performance. Anyway, there are many variations of chamber configuration with major differences in electrode structure.

11.1   Chemistry

Care must be taken to ensure that gas handling apparatus and vacuum systems are consistent with the application. Much of contemporary etch technology uses gases that are potentially hazardous, so safety is a principal concern. Not surprisingly, there are no universal etching methods that satisfy all needs. Although general principles can be established, etching is an application-specific process. What is suitable for one device or facility may be unsuitable for another. Process acceptability will vary with device, device level, time, and management philosophy.

11.2   Frequency

The exciting source of an electrical discharge is said to be high or low frequency, depending on its relationship to the collision rate between the particles in the gas. In this section we outline the characteristics of the different frequency regimes insofar as they alter etching characteristics and equipment requirements: beams or direct current (dc) sources, alternating current (ac), or low-frequency (lf) sources between 10 kHz and 1 MHz, high or radio frequency (rf) sources at 13.56 MHz and 27 MHz, and microwave (μw) sources at 2.45 GHz. The radio frequency discharge especially is widely used in applications for plasma etching, because it is able to produce a large volume of stable plasma. Reinberg [2] described the different sources as follows.

- dc sources: *"dc discharges are limited to those cases where there are conducting electrodes, free of insulating films, in contact with the plasma. An externally supported dc bias can be applied to conducting material in conjunction with a high frequency discharge [3]."*

- lf sources: *"If voltages are sufficiently high, 'electrodeless' excitation down to very low frequency is possible. At sufficiently low frequency, ions are able to follow the changes in electric field so that discharges behave similarly to dc discharges with alternating cathodes and anodes. Selection of an operating frequency depends on many factors. An important consideration has been the availability of suitable generators. Commercial solid state devices are obtainable with power outputs up to several kiloWatts. Matching circuits are needed to adapt the impedance of the source to that of the plasma. At high frequency ordinary L-C circuits are most common. Ferrite core transformers are useful up to about 0.5 MHz. The frequency band between this and 10 MHz presents special problems."* If low frequency is used, care must be taken to prevent sputtering of electrode material at the higher voltages needed to maintain the discharge.

- rf sources: *"The common radio frequency designated for experimental devices is 13.56 MHz and its low harmonics. They are frequently used when an alternating field is needed for excitation without the use of conducting electrodes in contact with the discharge. Generators operating in this range readily couple into a gas chamber through a dielectric wall such as a glass or quartz tube without the need for excessively high applied voltage."*

- μw sources: *"Microwaves are convenient source of excitation for several applications. The electric fields associated with a microwave discharge system are very complex so that they are used principally as free radical generators, or as a source of ions in conjunction with an acceleration system. The usual frequency is 2.45 GHz, identical to that used in microwave ovens. Most microwave discharge systems are non-resonant, using a wave guide with the discharge tube passing through it, and an appropriate matching device such as a stub tuner. Discharge volumes are usually small and require adequate cooling if high powers are used. Special techniques are needed to get large volume discharges."*

11.3 Electrode arrangement

Konuma makes a classification of the electrode arrangement into capacitive coupled versus inductive coupled and into internal electrodes versus external electrodes [1]:

"Some electrode arrangements are depicted in Fig. 11.1. The internal electrode mode, as shown in Fig. 11.1 (left), can be applied in the whole range of frequencies. If the frequency exceeds about 300 kHz, the high-frequency power can be supplied from an external electrode to the load. This can be done through a discharge tube which is made of glass (common glass or borosilicate glass) because of its heat resisting property and low dielectric loss. The high-frequency discharge (by means of external electrodes) is also called the electrodeless discharge. This mode can make it possible to reduce the effect of electrode materials, such as metal impurities on the plasma process. The high-frequency discharges have been classified into two types according to the method of coupling the high-frequency power with the load. They are the capacitive coupling (Fig. 11.1 top) and the inductive coupling (Fig. 11.1 (bottom)). In Fig. 11.1 the chamber itself, working as one of the electrodes, is grounded only in the case of the internal electrode mode. It is possible either to ground one of the electrodes or to electrically float it for the generation of a discharge regardless of the discharge mode and type."

So, in the case of the internal electrodes, there is free exchange of charge between the plasma and a possible conducting surface. The capacitive coupling has a solid dielectric barrier between the electrode and the plasma and the inductive coupling uses a changing magnetic field to induce an alternating voltage that sustains the discharge. Often, coupling

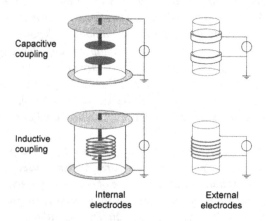

Fig. 11.1. Arrangement of electrodes in high-frequency discharges.

consists of a combination of these methods, even when it is not intentional. For example, sometimes barrel reactors are driven by winding a coil around a Pyrex tube but, except at very high-power density, coupling with such a coil is mainly capacitive between the winding and the plasma and the magnetic field of such an inductor has very little effect on the discharge.

11.4   Load capacity and technique

When plasma etching first found widespread use in the 1970s, barrel etchers were the most common type of batch reactors. In batch reactors, many wafers are simultaneously etched in the same chamber as opposed to one at a time or single-wafer reactors in which only one wafer is etched at a time. The capacity of batch systems varies considerably with reactor type and wafer size. Bulk, barrel-type reactors may hold several hundred wafers or equivalent material. Surface-loaded reactors range from a few wafers (= mini batch) to large parallel plates or hex-type systems that accommodate from 10 to 50 wafers, depending on wafer diameter. Load size is frequently limited by factors such as uniformity, rather than by the dimension of the chamber. Batch reactors generally have lower etch rates and work at lower pressures and lower power density than single-wafer reactors, therefore high throughput is achieved with large batch sizes. However, uniformity and wafer-to-wafer reproducibility are better when every wafer is exposed sequentially, as in case of a single-wafer etcher. Moreover, increases in wafer size and the demand for improved process control, such as individual end point detection, have made single-wafer etching reactors more desirable for many etching applications. But, in order to achieve adequate throughput, high etch rates are required. Single-wafer reactors must operate at 10 to 20 times the etch rate of batch reactors in order to have similar throughput. Since the arrival rate of reactive species at the wafer surface controls the etch rate, a high reactive species generation is a prerequisite (e.g., by way of a high density plasma or a high gas pressure). These harsher conditions usually lead to higher temperatures, less selectivity, and more interface damage from ion bombardment and sputtering.

Surface-loaded reactors place one side of the material being processed in direct contact with a surface, wall, electrode, or other part of the reactor. In bulk-loaded reactors, material is stacked so that only a small portion of each wafer is supported, with all surfaces in contact with the gas or vacuum space. The back (inactive) side of the material may be partially protected by loading wafers back to back. Surface loading usually results in better thermal contact providing better wafer cooling.

11.5   Plasma-sample distance

The plasma-sample distance classification of plasma equipment is rarely found in literature. Nevertheless, this subject is too important just to be neglected. Wafers are most frequently located directly in the plasma, separated from it only by the natural boundaries of the sheath, where they are bombarded by all of the energetic species of the discharge. These include positive and negative ions, electrons, photons from decaying excited atoms or radicals, photons created by electron bombardment of surfaces, short-lived radicals, and hot and metastable atoms. It is possible to protect the wafers from some of these particles by appropriate shields or by removing the wafers from the region where the plasma is created. So, while increasing the distance between the sample and the plasma, particles which are not wanted in the etch process can be filtered out from the plasma, and it is possible to operate at

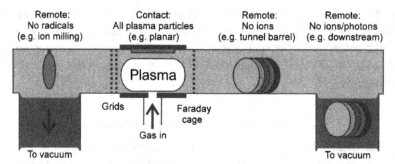

Fig. 11.2. Plasma etching at different distances between the plasma and the sample to filter out specific particles from reaching the surface of the sample: contact and remote plasma etching.

much lower pressures than when the sample would be mounted directly in the plasma glow region. Therefore, an important classification is to distinguish between a contact plasma, where the sample is in direct contact with the plasma, and a remote plasma, where the sample is located at a distance from the plasma, as found in Fig. 11.2.

11.5.1   Contact plasma etching

In contact plasma etching, samples are directly facing the plasma glow. Samples mounted in such reactors will be bombarded by all kinds of plasma particles. Members of this important group are the barrel reactors and parallel plate or planar reactors. This latter group can be split up into (1) *diode reactors* with horizontal target plates for single-wafer processing, (2) *triodes* with an extra electrode used for increased ion bombardment, and (3) *hexodes* designed for batch processing in which the vertically arranged cathode has the shape of a hexagon surrounded by the cylindrical chamber walls forming the anode.

Barrel reactor

In barrel etchers or volume-loaded reactors, wafers are simply stacked in a notched quartz support stand that is inserted into a reactor without any attempt to specify the directions of the exciting fields. External electrodes in either a capacitively or inductively coupled configuration sustain the plasma, which is usually fed by a 13.56-MHz rf source. Typical maximum power density is about 50 watts per liter of chamber volume. Most barrel systems do not provide for uniform passage of gases over the wafer surface and rely mostly on diffusion to obtain uniformity. Because the production rate of chemical species is limited, loading effects are common. Processing proceeds from the periphery inward with a constantly increasing rate with time. This is due to a combination of decreased load and wafer heating. Several attempts have been made to calculate optimum wafer spacing to minimize nonuniformity. So, barrel reactors are inexpensive, but uniformity is difficult to achieve and there is no temperature control: The etch rate can change irregularly due to reactor heating from the rf power source and heat produced by the etching. Therefore, some systems come equipped with special programmed preheat cycles. Nevertheless, barrel reactors are still common for noncritical processes such as stripping and cleaning, surface modifications such as changing adhesion and wettability, and large feature isotropic silicon etching due to their large batch capacity and low cost.

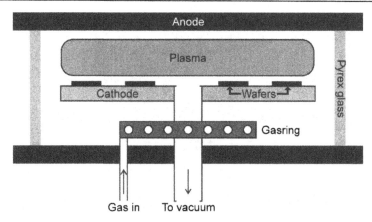

Fig. 11.3. The Reinberg reactor is a radial flow parallel plate system with wafers placed on the grounded lower electrode and the electrically isolated upper electrode rf driven from a matching network. Gas can flow from the outside-in (as shown) or inside-out. This configuration is sometimes referred to as plasma etching (PE) mode; by contrast, operation with the wafer-loaded electrode driven by rf has been called a reactive ion etch (RIE) mode (after Reinberg [2]).

Diode reactor

The planar "Reinberg" style reactor, originally introduced in the early 1970s for chemical vapor deposition, is shown in Fig. 11.3. In this reactor, the lower electrode supports the wafers and can be heated or cooled, while radial gas flow toward or away from the center helps uniformity. Diode reactors are available in both single-wafer and batch sizes, and many times it is possible for the user to switch manually between the PE and IBARE mode. Sometimes, one of the electrodes contains a perforated showerhead gas feed, which is used instead of the radial flow design. The external circuit usually has a blocking capacitor so that a negative dc self-bias voltage appears between the small electrode and the ground. At low pressure, these reactors develop still higher potentials and bias.

Hardware variations are usually determined by the equipment vendor, although minor modifications by the user can lead to significant improvements. Some hardware characteristics that may be candidates for modification are materials in contact with the plasma, electrode design and construction, gas distribution system like a showerhead (this may be part of the electrode design), gas metering system (primarily size change), plasma power supply, and endpoint monitoring system to satisfy special needs.

Triode reactor

Triodes, as their name implies, are three-element discharge systems. The purpose of the third element is to create two zones with distinct discharge properties. In one such arrangement, a grounded grid separates the two regions forming what are, in fact, two planar etch chambers. The wafer is placed in one chamber configured like an IBARE system. This chamber is downstream of, and shielded from, the other portion. There are two separate power supplies, which can operate at different power levels and need not be at the same frequency. By adjusting the power level of each part, the amount of chemical etching from radicals produced in the upper section can be varied with respect to the ion-enhanced and

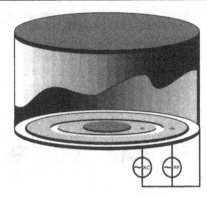

Fig. 11.4. The triode system has connections for two power supply units.

presumably anisotropic portion generated in the bottom chamber. For example, some reactors use 13.56-MHz rf power for the top chamber, which is higher than the ion plasma
frequency. Consequently, ions cannot follow the applied field and respond only to the time
average voltage between plasma and electrodes. The bottom chamber is supplied with a
source at a frequency of 100 kHz in which the ions can follow the field and will bombard
the cathode more heavily. In the triode design of Fig. 11.4, a 13.56-MHz power source (or
signal) is supplied to an annular electrode surrounding the chamber to produce reactants and
a 100-kHz power source is supplied to an additional electrode to maximize bombardment.

Hexode reactor

In the mid-1980s, the hexode or hex reactor was designed at Bell Laboratories. In this
design, wafer capacity is increased by folding the electrode surface into concentric cylin-
ders and expanding it vertically in the third dimension creating hexagonal right prisms.
Hexodes are diode systems having six-sided cathodes on which wafers are placed. They
have been particularly popular in both manual and automatic versions and are used in many
manufacturing operations. Figure 11.5 shows that the cathode is now surrounded by the
anode (bell jar) and that six wafers can be loaded onto each face on a vertical level. Wafer
loading is done in two steps. The first is placing wafers on special rectangular pallets using
a mechanical loader/unloader. The six pallets are then placed on the central core either by
hand or by a robot. A considerable amount of expertise has developed in using this etcher,
which is, however, not available in the open literature. A typical batch size for a hex system
is 24 four-inch wafers.

11.5.2   *Remote plasma etching*

A disadvantage of the contact plasma reactors is that the flux of all the impinging species
at the sample surface is difficult to control. A way to improve this is by way of the remote
plasma reactors, such as the tunnel-barrel, downstream, and ion mill reactors.

Tunnel-barrel reactor

One notable variation on the barrel reactor is the incorporation of a perforated metal insert,
a tunnel or Faraday cage. This Faraday cage is used to shield ions from reaching the surface

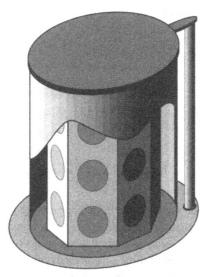

Fig. 11.5. The hexode reactor consists of connected electrode surfaces folded into a right hexagonal cylinder, with the larger outside vacuum vessel serving as the anode. Wafer capacity is large because the wafers are stacked in the vertical direction.

so only long-lived radicals and photons determine the etch rate. Chambers with tunnels or shields usually produce more uniform processing than those in which the material is actually in contact with the plasma. Lowering the pressure, however, permits diffusion of charged particles into the shielded zone. A disadvantage of this reactor type is that material might be sputtered from the cage on top of the samples.

Downstream reactor

In downstream, outside, or after glow etching, a sample is mounted far away from a remote plasma. Ions and even photons are not able to reach the surface of the sample, and only the longer-lived radicals are transported to the wafers by forced convection created by fast-flowing vacuum pumping. In this way, surface damage due to ion impact and radiation damage is minimized. However, the lifetime of radicals can be greatly reduced due to collisions of the radicals with the reactor walls between the plasma glow and sample. This could cause radicals to recombine, which lowers the etch rate. The absence of ions, electrons, or other directed species results in chemical reactions providing primarily isotropic etching. Microwave excitation is a popular choice for downstream reactors.

Ion mill reactor

To control the energy and flux of ionic species, in 1961 H. R. Kaufman invented the ion source. In first instance, the source was used as an ion thruster for space propulsion, but later on it was used for milling purposes too. It can generate a collimated ion beam from a wide choice of gas species. The low-pressure (<1 mTorr), line-of-sight nature of beam techniques prevents ion scattering and redeposition of sputtered material. This possibility of separately controlling ion energy and flux provides a flexibility of directional bombardment not available in other plasma processes. Because it is difficult to start a plasma at such

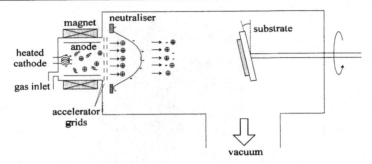

Fig. 11.6. Basic schematic for ion beam etching.

low pressure, it is created in a remote small, higher-pressure chamber (Fig. 11.6). In this so-called Kaufman-source chamber, a high density of electrons is created by way of a heated cathode wire. These electrons are forced to the anode by way of an external voltage source (ca. 40 V). Moreover, their lifetime is increased with the help of permanent magnets surrounding the chamber. The ions are extracted from this plasma with electrostatically controlled grids (up to 1000 V and more) and are directed to the substrate to be etched. A neutralizer is used to prevent charging of the (insulating) substrate, which could cause a nondesirable deflection of incoming ions.

11.6   Operating pressure

Classification of pressure into high, medium, and low is convenient to differentiate among the various types of pumping systems and to illustrate differences in the plasma and neutral background. The usual gas kinetic conditions apply. Lower pressure means decreasing (neutral) particle density, longer mean free path, fewer collisions, etc. Generally, at lower pressure the flux of ions toward the surface is much more directed and their kinetic energy is higher, which improves the anisotropy. But the concentration of radicals is lowered, which decreases the etch rate. We call pressures below about 20 Pa ($\sim$150 mTorr) low, since above this, backscattering of sputtered material becomes significant. High-pressure etching usually refers to considerable higher pressure, about 100 Pa and above. Most etching is performed at pressures less than 1000 Pa. The various pressure regimes, their characteristics calculated with the formulas from Chapter 10, and the methods employed to establish them are outlined in Table 11.1 [2].

11.7   Type of etching species

In Fig. 11.7 the results of the interaction between ionic and radical species with a surface facing a plasma are shown, i.e., (1) the *chemical* radical etching (RE), which uses predominantly radicals, (2) the *synergetic* ion beam assisted radical etching (IBARE), which uses both radicals and ions, and (3) the *physical* ion beam etching (IBE), which uses only ions. Generally, IBE shows only positively tapered profiles, low selectivity, and low etch rates, whereas RE gives rise to isotropic profiles, high etch rates, and high selectivity. IBARE enables the achievement of profile control due to the synergetic combination of physical sputtering with chemical activity of reactive species with high etch rates and high selectivity.

Table 11.1. *Pressure ranges used in plasma etching and important gas parameters.*

Pressure [Pa]	Density [m^{-3}]	Mean free path [mm]	Particle flux [1/m^{2}s]	Pumping
High: >100	>2 × 10^{22}	<0.01	>10^{24}	Mechanical Blower
Medium: 10–100	2 × 10^{21}–2×10^{22}	0.01–0.1	10^{23}–10^{24}	Blower Diffusion Turbo
Low: 1–10	2 × 10^{20}–2 × 10^{21}	0.1–1	10^{22}–10^{23}	Blower Diffusion Turbo
Very low: 0.1–1	2 × 10^{19}–2 × 10^{20}	1–10	10^{21}–10^{22}	Diffusion Turbo Cryogenic
Ultra low: <0.1	<2 × 10^{19}	>10	<10^{21}	Diffusion Turbo Cryogenic

11.7.1   Radical etching (RE) or plasma etching (PE)

Radical etching (RE) is characterized by purely chemical etching and a minimum of ion bombardment. Unfortunately, these chemical plasma systems are also called plasma etchers (PE). Of course, the IBARE and IBE systems are plasma etchers as well. Typical reactor types for RE include the barrel etchers (no ions) and downstream etchers (no ions and photons) in which, generally, the plasma is excited using microwaves (Fig. 11.2). Such systems are often used for photoresist stripping and other applications where high selectivity and low radiation damage are key requirements and the isotropic nature of the etch is not a problem or even an advantage (think about plasma releasing).

Sometimes symmetrical capacitive planar reactors are used. Because of the equal area of both electrodes, there is no bias created between the two electrodes. When the pressure in the reactor is high (say 1 Torr), then the plasma potential and bombardment of ions will be low. Generally, it is more convenient to use an asymmetrical system in which the rf powered electrode (where the wafers are situated) is large relative to the grounded surface area. This is the so-called PE mode found in Fig. 11.3. In these cases the plasma potential will be high with respect to the grounded electrode, but will almost equal the target electrode potential. Hence, no highly energetic ion bombardment will occur at the substrates. Unfortunately, the wafer can easily be contaminated by material sputtered off the grounded counterelectrode. Because the grounded surface is more negatively charged as the target electrode, this type of system is also referred to as an "anode-loaded PE system," as opposed to the cathode-loaded IBARE systems.

11.7.2   Ion beam assisted radical etching (IBARE) or reactive ion etching (RIE)

In applications where, besides radicals, ion bombardment is required, planar reactors can be employed. As for the chemical plasma etching, the name for this synergetical plasma

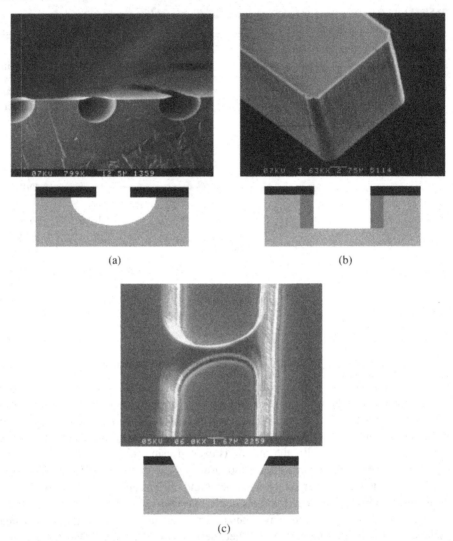

Fig. 11.7. Typical products of the three basic mechanisms of dry plasma etching.
(a) Chemical radical etching (RE); (b) synergetic ion beam assisted radical etching
(IBARE); (c) physical ion beam etching (IBE).

etching system is not always straightforward. Trade jargon often refers to low-pressure
asymmetric planar reactors as reactive ion etchers (RIE), distinguishing higher-pressure
operation (>100 mTorr) with the term plasma etchers (PE). Of course plasma etching is
taking place in both regimes and ions are almost never the primary etchants. We prefer
the name of ion beam assisted radical etching (IBARE) instead of reactive ion etching. In
IBARE it is difficult to independently control the temperature and fluxes of species. This
is one of the reasons for the lack of consistency in the reports concerning the fluor-based
IBARE of silicon. Nevertheless, IBARE is the most important plasma technique in use
today. Popular types of IBARE systems are the diode reactors. As in the case of the radical
etch equipment, planar reactors for IBARE processes can be either symmetric, where the
cathode area equals the anode area, or asymmetric, where these areas are different in size.

In symmetrical *low-pressure* systems, the plasma potential is high and both electrodes are bombarded by energetic ions. Due to the zero bias caused by the equal electrode areas, this type of etcher is often confused with real RE (i.e., no bombardment). More commonly, in asymmetrical systems, the wafer-loaded rf-powered electrode is small relative to the grounded surface area. These systems are characterized by a low plasma potential (10–50 V) and a large bias on the rf-driven electrode (10–1000 V), which increases ion energy and improves the directionality of the etch process. Because of the low plasma potential, relatively little sputter contamination from the grounded surface occurs. Moreover, because the target electrode is more negatively charged as the grounded surface, this type of system is also referred to as a "cathode-loaded IBARE system," as opposed the anode-loaded PE systems.

11.7.3 Ion beam etching

It is often hardly possible to etch a specific substrate material because there is no chemical etchant available. In these cases the patterning is fulfilled by sputtering the layer with a directed energetic ion flux. When the ion bombardment takes place at higher pressures in contact with the plasma, it is called sputter etching. In contradiction, at low pressures in remote plasmas, it is known as ion milling.

In sputter etching (SE), a sample is mounted on top of the cathode electrode of, e.g., a planar reactor. The sample is in contact with the plasma where it is bombarded by highly energetic ions from the plasma glow region which pass the plasma sheath region. In sputter processes, radicals and photons are not important enough to influence the overall etch rate, and the ions control the etch rate by way of their impact on the sample surface. However, due to the relatively high pressure, redeposition is often found because of gaseous particles, just sputtered from the surface, which are reflected back from the plasma due to collisons with plasma particles. Moreover, due to collisions of the ions with plasma particles while passing the sheath, i.e., ion scattering, the impact angle of the ions with respect to the surface is not fully normal anymore, resulting in an undercut.

The collisional problems of sputter etching are less pronounced in the electron bombardment ion source used in the ion millers. This latter technique is known as ion milling (IM) or ion beam etching (IBE). The inert (IBE) or reactive (RIBE) ions are accelerated toward the sample with the help of a high-voltage source. Because physical etching is a slow process, typically 1–30 nm/min, sometimes an extra reactive feed gas is led into the reactor. Such processes are known as chemical-assisted (reactive) ion beam etching (CAIBE, CARIBE). In RIBE, and more in particular CAIBE, it is possible to manipulate important parameters in plasma-assisted etching, e.g., temperature, and electron or photon impact. For this reason, basic studies of the surface science aspects of plasma etching were initiated using CAIBE in the beginnings of the 1980s in laboratories all over the world.

References

[1] M. Konuma, *Film Deposition by Plasma Techniques* (Springer-Verlag, Berlin Heidelberg, Germany, 1992).
[2] A. R. Reinberg, *Plasma Etch Equipment and Technology: Plasma Etching: An Introduction*, eds. D. M. Manos and D. L. Flamm (Academic Press, Inc., San Diego, 1989).
[3] R. Bruce and A. Reinberg, J. Electrochem. Soc. **129**, 393 (1982).

12

Contact plasma etching

This chapter is a brief review of ion beam assisted radical etching (IBARE) as applied to pattern transfer, primarily in silicon technology. It focuses on concepts and topics for etching materials of interest in micro system technology. Although many reviews on plasma systems have been published, generally they deal only with subjects concerning the integrated technology [1–12]. Nevertheless, it is convenient to use these reviews as a blueprint, and therefore the author has chosen the well-organized review *Reactive Ion Etching* by Oehrlein to treat many plasma concepts [7].

An important credit of IBARE is its ability to achieve etch directionality. The mechanism behind this directionality and various plasma chemistries to fulfill this task will be explained. Multistep plasma chemistries are found to be useful to successfully etch, release, and passivate micromechanical structures in one run. Plasma etching is extremely sensitive to many variables, making etch results inconsistent and irreproducible. Therefore, important plasma parameters, mask materials, and their influences will be treated. Moreover, IBARE has its own specific problems, and solutions will be formulated. The result of an IBARE process depends in a nonlinear way on a great number of parameters. Therefore, a careful data acquisition is necessary. Also, plasma monitoring is needed for the determination of the etch end point for a given process. This chapter is ended with some promising current trends in plasma etching.

12.1 Etch directionality in IBARE

Etch directionality is due to directed energy input into an etching reaction at a surface and can be accomplished by neutral, ion, electron, or photon bombardment of a surface exposed to a chemical etchant. In the case of IBARE, the achievement of etch directionality is due to energetic ion bombardment. An important clarifying experiment performed by Winters and Coburn was the exposure of a silicon surface to a well-defined dosage of chemical etchants, XeF_2, and energetic ions, argon [9]. They found that the etch rate obtained for a silicon surface exposed to both particle fluxes simultaneously is much greater than the sum of the etch rates for exposure to the ion beam and chemical etchant separately.

Ion-enhanced or ion-assisted etching can be divided in two main groups: ion-induced (reaction controlled etching) and ion-inhibitor (desorption controlled etching) IBARE.

12.1.1 *Ion-induced IBARE*

This technique is used when the surface of the substrate is not reacting spontaneously, such as in the chlorine-silicon or oxygen-polymer system. Ions do modify the surface reactions

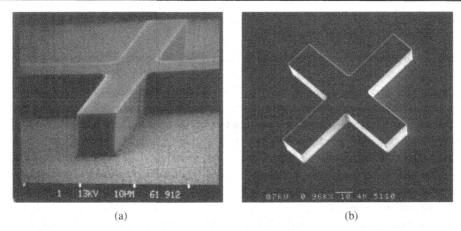

Fig. 12.1. (a) Ion-induced polymer etching with the help of an oxygen-based IBARE plasma system. (b) Ion-inhibitor silicon etching with a siliconoxyfluoride film at the beam sidewalls using a fluor-based IBARE plasma system.

in one or another way and make it possible for radicals to react with the substrate [7]. There are many models and, depending on the particular etchant-substrate under consideration, a different ion-induced mechanism may be dominant. For instance, it appears that the lattice damage model is important for the oxygen-polymer, fluorine-tungsten, and chlorine-silicon systems but not for the fluorine-silicon system.

Lattice damage produced by ion bombardment increases the reaction rate of etchant species with the substrate relative to undamaged material [4]. For example, when an organic compound like a polymer is exposed to reactive oxygen, the radicals may react spontaneously with the backbones of the long-stretched nonvolatile molecules. However, it is difficult for the oxygen to break such molecules into smaller volatile pieces and thus the etch rate is low. Therefore, an extra energetic ion beam may help to crack these big molecules and will increase the etch rate, because now the oxygen will remove the smaller parts. In Fig. 12.1(a), a typical result of this technique is shown: the directional etching of a polymer with the help of an oxygen IBARE sytem. Note that, when the temperature is increased above the so-called glass temperature of the polymer, the reaction of the polymer with the oxygen may become spontaneous, i.e., no ion bombardment is needed anymore, and the profile will become isotropic. This is not an IBARE process anymore, but a thermally assisted radical etching (TARE) process.

12.1.2   Ion-inhibitor IBARE

In this technique the substrate is etched spontaneously and therefore an inhibiting layer is needed to achieve directionality. Sidewalls of trenches are not exposed to ion bombardment and will be covered by the film. However, the bottom of the trench is exposed to ion bombardment and thus free from this deposit and etching can proceed. The passivating layer can be grown by [15]

(a)   inserting gases which act as a silicon oxidant forming siliconoxyhalogens, as shown in Fig. 12.1b,

(b)   inserting gases which act as polymer precursors forming carbonhalogens [14],

(c)   eroding and redepositioning mask material such as metalhalogens, or

(d)   freezing the normally volatile reaction products of silicon with radicals at the trench
       walls, such as siliconhalogens [18].

The deposition of a polymeric carbonhalogen film has the disadvantage that this film
is thermally less stable than an inorganic siliconoxyhalogen film. The freezing of reaction
products uses expensive cryogenic coolers, and the redeposition of mask material is not
acceptable because areas which should stay clean are also contaminated.

Because the passivating film is very thin, the incoming ions should not be highly energetic,
so the mask to silicon selectivity will be very high and the substrate damage will be low.
Also, because of the low energy of the ions, trenching and faceting (typical IBE problems)
are not found, and it is very easy to change the direction of the impinging ions, thus changing
the etched profile [15]. It is exactly these characteristics which make the ion-inhibitor such
a popular plasma technique.

With respect to the "freezing technique," McFeely et al. [53] have shed some light on the
mechanism of ion-enhanced etching for the XeF_2/Ar-silicon system: It was observed that a
silicon-fluorine reaction layer much thicker than a monolayer was formed. Volatile SiF_4 was
seen trapped in this nonvolatile reaction layer. A preponderance of SiF_3 was observed which
suggested that the reaction to form SiF_4 from SiF_3 was the rate-limiting step of the etching
reaction. This inhibiting layer is blocking the XeF_x somehow, which decreases the etch
rate. It was observed that when the reaction layer was bombarded with argon ions the argon
ion beam tended to drive a disproportionate reaction in which involatile SiF_3 molecules on
the silicon surface were converted into SiF_2 and SiF_4 etch product. Nevertheless there are
more ways to explain the increased desorption rate. For example, the SiF_x species are less
tightly bound than silicon and have a greater sputtering yield. Therefore, an extra energetic
ion beam may help to remove this inhibiting layer. Alternatively, we could say that the ion
bombardment increases the surface temperature of the silicon with its SiF_x layer. The higher
temperature removes this layer directly without the SiF_4 reaction step being necessary.

12.2   Pure plasma chemistries

Because of their high etch rates, normally hydrogen- and halogen-based (i.e., F, Cl, and
Br) plasmas are used for the IBARE of Si and the etch products are volatile SiH_4, SiF_4,
$SiCl_4$, and $SiBr_4$ (Table 12.1). Whereas F-based plasmas are generally used for isotropic
etching, Cl- and Br-based plasmas such as Cl_2 are primarily used to achieve anisotropic
etch profiles. Except for F-based mixes, these gases are particularly hazardous (Br_2 or Cl
compounds) and special precautions are recommended.

12.2.1   Hydrogen based

There has been extensive surface science activity in the H-Si system, most of which has
been directed toward developing an understanding of the surface structure resulting from
an exposure of single-crystal silicon to H atoms. Because of the small size of the H atom,
ion-induced IBARE might be impossible (see Section 12.2.3).

12.2.2   Fluorine based

Mogab studied the etching of Si in a CF_4 plasma and found a linear relation between the etch
rate and F-atom density, showing that F atoms are directly involved in the etching process

Table 12.1. *Important gases for Si trench etching with their main plasma radicals, products, and inhibitor. *means: Only with cryogenic cooling.*

Gas	Radicals	Products	Inhibitor	Gas	Radicals	Products	Inhibitor
H_2	H	SiH_4	$Si_xH_y^*$	CHF_3	CF_2	HF, (SiF_4)	$Si_xC_yF_z$
CH_4	H, CH_3, CH_2	SiH_4, H_2	$Si_xC_yH_z$	CH_2F_2	CFH, C	HF	$Si_wC_xF_yH_z$
F_2	F	SiF_4	$Si_xF_y^*$	CH_3F	CH_2, CFH	HF, H_2	$Si_wC_xF_yH_z$
NF_3	F, NF_2	SiF_4	$Si_xN_yF_z^*$	CF_4/O_2	CF_3, F, O	SiF_4, F_2, OF, O_2F, COF_2	$Si_xO_yF_z$
SiF_4	F, SiF_3	SiF_4	$Si_xF_y^*$	CF_4/H_2	CF_3, F, H	SiF_4, HF, CHF_3	$Si_xC_yF_z$
CF_4	F, CF_3	SiF_4	$Si_xC_yF_z$	SF_6/O_2	SF_5, F, O	SiF_4, SOF_4	$Si_xO_yF_z$
SF_6	F, SF_5	SiF_4	$Si_xS_yF_z^*$	SF_6/H_2	SF_5, F, H	SiF_4, HF	$Si_xS_yF_z^*$
S_2F_2	F, S_2F	SiF_4	$Si_xS_yF_z^*$	SF_6/N_2	SF_5, F, N_2	SiF_4	$Si_xS_yF_z^*$
Cl_2	Cl	$SiCl_4$	Cl	SF_6/CHF_3	SF_5, F, CF_2	SiF_4, HF	$Si_xC_yF_z$
Br_2	Br	$SiBr_4$	Br	$CBrF_3$	forbidden	greenhouse	ozone
CBr_4	Br, CBr_3	$SiBr_4$	Br, $Si_xC_yBr_z$	CCl_4	forbidden	greenhouse	ozone

[17]. Etching of Si in F-based plasmas, e.g., SF_6, CF_4, SiF_4, NF_3, XeF_2, or F_2, normally results in a large undercut of the mask. However, Tachi and his coworkers showed that the horizontal Si etch rate using an SF_6 plasma can be reduced dramatically by cooling the substrate to $-120°C$ and near-ideal etch profiles can be obtained [18]. If anisotropy is due to an inhibiting Si_xF_y layer, the same effect should occur for the other gases. Ion-assisted etching – dominating at the bottom of the trench – was not affected by the low substrate temperature, although the vertical Si etch rate increased as the temperature was decreased, possibly due to condensation of etchant.

12.2.3   Chlorine based

Etch directionality for Cl-based plasmas (Cl_2 or $SiCl_4$) may be explained by the observation that Si and SiO_2 are not etched spontaneously at room temperature by Cl atoms, making only ion-induced etching possible [19]. Cl atoms chemisorb on Si and form an ordered Cl monolayer. In contrast to the Si_xF_y layer in F-based etching, this layer is barely influenced by temperature and will "inhibit" etching almost completely.

   Cluster calculations have shown that Cl atoms on an Si surface have to overcome a Van der Waals energy barrier of ~ 10 eV to attack the backbones of Si surface atoms to form $SiCl_4$ [20]. No energy barrier for F atoms to the penetration of the Si surface was found, indicating that subsurface SiF_x species will form spontaneously. The size of the halogen relative to the Si-Si interatomic distance plays an important role.

12.2.4   Bromine based

Recently, the use of Br chemistry has received considerable interest in IBARE process development because of the low spontaneous etch rate of Si and SiO_2 with Br atoms (Br monolayer) [21]. Unfortunately, very little surface science activity has been reported for this system.

12.2.5   Oxygen based

Oxygen plasmas are mainly used for polymer etching and are important in Si trench etching because they are able to remove polymeric residues afterward. At elevated temperatures, spontaneous etching (ashing) occurs. However, the IBARE etching of polymers below the glass temperature is ion-induced, thus the film is not removed at the sidewalls of a trench. This principle is used to release movable structures in MEMS applications (see Section 12.4).

12.3   Mixed plasma chemistries

Mixed molecules (e.g., CCl_2F_2) and gases (e.g., SF_6/Cl_2) containing halogens are often used for anisotropic etching (Table 12.1). If the plasma chemistry is chosen such that etch-inhibiting films can form at the sidewalls of a trench, directional etching is possible. By changing the relative atom density, e.g., the F/Cl ratio in the feed gas, it is possible to vary the trench profile.

12.3.1   Mixed molecules

The strongest halogen-carbon bond is the F—C bond. Thus, electron impact dissociation of mixed halocarbons produces primarily Cl or Br atoms, and the etch characteristics using these plasmas are Cl- or Br-like. For example, Matsuo used a 30-mTorr $CBrF_3$ plasma to etch Si directionally, whereas a CF_4 plasma using similar conditions resulted in nearly isotropic etching [22]. Directionality is due to an inhibiting Br monolayer. The most widely used mixed molecule is CHF_3. It dissociates primarily into CF_2 and HF species. The CF_2 intermediates are responsible for polymer formation, C_xF_y, on surfaces (Fig. 12.3).

On March 22, 1985, in Vienna and September 16 in Montreal, there were international commitments to reduce the use of chloro- and bromofluorocarbons in order to protect the ozone layer and to prevent the greenhouse effect. On the "forbidden list" are $CBr_xCl_yF_z$, $CCl_xF_{(4-x)}$, and $C_2Cl_xF_{(8-x)}$ for $(x > 0)$.

12.3.2   Mixed gases

A large variety of F-, Cl-, Br-, and O-based plasmas with profusion of gas additives are used in micromachining. In most cases a specific gas mixture or "recipe" is based on a great deal of empirical evidence obtained for a particular application rather than a real basic understanding of the relevant plasma chemistry. However, certain insights have proven to be helpful in formulating gas mixtures, as is shown below.

Effect of oxygen addition

The addition of small amounts of O_2 (<5%) to a CF_4 plasma is known to increase dramatically the F atom density, thus etch rate, in the discharge [23]. The CF_4/O_2 dissociates primarily in CF_3, F, O, and, in minority, CF_2 radicals. The increase in F atoms is due to reactions of O_2 with CF_x radicals forming CO_2 and COF_2, and CF_4 with O atoms forming COF_2. This effect reduces also the recombination of F atoms with CF_3, again increasing the F-atom density, although this effect is probably less important because the chance for two radicals to collide is rather small (note that in a plasma there is no thermodynamic equilibrium; many reactions are forced by electron impact). The consumption of unsaturated CF_x

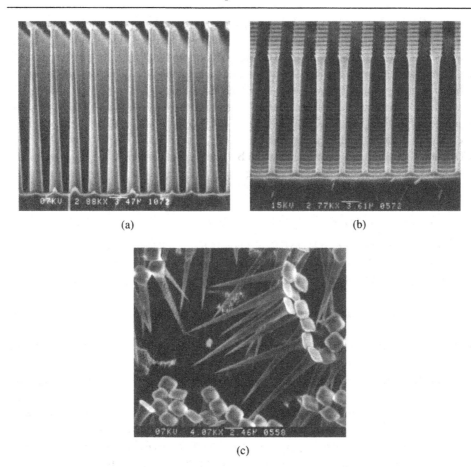

(a)   (b)

(c)

Fig. 12.2. (a) The effect of 30% oxygen addition in an SF_6/O_2 plasma. Due to the strong passivation caused by the oxygen content, the profile is IBE-like, making positively tapered spikes. (b) The effect of 25% oxygen addition in an SF_6/O_2 plasma. Due to the decreasing oxygen content with respect to (a), the profiles are less positively tapered, creating almost vertical pillars. (c) The effect of 20% oxygen addition in an SF_6/O_2 plasma. The passivation is even lower as in case of (b). The result is a negatively tapered profile which makes nails of the pillars.

species by oxygen has the additional effect of suppressing polymer formation. At higher O_2 content ($>7.5\%$) a passivating inorganic $Si_xO_yF_z$ film is formed on the Si surface, and the etch rate is controlled primarily by the thickness of this layer rather than the F-atom density, which increases up to ca. 15% O_2 addition. Too much O_2 will decrease the F-atom density due to dilution. In summary, the Si surface changes from primarily Si-C to Si-F to Si-O bonding as the O_2 percentage is increased.

The effect of the addition of O_2 to an SF_6 plasma is practically the same as its addition to a CF_4 plasma. The mix primarily dissociates in SF_5 (little SF_4), F, and O radicals, and the higher F-atom density is due to the reaction of O_2 with SF_x forming SO_2 and SOF_4, and SF_6 with O forming SOF_4. The vertical etch rate first increases because of the higher F-atom density and subsequently decreases due to a growing $Si_xO_yF_z$ film and F-atom dilution. A big difference is that there is no formation of a passivating S_xF_y film due to its high volatility at room temperature in contrast to the C_xF_y film in the CF_4 chemistry. In Fig. 12.2 SEM images of Si trenches using various percentages of O_2 are displayed. The horizontal etching depends on the thickness of the passivating $Si_xO_yF_z$ layer and the F-atom density trying to etch the Si by penetrating this layer. The thickness of the $Si_xO_yF_z$ layer is a function of, e.g., the O-atom density, the ion impact, and the local temperature. The F-atom density is a function of, e.g., the SF_6 flow, power, and (micro)loading.

Adding O_2 to mixed halocarbon gases will increase the halogen atom density also. For example, a $CBrF_3/O_2$ mixture will produce CF_3 (CF_2), Br, and O radicals. Subsequently, the reaction of $CBrF_3$ with O atoms will form COF_2. Additionally, O_2 reacts with CF_x into COF_2 and CO_2. Both processes increase the Br- and F-atom density. The extra F atoms are responsible for an increase in lateral etching. So, it will change the trench profile and may prevent the forming of grass (see Section 12.7.2).

Effect of hydrogen addition

The effect of small amounts of H_2 to CF_4 plasmas (CF_3, F, and H) is twofold. (1) H_2 reduces the F-atom density because of relatively inert HF formation, and the Si etch rate is consequently reduced. (2) More important, H_2 reacts with CF_3 forming polymeric precursors, such as CF. As a result, a C_xF_y film will form on surfaces where ion bombardment fails, such as the trench sidewalls. If the H_2 concentration is high ($> 30\%$), polymerization occurs on all surfaces and etching stops. Figure 12.3 shows a typical example of a deposited C_xF_y film covering an Al beam. At even higher H_2 content, the plasma becomes H based, and again etching is observed.

As a useful indicator of the predominance of etching (F) over deposition (CF_x), the F/C ratio of the discharge is often used [24]. The F/C ratio is 4 for CF_4, 3 for C_2F_6, etc. The ratio is lowered when extra Si, which consumes F atoms, is added or when CF_4 is mixed with H_2 or CH_4. The turnover from deposition to etching is stimulated by ion bombardment. For a Si substrate, deposition is observed for small F/C ratios (<2). However, it is known that H-based plasmas (e.g., H_2) are etching Si and, identically, C_xH_y film formation might occur. The model is not accounting for this and should be modified.

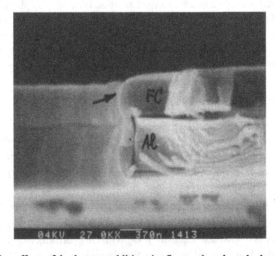

Fig. 12.3. The effect of hydrogen addition in fluorcarbon-based plasmas. A CHF_3 plasma is depositing a Teflon-like film which covers an aluminium beam. The chemistry is practically identical to a CF_4/H_2 gas mixture.

When H_2 is added into SF_6, no $S_xF_yH_z$ film formation occurs at room temperature, so directionality is not possible this way. As for CF_4 plasmas, SF_6 and H_2 react with radicals into "inert" HF, thus decreasing the etch rate.

The effect of H_2 addition on a mixed halocarbon plasma is similar to effects observed with CF_4/H_2; e.g., in a $CClF_3/H_2$ plasma the deposition of a $C_xCl_yF_z$ film is observed [25]. At the same time, the Cl atom density decreases due to the stable HCl molecule, thus decreasing the etch rate.

Effect of nitrogen addition

The addition of N_2 gas into a CF_4 or SF_6 plasma is another important mixture in etching Si because it increases the F-atom density. In contrast to most other gases, nitrogen doesn't dissociate on excitation [26]. Instead it is found in bound excited electronic states and not as an atom or an ion. These excited molecules are more effective in splitting SF_6 into SF_5 and F radicals than the light electrons (in general, the electronic and thermal dissociation of species is not the same [10]). Again, at higher N_2 content the etching will decrease due to dilution. SF_6/N_2 differs from SF_6/O_2 etching because relatively more SF_5 ions can respond to the bias and sputtering might increase. Adding N_2 into a CF_4 plasma has the additional effect that polymer forming is decreased because of volatile CN species. In the same way F or O atoms react with bare Si into $Si_xO_yF_z$, F or N_2 radicals might turn Si into $Si_xN_yF_z$. This film is weakly passivating and thus is never used in ion-inhibitor processes. Instead, these F-rich plasmas are effective in fast isotropic etching of Si.

Effect of CHF₃ addition

The addition of CHF_3 gas does not play a special role in CF_4 mixes because the CF_4/CHF_3 system is almost identical to CF_4/H_2 mixes (Fig. 12.3). $CF_4 + H_2$ has a higher internal energy than $CHF_3 + HF$, and when CF_4 is mixed with H_2, CHF_3 and HF might be created spontaneously. In contrast, SF_6/CHF_3 and SF_6/H_2 plasmas are different. Unlike H_2 addition, when adding CHF_3, profile control is possible because CF_2 radicals, a product of the CHF_3 plasma, will form a blocking C_xF_y layer on the Si surface [14, 27]. The film decomposes at much lower temperatures than, e.g., an inorganic $Si_xO_yF_z$ film. Thus, for a vertical sidewall, ion bombardment, exothermic reactions, etc. should be sufficiently low to ensure the growing of a polymer film. The SF_6/CHF_3 mixture differs from the SF_6/O_2 and SF_6/N_2 mixture because the F-atom density is barely increased, resulting in a lower etch rate.

Other gas additives

Noble gases such as argon and helium are often added to stabilize plasmas or for cooling purposes (He in high-pressure plasmas). Ar addition can also cause inert ion bombardment of a surface and results in enhanced anisotropic etching (e.g., Cl_2/Ar IBARE of Si [16]). The consequences of diluting a reactive gas with a noble gas are not easily understood. The addition of a chemically inert gas may significantly change the electron energy distribution in a plasma and may alter the reactive species population in the discharge. This effect is observed when the ionization potential of the chemically inert additive is very different from the ionization potentials of the plasma species of the primary gas. An altered reactive species makeup of the discharge, e.g., enhanced dissociation, can also be due to more complex effects. Schwartz and Gottscho examined the mixing of BCl_3 with He, Ar, and

Kr and found that energy transfer from noble gas metastable states to BCl_3 states causes enhanced dissociation of BCl_3 [28].

Special mixes

In an $SF_6/O_2/CHF_3$ plasma, each gas has a known specific function and influence, so the etched profile is easily controlled just by changing the flow rate of one of these gases [15, 34]. In such a plasma, SF_6 produces the F radicals for the chemical etching of the Si forming volatile SiF_4, oxygen creates the O radicals to passivate the Si surface with $Si_xO_yF_z$, and CHF_3 (or SF_6) is the source of CF_x^+(SF_x^+) ions, responsible for the removal of the $Si_xO_yF_z$ layer at the trench bottom forming the volatile CO_xF_y (or SO_xF_y).

12.4   Multistep plasma chemistries

After the successful (anisotropic) etching of micromechanical structures, they often have to be released. In bulk micromachining some very useful dry release techniques have been proposed, such as the SIMPLE and SCREAM processes [30, 31]. The BSM one-run multistep process is a more sophisticated dry release technique [29]. The technique starts with commercially available silicon-on-insulator (SOI) wafers. After the deposition of a (lift-off) mask for the pattern definition, the movable structures can be fabricated in only one IBARE run with four individual steps. These are (Fig. 12.4)

(1)   the (an)isotropic IBARE ($SF_6/O_2/CHF_3$) of the top silicon,
(2)   the IBARE (CHF_3) of the insulator together with the passivation (C_xF_y film) of the sidewalls of the structures,
(3)   the IBARE ($SF_6/O_2/CHF_3$) of the floor, and
(4)   the isotropic IBARE (SF_6) of the bulk silicon.

Eventually, the process can be finished with a conformal step coverage of a C_xF_y film to protect the released structures from the environment [14]. For instance, these fluorocarbon films have an extremely low surface tension and therefore they repel water and other liquids. With this technique it is possible to release very long thin Si beams successfully.

12.5   Plasma parameters/influences

Probably the biggest disadvantage of plasma etching is its extreme sensitivity to many variables. Partly, these parameters are well-known, such as pressure, power, and flow. However, more often, influences such as target/reactor materials and cleanliness are unintentionally disregarded. Surprisingly, number one on this list, temperature, is almost never accounted for enough. Therefore, many times etch results are, not surprisingly, inconsistent and irreproducible.

12.5.1   Doping

In contrast to undoped Si, highly doped Si etches spontaneously in a Cl_2 discharge. N-type Si (e.g., P- or As-doped) etches faster than intrinsic Si, which etches faster than p-type Si (e.g., B- or Ga-doped), and this effect is not chemical in nature since it is absent if the dopants are not electrically activated [32]. Thus, the etch rate depends on the electronic

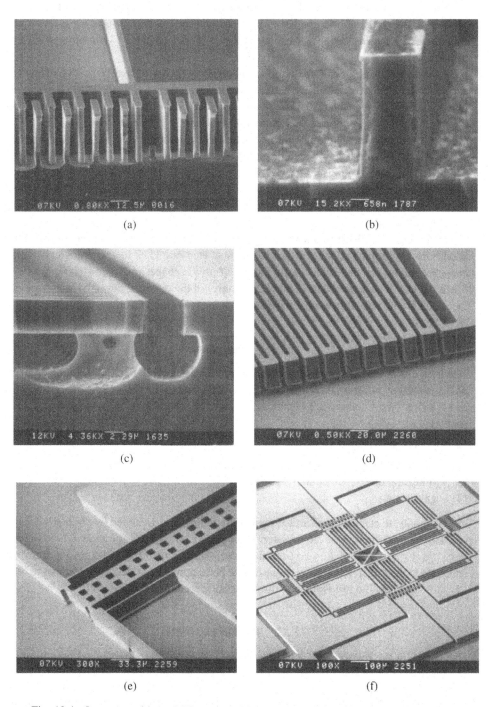

Fig. 12.4. One-run multistep RIE process [29]. (a) After anisotropic etching the top Si of an SOI wafer; (b) after etching the insulator and sidewall passivation; (c) during isotropic etching of the base Si; (d) after isotropic etching the base Si; (e) and (f) typical finished MEMS products.

properties of the substrate, and this has been explained by band-bending effects at the Si surface [33]: Coulomb attraction between uncompensated donors (n-type), e.g., As^+, and chemisorbed halogens, e.g., Cl^-, enhances the etch rate, whereas Coulomb repulsion in p-type Si inhibits the etch rate.

Schwartz and Schaible observed horizontal etching of a buried highly As-doped layer in a low-pressure (10 mTorr) Cl_2 discharge [16]. However, for a CCl_4 discharge (C_xCl_y film) or intrinsic Si (Cl monolayer), the etching was perfectly directional. Mogab and Levinstein observed that etching of doped poly-Si in a 300-mTorr Cl_2 plasma resulted in an isotropic profile [32]. Directional etching could be achieved by adding C_2F_6 to Cl_2, which formed a sidewall passivation layer. A detailed study of the doping effect for Si has recently been completed by Winters and Haarer [35].

The etching of poly-Si by Br atoms has been measured in an afterglow experiment [36]. No etching at room temperature was observed, even for n^+ poly-Si, and a very large doping effect was seen.

The doping effect decreases with ion bombardment and is difficult to observe for IBARE conditions as a doping dependence of the vertical etch rate [37]. Its technological significance lies in the fact that it makes the control of profile shapes in trench etching possible. Since the lateral etch rates (chemical etching only) of the different doped Si layers are not the same, dry release of free-standing structures for MEMS applications is possible [31].

12.5.2   Temperature

Needless to say, temperature is the most important parameter in IBARE etching. Together with the entropy and enthalpy, it rules every energy step in the reactor, such as adsorption and reaction. Many sources are known to increase the temperature at the substrate surface, such as

(1)   ion bombardment,
(2)   exothermic reactions at the substrate surface,
(3)   rf heating due to eddy currents, and
(4)   gas heating.

Commonly, to stabilize the surface temperature, the target is cooled by circulating water (or other liquids) through the target platen. Of course, the wafer has to be clamped sufficiently (e.g., mechanical, electrical, or vacuum grease) to maximize the heat transfer from the substrate to the target. Alternatively, gases like helium may be added to the plasma to cool the substrate from the frontside, or helium backside cooling can be utilized.

12.5.3   Reactor materials

The choice of the reactor and target materials is of critical importance and may result in (un)desirable etch characteristics, such as

(1)   depletion of reactant, e.g., graphite (an Si or quartz target consumes F atoms, graphite or Teflon consumes O atoms, and Al consumes Cl atoms);
(2)   generation of active species, both *indirectly* (e.g., Al may increase the F-atom density due to catalytic reactions) and *directly* (e.g., Teflon produces F atoms, $CF_2 + O \rightarrow 2F + CO$, and quartz produces O atoms which may prevent polymer building);

(3)   generation of polymer precursors, e.g., graphite or Teflon produces C_xF_y;
(4)   micromasking due to redeposition resulting in surface roughening during processing, e.g., SF_x^+ ions may sputter Al, forming involatile AlF_3 particles; and
(5)   secondary electron emission coefficient of electrode surfaces influencing the nature of the discharge.

12.5.4   Reactor cleanliness

The addition of small amounts of contaminants to a plasma may alter an etch result significantly. For example:

(1)   The etching of native oxide is made irreproducible if small amounts of water are present in the chamber; the water will react with oxygen scavengers or oxidize the substrate, e.g., $Al \rightarrow Al_2O_3$ in Cl-based and $Si \rightarrow SiO_2$ in F-based etching. Since the presence of water is primarily due to exposure of the chamber to room environment ambient, load locks eliminate this problem.
(2)   Small concentrations of N_2 or O_2 gas due to leakage can noticeably change the plasma chemistry. It can be caused when particles are trapped in the rubber seal during closure of the reactor or because seals are etched by long-living reactive atoms like fluorine. A solution for this problem is to check the base pressure periodically.
(3)   The etching process itself can lead to chamber contamination, e.g., in CF_4 etching a C_xF_y film is deposited at the reactor walls which may lose adhesion and cause particles. The film is recycled or will alter the F or O atom concentration in a next run. The best procedure appears to be to clean, e.g., with an O_2 plasma, and "condition" the chamber by running the process until equilibrium is reached.

12.5.5   Loading

Loading occurs whenever the reactant density is depleted due to an excessive substrate load. As a result, the etch rate will decrease inversely proportionally to the Si area which is exposed to the plasma glow [38]. Moreover, the etch rate/volume depends on the Si shape; a long, small structure etches faster than a square. The importance of this effect depends strongly on the radical lifetime. At the same time, the etched profile in, e.g., SF_6/O_2 etching will change while increasing the loading because the F/O density ratio and ion impact (or dc bias) is decreasing. The bias decreases because of the increase in reaction products which increases the plasma impedance. At higher loading there will be less underetching due to the smaller F-atom density.

Depending on the gas phase mean free path and the number and structure of specimens being etched, the loading effect may be both global, i.e., the reactant concentration in the reactor is uniformly lowered, and local, i.e., microloading. Thus, microloading is formally equivalent to loading, and it describes the etching rate dependency on pattern density. Structures in the locality of big Si areas are etched at a slower speed than those situated in nonetching areas. In Fig. 12.5 an example of microloading is given. In this picture, four poly-Si combfingers on top of a SiO_2 layer are etched. Because of RIE lag (see Section 12.7.3), the open area at the right is etched faster than the areas between the fingers where there is still some Si left. At the moment that the SiO_2 surface is reached for the open area, the local F-atom density increases, resulting in an enhanced chemical underetching.

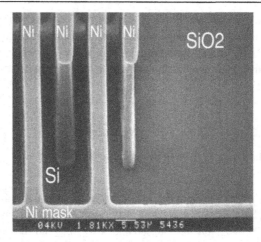

Fig. 12.5. The influence of microloading. Polysilicon fingers (2 × 2) on top of an insulating layer (SiO₂) are etched. Because of RIE lag (Section 12.7.3), the open area at the
right is etched faster than the areas between the fingers, where there is still some silicon
left. At the moment that the insulating surface is reached for the open area, the local fluor
atom density increases. This increase results in an enhanced chemical underetching, of
the most-right beam.

The importance of the (global) loading effect is decreased by consuming etchant species
through processes other than reaction with the wafer load, e.g., rapid pumping or an Si
target. Another possibility is making the synergy such that ions – and not radicals – control
the etching rate. In other words, ion-induced etching (e.g., Cl₂-Si) is less sensitive for
loading than ion-inhibitor etching (e.g., SF₆-Si). Additionally, the shorter lifetime of Cl
atoms with respect to F-atoms will decrease the loading effect in Cl-based etching. The
effect of microloading is minimized by changing the original mask pattern density into a
more uniform pattern density.

12.6 Mask materials/influences

In order to copy a pattern into an Si substrate, a mask is needed. It is therefore important to
know the influence of this mask on the etch result, such as the etchability.

12.6.1 Etchability

In general, an etching mask will influence the Si trench profile, because the mask is re-
tarding when its profile is not fully vertical. Almost infinite selectivity is possible in using metal(oxide) masks (except, e.g., Ti, Mo, W, Nb, and Ta for F-based plasmas and
Al or Cr for Cl-based plasmas) as long as the ion bombardment is sufficiently low (e.g.,
the C_xF_y deposit at the sidewalls of Si trenches in a CHF₃ plasma is not attacked by
F radicals). Important parameters to consider are the sputtering threshold (generally between 10 and 40 eV) and yield (0.01–10 atoms/ion, Fig. 12.6) [39]. Sometimes a layer of
10 nm is enough to etch through an Si wafer due to the low volatility of the metal(oxy)-
fluorides.

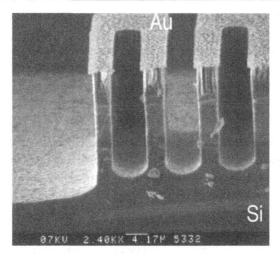

Fig. 12.6. Mask erosion for a gold mask layer at 40 eV. The silicon sample is etched in an SF_6/O_2 chemistry.

12.6.2 Film formation

Sometimes, redeposition of mask material will prevent spontaneous etching. This inhibitor may be a metal fluoride (e.g., AlF_3 from an Al mask) or a polymer (e.g., $C_xH_yF_z$ from a resist mask).

12.6.3 Catalytic reactions

Fedynyshyn and coworkers found an increase in the Si etch rate when using specific mask materials (e.g., Al or Ag) in F plasmas [40]. They proposed that catalytic reactions at the mask surface are responsible for the higher etch rate. For instance, Al would dissociate SF_6, forming more F-atoms and leaving the Al unaffected. However, at this moment we believe that an increase in substrate temperature, and not these catalytic reactions, is more likely to be responsible for increasing the etch rate.

12.6.4 Selectivity

Etch selectivity, i.e., the etch rate ratio between two materials, is required when a film is being etched with respect to an etch mask or stop layer. This is due to the following:

(1) Selective formation of an etch-inhibiting layer, e.g., C_xF_y on top of Si in a CF_4-based plasma. This film will not grow on SiO_2 because its oxygen produces volatile CO_x and COF_2.

(2) Nonreactivity of one of the materials, e.g., Si etching with an AlF_3 mask in an F plasma. The AlF_3 layer is not attacked by F-atoms, but Si will volatilize by forming SiF_4 species.

(3) Nonvolatility of reaction product, e.g., Si etching with an Al mask in an F plasma. Now the Al reacts into involatile AlF_3.

(4) Selective formation of an electrostatic screen at the mask surface, e.g., Al_2O_3 on top of Si. When an insulating layer is bombarded with impinging ions, the layer will charge up and, subsequently, this charge will repel new incoming ions, thus decreasing

synergetic etching. Si cannot be charged to a different potential, and therefore maximal ion bombardment will proceed.

(5) Loading, e.g., in F-based etching the Si etch rate decreases with loading whereas the SiO_2 etching is barely loading dependent. In other words, the selectivity is increased while decreasing the Si loading.

12.6.5 Materials

The choice of a mask material will depend on a lot of arguments, such as availability, IC compatibility, easiness, familiarbility, etc. In microtechnology, almost any element from the periodic sytem together with their combination can be used (e.g., C. Ti, SiC, Al_2O_3, ...). Therefore, in this section a variety of materials, and how they behave in silicon etching with fluor chemistry, are summarized.

- *Photoresist* (PR) is the most straightforward mask material. Unfortunately, the PR/Si selectivity is never very high and is difficult to control, especially when there are O and F-atoms present in the plasma. F-atoms react with polymer into HF and leave a reactive polymer surface behind, which may react with oxygen gas. During etching PR and Si, a lot of heat is produced and therefore the temperature rises and the selectivity decreases (this has its origin in the low glass temperature of PR, making the etching thermosynergetic). When the temperature is low enough ($20°C$), the only etch mechanism is due to impinging ions; a typical ion-induced (synergetic) etch mode. The etching of PR can be suppressed when, e.g., CF_x monomers from a CHF_3 additive are allowed to adsorb at the PR mask. In short, the etching of PR can be suppressed by
  (1) cooling the substrate by clamping, cryogenically, or by the addition of a cooling gas (e.g., He) in the gas mixture,
  (2) lowering the bias voltage by changing the reactor geometry (shower head), adding an insulating target, or adding an extra dummy Si in the chamber, or
  (3) using additives such as CF_x which don't etch but only compete with the other radicals.
- *Silicon oxide* is etched synergetically and therefore the selectivity is limited. The etching is ion-induced and the selectivity to Si can be increased by suppressing the bias. When hydrogen is added to a CF_4 plasma, a decrease in SiO_2/Si etch selectivity is found as a result of the growing of a C_xF_y film on the Si surface. The addition of too much oxygen will form an $Si_xO_yF_z$ layer and thus decrease the selectivity. The highest SiO_2/Si selectivity is reached when there is no passivating film grown at the Si surface.
- *Silicon nitride* is even less attractive as a mask than SiO_x, especially when it is not stoichiometric (i.e., Si_3N_4), mainly because it can be etched chemically. Thus nitride takes an intermediate position between Si and SiO_2. Nitride reacts faster with F-atoms than oxide because of the intrinsically weaker Si-N bond, although not as fast as Si. It also appears to react with CF_2 precursors into volatile CN species, as it is etched in CF_4/H_2 at roughly the same rates as SiO_2.
- *Aluminium* is IC compatible, but there is mask erosion (sputtering) visible after greater etch depth giving rise to roughening of Si adjacent to the Al-protected regions, even at relatively low ion energies ($<40\,eV$). The high erosion rate may be explained by the low sputtering threshold of Al ($13\,eV$) together with the existence of eddy currents in

Al. Nevertheless, the Al/Si etch selectivity is extremely high for low ion energy and can easily exceed 100,000.

- *Chromium and nickel* seem to be perfect mask materials. They are minimally sputtered at bias energies up to 200 eV.
- *Others: Platinum* – a noble metal – is strongly sputtered at low bias voltage and the soft noble *gold* is even worse (Fig. 12.6). Their use therefore is limited to extremely low bias voltages (<20 eV). *Yttrium* is easily oxidized into the strong insulator Y_2O_3 giving rise to mask undercut, thus it should not be used in plasmas containing oxide atoms. *Zinc oxide* is a semiconductor which is etched, although not heavily, so its use is limited to low voltages. *Palladium*, an expensive rare earth, seems to be okay. *Copper*, although noble, is easily oxidized but gives good results.

12.6.6   Conductivity

An insulating mask might be charged to the plasma potential [38], thus creating strong local electrical fields at the edge of a mask giving rise to ion bowing and therefore an undercut directly beneath the mask surface (Fig. 12.7). Additionally, there will be a lower mask erosion due to this charge.

12.6.7   Temperature

The local temperature of sidewalls should be constant when etching deep Si trenches, because surface reactions (adsorption or desorption) are temperature dependent. Eddy currents in the mask and ion bombardment increase the local temperature of sidewalls. This alters the adsorbility of oxygen, thus changing the thickness of the inhibitor, and this will change the profile. To demonstrate the influence of the mask (and temperature), we examined the etched profile for three different materials during the same IBARE run (Fig. 12.8). A Cu mask resulted in a nearly anisotropic profile, a Pt mask showed a negatively tapered profile, and an Al mask, a well-known material giving rise to eddy currents, resulted in trench-opening-depending profiles. By cooling the substrate with grease clamping, this difference almost vanishes. More experiments have to follow.

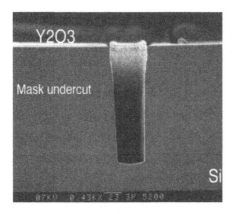

Fig. 12.7. Undercutting of an insulating Y_2O_3 mask in a silicon sample due to ion bowing. The insulator has been charged by incoming ions or electrons from an SF_6/O_2 plasma.

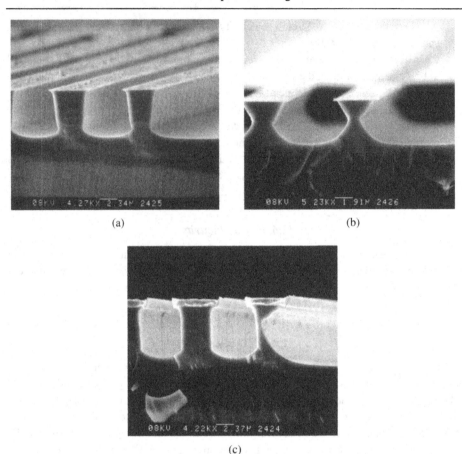

Fig. 12.8. The influence of a (a) copper, (b) platinum, and (c) aluminum mask material
on the profile.

12.7   Problems and solutions

It should now be obvious that IBARE is an incredibly complex technique and it takes quite
some time before one is familiar with it. Unfortunately, this is not all; IBARE has its own
specific problems, and this section will examine a few of them.

12.7.1   Uniformity

The uniformity, global as well as local, of an etch result depends completely on the loading
and the aspect ratio dependent etching (ARDE) effect, and we treat the solution for these
problems in the corresponding sections.

12.7.2   Roughness

A major problem during etching silicon vertically is the forming of "grass" or "black silicon"
on the surface as a consequence of all kinds of micromasks deposited or grown on the Si
(Fig. 12.9), e.g., native oxide, dust, etc., which is already on the wafer before etching. But,

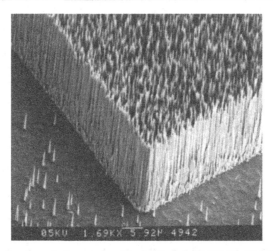

Fig. 12.9. The forming of micrograss or black silicon in an SF_6/O_2 IBARE plasma. The grass is caused by micromasks (sputtered) on top of a silicon sample together with the highly anisotropic etch behavior of the plasma chemistry.

it is also formed during etching due to redeposition of mask material from imparting ions or passivation of the surface together with angle-dependent ion etching of this inhibitor.

Spikes formed due to dirty wafers before etching are easily prevented by giving the wafer a precleaning step. The redeposition of mask material decreases for low ion energies (thus low self-bias), and the growing of inhibitor particles is excellently controlled when a slightly lateral etching is allowed.

12.7.3 ARDE

Currently, in micromechanics the etch depth of trenches increases while the trench width (or opening) stays the same or will become even smaller. The aspect ratio (depth/width) therefore increases and aspect ratio dependent etching (ARDE) will become important. ARDE is a collective noun for

(1)   sidewall bowing, i.e, the deflection of ions to sidewalls during their trajectory along these walls,

(2)   feature size dependency of profiles, i.e., different tapered profiles observed for different trench openings, and

(3)   RIE lag, i.e., the effect that smaller trenches are etched slower, positive lag, or faster, negative lag, than wider trenches.

These are well-known phenomena observed during etching trenches into a conducting substrate, as shown in Figs. 12.8 and 12.10, and seem to be strongly correlated by the effect of ion bowing. Ion bowing is caused by the diffraction of ions while entering a trench or by the negative potential of trench walls with respect to the plasma glow resulting in a deflection of ions to the walls. Ions are the main etching specimen of the passivating $Si_xO_yF_z$ layer in an SF_6/O_2 plasma and are controlling the etched profile by their direction. The etch rate decreases almost linearly as the aspect ratio increases and is determined by the aspect ratio, regardless of the opening size. This phenomenon has been attributed to a diverging electric field in the trench [41], diffusion effects on the supply of reactant to

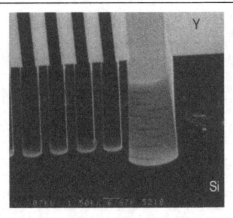

Fig. 12.10. The influence of ion bowing on a silicon profile. The wider trench at the right of the silicon sample is more negatively tapered due to ion bowing. This bowing is caused by the diffraction of ions due to the image force while entering the trench. The image force attacks charged particles until they collapse. This effect is less pronounced in the smaller trenches because the opposite wall is farther away.

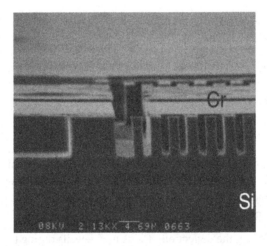

Fig. 12.11. The black surface methodology (BSM) for suppressing RIE lag. The sample is etched with the help of $SF_6/O_2/CHF_3$ chemistry. The sample is sufficiently clamped (vacuum grease) and cooled (10°C) with the target platen to ensure polymer deposition at the silcon surface from the CHF_3 feed. The depletion of CF_x radicals in the smaller trenches counterbalances the depletion of ions.

the bottom of the trench, and consumption of reactant at the trench sidewalls. However, in Ref. [38], the authors have made a plausible explanation that the physical ion depletion and bowing are responsible for the ARDE phenomena observed. It was demonstrated that the effect of RIE lag could be suppressed by changing the plasma chemistry. Figure 12.11 gives an example of an "RIE-lag-free" etched structure. The sample was sufficiently clamped (vacuum grease) and cooled (10°C) with the target platen to ensure polymer deposition at the silcon surface from the CHF_3 feed. The depletion of CF_x radicals in the smaller trenches counterbalance the depletion of ions, which explains the lack of RIE lag.

12.7.4 RIE damage

The impact of IBARE on properties of devices is thought to be due to IBARE-related surface contamination and substrate displacement damage [7]. The term "RIE damage" has been used for a variety of undesirable IBARE effects, such as the following

(1)   Surface residues, such as halocarbon films, which can be removed by an oxygen PE, or AlF_3 on top of Al in F-based etching which dissolves in, e.g., KOH but not in standard Al etch.
(2)   Impurity implantation or penetration, such as hydrogen diffusion.
(3)   Lattice damage due to energetic ions or radiation. Heat treatments can anneal out this damage.
(4)   Dopant loss due to, e.g., hydrogen–boron interactions.
(5)   Heavy metal contamination from, e.g., the reactor walls diffusing readily into Si. This effect reduces with the plasma potential.
(6)   Surface roughness as treated before.
(7)   Oxide breakdown as the rf power is turned off and the "oxide" capacitor discharges.
(8)   Mobile ion contamination, such as sodium from Teflon electrodes.
(9)   Post-IBARE corrosion, as in, e.g., Cl-based etching of Al. Upon exposure to atmosphere, HCl is formed and corrosion of the Al takes place. The chlorine-containing residues are removed by post-IBARE plasma/wet cleaning treatments.

The residual damage is a strong function of, e.g., the maximum ion energy or flux, but particularly of the Si etch rate. Upon exposure of a sample to an IBARE plasma, damage will be introduced into the substrate and will accumulate. At the same time, however, the etching will consume the damaged layer. Thus, for high etch rates, little residual damage should be observed.

12.8 Data acquisition

Despite its widespread use, plasma etching remains a poorly understood process. The plasma chemistry of even simple etch systems involves many chemical reactions and is complicated by the strong interaction between electrical, chemical, and physical effects, thus creating an operation space that is difficult to characterize. Moreover, the result of an IBARE process, such as the etch rate or profile, depends in a nonlinear way on a great number of parameters, e.g., power, pressure, flow, or residence time. This situation requires extensive "cooking" in the development of suitable etch processes, the process of trial and error. It is for this reason that people often refer to plasma chemistry as "an art more than a science." So, it is possible to find a recipe for the IBARE of silicon giving, e.g., a vertical wall profile. However, established recipes are practically never transferable between different etch chemistries and cannot be used in other laboratories, especially when they have a different reactor.

To encounter these problems, fundamental models based on an improved understanding of the science of rf discharges are increasingly applied. In fundamental models, all potentially significant plasma reactions are mathematically represented and the creation, transport, and loss of species can be modeled from a continuum approach using the mathematical expressions directly or by Monte Carlo simulation. It is expected that such computer models of plasma etching processes will aid significantly in optimizing their use for materials processing. However, many fundamental plasma parameters, such as the rate coefficients and cross sections for most reactions, are unknown. Moreover, due to the extremely complex nature of particle dynamics within a plasma, the connection between these microscopic models and macroscopic characteristics such as etch rate has yet to be cleared.

Since it is presently difficult to model plasma processes from a fundamental approach, parametric (empirical) modeling techniques are finding their way into plasma modeling. These statistical design-of-experiment models (DOEM) and data analysis methods such as response surface methodology (RSM) have been used to obtain statistical models of, for example, the etch rate or profile by directly looking at the output of the etch result.

Despite their succes, statistical techniques are limited because they make a linearization (matrix equations) of all the events in the plasma process, and therefore they are not able to track nonlinear behavior in plasmas. This limitation is almost never a disadvantage. However, during the anisotropic etching of materials like silicon or polymers, the appearance of a rough surface (micrograss or black silicon) has a strong nonlinear behavior. Since this roughness and anisotropy are firmly correlated and the roughness is easy to observe, it is a perfect tool for finding the vertical profile regime more quickly than a statistical technique would. This technique is known as the black silicon method or, alternatively, the black surface methodology (BSM).

In this section, both fundamental and statistical models and the BSM are explored to characterize and optimize plasmas.

12.8.1   Fundamental models

In the fundamental models, a threefold approach is being pursued in order to reach an optimized plasma result for, e.g., an IBARE system: experiments, modeling, and implementation.

- *Experiments*: First, nonintrusive, real-time, and in situ measurements on real IBARE systems are being performed. Significant progress has been made in determining species densities and their energy distribution to be used as basic variables in the fundamental theories or as input variables for computer simulators.
- *Modeling*: Second, IBARE models are studied since real glow discharges make well-controlled experiments difficult. Due to the coupling of most parameters, a controlled change in one quantity invariably, and often irreversibly, changes other quantities and it is difficult to asses the relative importance of the change of a specific quantity in producing a new result. The goal of the model system studies is to investigate the interaction of fluxes of species with well-specified surfaces.
- *Implementation*: The third component needed is numerical modeling. Values of the controllable plasma operating parameters and the results of the model system approach on cross sections, reaction rates, etc. are used as inputs of a computer model of a glow discharge for a specific application. The output of the numerical model can be compared to the results of measurements performed on real systems. For prototypical plasma processes, such as Si etching using SF_6, numerical models are already quite advanced and increasingly accurate.

12.8.2   Design-of-experiment models

Because of the large number of factors (input variables), plasma process optimization should be considered as a multidimensional problem, and a systematic procedure needs to be found. Design-of-experiment models (DOEMs), response surface methodology (RSM), and the black surface methodology (BSM) are valuable tools to explore unknown regions of parameter space, to obtain initial information on important variables, and to refine the information so that optimum conditions can be obtained.

The most simple DOE model is the one-dimensional search. In this approach, the factors are varied continuously only one at a time, and the result is monitored as a continuous function. Although this technique is accurate, it is also time consuming. A somewhat more sophisticated technique is the multidimensional approach. In this case, it is possible to change among the factors from experiment to experiment, which increases the optimiza-tion flexibility but does not forestall the time wasted very much. In order to decrease the time wasted for process optimization, a structural strategy of experimental design needs to be implemented to extract as much information as possible from a limited number of experiments. The unifying feature of such statistical designs is that all factors are varied simultaneously, in contrast to the multidimensional search. Statistical experimental designs include mainly factorial design and fractional factorial designs. Due to the discretization, this method is less accurate than the multidimensional technique and process optimization is only possible within the parameter space of the originally predetermined discrete level settings. To overcome this last problem, response surface methodology (RSM) has been developed which generally makes use of one of the fractional factorial design methods. In RSM, plasma-etching characteristics are expressed as continuous functions of instrumen-tal variables using polynomial mathematical models. Although the input factors are still discrete (caused by the fractional factorial design), the mathematically rearranged output is semicontinuous. The RSM is fast and accurate as long as the basic assumptions are met by the system. For example, for a three-level design, not more than a parabolic behavior can be obtained or fitted: The higher-order variations are thus lost. To be able to track the higher-order variations, in special cases the black surface methodology (BSM) can be used. The BSM is a nonstatistical interactive multidimensional search in which the input factors can be changed continuously and the continuous output data are directly monitored. In the BSM, a certain recognizable output (observable) is kept constant while changing the input factors one by one through the whole parameter space. This observable is defined as a recognizable result, which is used as a reference or marker, e.g., black silicon. Due to this interactive approach, this technique is accurate and fast. However, this technique is only favorable when the observable is easy to monitor or recognize.

In the next sections, some popular DOEMs in use for plasma characterization nowadays are given. It is a case study about statistical design-of-experiment methods and data analysis, as given in the thesis "Plasma Etching for Integrated Silicon Sensor Applications" by Yuan Yiong Li [52]. Some of these DOEMs are the one- and multidimensional search, factorial design, and fractional factorial design. The data analysis with the help of RSM is also found in this section, but the BSM will be treated more extensively in the next section.

One-dimensional search

A process is in practice often developed by varying one factor at a time, while holding all other factors at some constant level: the so-called one-dimensional search. This approach does not require a statistical design and can result in successful process development if an appropriate baseline recipe is already available. However, it is expensive and can yield incomplete and often misleading results. This type of experimentation requires testing at many factor levels, does not account for experimental errors, and ignores interactions among independent input variables.

Multidimensional search

In the multidimensional search it is allowed to "change from one variable into another" between subsequent experiments. In this way it is possible to trace more quickly certain trends and to optimise the final result much better than would be possible with the one-dimensional search.

Factorial design

Factorial design consists of statistical experiments using all combinations of level settings for all factors. The experimental results are analyzed to understand the correlations between these factors and the performance factors, including the effects of individual input factors and also the effect of interactions between the factors. Although factorial designs are thorough in a sense of covering all possibilities of input factor combinations, these contain a large number of experiments, which make this method still expensive with respect to the multidimensional search. The total number of experiments for k factors at m levels is m^k. For example, in the case of 4 factors at 3 levels, 81 runs are required. The exponential increase in the total number of experiments with the number of factors prevents this design to be practical when k is larger than 4, which is often the case in plasma process development.

Fractional factorial design

This DOEM is a subset of factorial design that reduces the number of experiments to be performed by exploring only a fraction of the input variable space in a systematic manner. A number of fractional factorial designs have been developed, including the orthogonal design, the face-centered cube design, the Box–Behnken design, and the central composite design. As an example, Fig. 12.12 portrays data points used for a central composite model

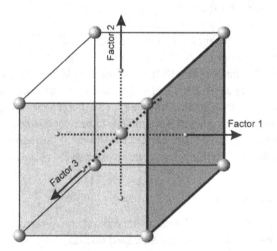

Fig. 12.12. Central composite model in three variables. One model of data selection for response surface analysis. Analysis of this type of data is readily accomplished with a personal computer.

in three variables. Data so obtained are sufficient to describe the result (say etch rate) over the entire space represented according to a complete second-degree equation including interactions of the form

$$\text{Response} = A_0 + A_1 X + A_2 Y + A_3 Z + A_4 XY + A_5 XZ$$

$$+ A_6 YZ + A_7 X^2 + A_8 Y^2 + A_9 Z^2. \tag{12.1}$$

A complete array, including replication to determine experimental error, would require 17 experiments. A minimum of 11 experiments are sufficient to solve for the coefficients in the equation.

Even if a fractional experimental design is applied, the required number of experiments can be overwhelming when many factors (say greater than 6) are considered. Therefore, a strategy to determine the relative importance of factors is necessary, which allows us to select the factors that influence each performance most. A physical understanding of the process can suggest which factors are important. However, the complexibility of plasma etching processes typically makes a priori decisions about the significance of variables difficult. Therefore, prior testing is often used to select the most important factors. A very effective method to facilitate this selection is based on screening designs. They are developed to test a large number of factors with few experiments and are analyzed by fitting the results in a linear model.

So, (fractional) factorial designs can be used to qualitively characterize a plasma process and to select an optimized level setting. However, the optimized level setting found is discrete as a result of the experimental trials and always one of the originally predetermined level settings. Therefore, important optimizing input level settings may be ignored.

Response surface methodology

To overcome the "discreteness" of the factorial designs, response surface methodology (RSM) can be utilized. In this approach, any optimized level setting between predetermined settings can be found and multiobjective optimization is possible. RSM is a statistical method in which data from suitable designed experiments are used, such as fractional factorial designs, to construct polynomial response models. The methodology consists of experimentally measuring the response to a mathematical function and statistically evaluating the quality of the experimental data and process representation by the mathematical models. Parametric models are fitted to each response and are used to numerically explore and optimize the process. The contour or three-dimensional plot of each response gives a visual representation of the effects of the variation of input variables. Generally, a full quadratic model, which includes linear two-factor interactions and quadratic terms for curvature, has been found to adequately represent the real world of plasma processes. Nevertheless, black silicon, which is an indication for the roughness of a silicon surface, is a strong nonlinear function and therefore almost impossible to optimize with RSM.

12.8.3   Black surface methodology

In ion-inhibitor IBARE the plasma is provided with a chemical etchant for the etching of the substrate, a passivator for blocking the etching at the sidewalls of a trench, and an ion source for the local removal of the passivation layer at the bottom of the etching trenches.

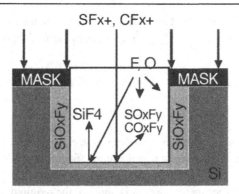

Fig. 12.13. The SF$_6$/O$_2$/CHF$_3$ chemistry.

When these processes are controlled in the correct manner, it is possible to create all kinds of trenches with excellent profile control, high etch rates, and selectivity.

As a typical example of a plasma making use of this ion-inhibitor process, the SF$_6$/O$_2$/CHF$_3$ chemistry is taken. After explaining the synergetic mechanism of the SF$_6$/O$_2$/CHF$_3$ chemistry, the origin of black silicon is clarified together with how it is prevented. The major target of this section, however, is to formulate a method to harness an IBARE system and not to explain just a specific plasma chemistry. This method, the so-called black surface methodology (BSM), should be treated as a kind of design-of-experiment method.

The synergetic mechanism of SF$_6$/O$_2$/CHF$_3$ plasmas

In an SF$_6$/O$_2$/CHF$_3$ plasma, each gas has a known specific function and influence, so the etched profile is easily controlled just by changing the flow rate of one of these gases (see Fig. 12.13). In such a plasma, SF$_6$ produces the F radicals for the chemical etching of the silicon forming the volatile SiF$_4$, O$_2$ creates the radicals to passivate the silicon surface with SiO$_x$F$_y$, and CHF$_3$ (or SF$_6$) is the source of CF$_x^+$ (or SF$_x^+$) ions, responsible for the removal of the SiO$_x$F$_y$ layer at the bottom of the etching trenches forming the volatile CO$_x$F$_y$ or (SO$_x$F$_y$).

The origin of black silicon

As stated in the above, there is a constant competition between the fluorine radicals that etch and the oxygen radicals that passivate the silicon. At a certain oxygen content there is such a balance between the etching and the passivation that a nearly vertical wall results. At the same moment, native oxide, dust, etc. will act as micromasks and, because of the directional etching, spikes will appear, as shown in Fig. 12.14. These spikes consist of a silicon body with a thin passivating siliconoxyfluoride skin. They will become higher in time and, depending on the etch rate, they will exceed the wavelength of incoming light after some time. This light will be "caught" in the areas between the spikes and can't leave the silicon surface anymore. So, all the light is collected by the etching surface and turned into black.

The origin of micromasks is caused by native oxide, dust, and so on which is already on the wafer before etching. But, it is also formed during the etching because siliconoxide particles

Fig. 12.14. The forming of grass used as an "observable" for the black surface methodology (BSM).

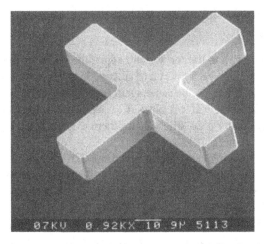

Fig. 12.15. A structure etched close to the black silicon line with the help of the BSM.

coming from the plasma are adsorbing at the silicon·surface or because of the oxidation of the silicon surface together with the angle-dependent ion etching of this oxide layer. Another source of particles during etching which will act as micromasks is the resputtering of mask material due to imparting ions.

Preventing black silicon

It is possible to forbid spikes from forming by constantly underetching the micromasks isotropically or by etching the features with a slightly negative undercut. The isotropic solution makes sense only when it is used as a post etch, because otherwise the feature density is limited. On the other hand, the negative underetching is an excellent way to control the smoothness of the substrate surface, barely limiting the feature size density. In this case, the addition of CHF_3 to an SF_6/O_2 plasma and its ability to prevent grass are described. In Fig. 12.15 a typical result is shown from a structure etched quite near the regime of black silicon but with a slightly negatively tapered profile, which prevents roughness and creates very smooth surfaces.

The formulation of the black silicon method

In this section, an easy way to find the vertical wall regime is described with the help of the information already given. A more or less general tool is reached in which the recipe for any IBARE system can be found just by fulfilling the sequence written in the box. As can be concluded from point 3 of this sequence, purely vertical walls can be achieved for any pressure, power, or O_2, CHF_3, or SF_6 flow. This is an important conclusion, because now we are able to create any dc self-bias we want without changing the profile. For instance, it is possible to develop very low bias voltages (<20 eV) at the higher pressures giving very high mask selectivity, maintaining the profile. In such cases the etched silicon bottom and the sidewalls are nearly perfect. It is also observed in Fig. 12.16a that a vertical wall profile is found for zero CHF_3 flow. This means that the passivation with siliconoxyfluorides at the sidewalls is more likely than the passivation with a fluorocarbon layer.

The BSM sequence

1. Place a piece of silicon in the reactor and adjust the preferred power and pressure for an SF_6/O_2 plasma. Etch ca. 1 micron of silicon, open the process chamber, and see if the silicon is black. If not, do the same again but increase the oxygen flow. Proceed with this sequence until the wafer is black. Increasing the oxygen too much still will give rise to black, or better grey, silicon since there exists a positively tapered profile without any underetching (Fig. 12.16a).

2. After the black silicon regime is found, add some CHF_3 to the mixture and increase this flow until the wafer is clean again. Too much CHF_3 will make the profiles isotropic (and smooth) because the CF_x species are scavenging the oxygen radicals which are needed for the blocking layer.

3. Now a wafer with the mask pattern of interest is inserted in the reactor and the etched profile is checked. If necessary, add some extra silicon in the reactor chamber until the exposed silicon area is the same as in steps 1 and 2. Increasing the SF_6 content will create a more isotropic profile. Adding more oxygen will make the profile positively tapered and extra CHF_3 will make it more negatively tapered (Fig. 12.16a). Adding at the same time O_2 and CHF_3 with the correct balance will create very smooth and nearly vertical walls. Increasing the pressure or decreasing the power will make the profile more positively tapered. In Figs. 12.16a and b the influence of the O_2/CHF_3 flow and the pressure/power on the profile is given. Increasing at the same time the O_2 and CHF_3 flow or power increasing the O_2 flow while decreasing the pressure, decreasing the pressure and power or CHF_3 flow, or decreasing the CHF_3 flow while increasing the power will hardly change the profile. However, such a change will increase the dc self-bias and a higher dc self-bias will give the off-normal ions enough energy to etch the sidewalls, thus changing the profile a little.

So, the plots belonging to the BSM give a qualitative behavior of the plasma reactor and its chemistry. After the plots have been obtained, they can be used as a guideline to fix the parameter setting of the reactor for a specific profile.

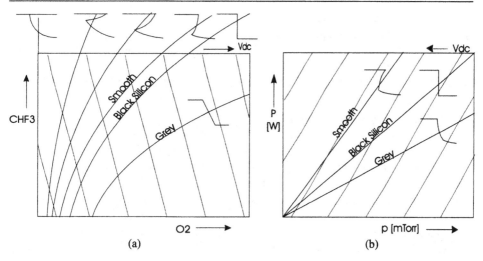

Fig. 12.16. The influence of the (a) flow and the (b) power and pressure on the profile.

General remarks concerning the BSM

After the introduction of the BSM in the literature it was experienced that some groups were not able to etch high aspect ratio trenches (HARTs) with their IBARE while using the BSM. Therefore, some general remarks will be given here which should be noted before starting the BSM. In this section, three different conventional IBARE machines at different locations were prepared for the BSM. The first is located at the MESA laboratory of the University of Twente: the Plasmafab 310/340 model from the STS company. The second is situated at the AMP Inc. Laboratory in Harrisburg: the model Plasma-Term 700 from Plasma-Term Inc. The last one is placed at the Carnegie Mellon University laboratory in Pittsburgh: a Plasma-Term 790 model from Plasma-Term Inc.

The *Plasmafab 310/340* is the IBARE reactor where the BSM was introduced. Working with this reactor, the most critical influence has been found to be the temperature. Together with entropy and enthalpy, it rules every energy step in the reactor, such as adsorption and reaction. When the temperature is not regarded, profile control is almost impossible. Therefore, the temperature should always be stabilized with the help of a proper heat sink.

A different problem is found when opening the throttle valve completely. This valve controls the pressure in the reactor during etching. Although the pressure sensor, which is placed between the reactor and the turbopump, shows a pressure lower than 5 mTorr, it is easy to create a plasma. It is found that the pipeline between the reactor and the sensor has a major flow resistance of approximately 2 mTorr/sccm when the plasma is on. Thus, using a flow of 45 sccm SF_6 with 5 sccm oxygen results in a pressure difference between the reactor and the pressure sensor of 100 mTorr. In other words, the minimum pressure of the reactor is controlled by the amount of gas flow. This situation can be changed only when the reactor geometry is changed. Nevertheless, this problem isn't hampering the use of the BSM, although care should be taken in interpreting the diagrams which go with this technique. The pressure setting given in this book and in all the papers presented by the authors correspond to the noncorrected output of the pressure sensor mounted downstream from the vacuum system. The sensor isn't relocated because (i) the set of papers would become confusing, (ii) a sensor mounted near the reactor is easily destroyed by the aggressive plasma environment, and (iii) the real reactor pressure can be calculated easily knowing the flow resistance.

Working with the *Plasma-Term 700* machine, it became clear how important clean gases are. At first, it was almost impossible to produce HARTs in silicon with this machine, although a certain degree of anisotropy could be achieved (approximately 1-μm undercut for a 5-μm trench depth). Also, the results were quite irreproducible and depend and on the time working with this reactor. Giving the IBARE its own oxygen cylinder, the problem vanished. It was found that a different reactor was polluting the oxygen line with an interfering gas, probably nitrogen.

Starting to etch Si with the *Plasma-Term 790* machine, it was noticed that the color of the plasma was not right. Normally, the color should be purple-blue and the color becomes more blue when the amount of etched Si in the reactor is increased. In fact, looking through the viewport of the reactor, sometimes it is possible to see a bright blue aura around the Si wafer. The blue color is produced by excited states of SiF_4 and the purple color comes from the SF_6. To find the problem, a pure oxygen plasma was flashed without any material other than the aluminium from the reactor walls. The color turned out to be pink/blue. Normally, the pink color is indicative of a leakage of nitrogen or air in the reactor, but a careful checking of the valves and cleaning of the vacuum seal could not end this problem. The blue color is indicative for excited fluor atoms, and therefore an oxygen plasma was started at high power in order to clean the reactor walls. After a few hours the color of the plasma changed slowly into grey/white, which is normal for oxygen plasmas at 100 mTorr. After this, the BSM worked perfectly and HARTs could be created easily.

12.9   End point detection and plasma diagnostics

The most direct need for plasma diagnostic techniques arises in the determination of the etch end point for a given process. In addition, plasma diagnostic techniques are employed for process monitoring and provide information on the types of species present in an IBARE plasma, its energy content, concentration, and so forth. The most commonly used techniques for etch end point and plasma diagnostics are laser, spectroscopic, and probe measurements [7, 10]. Less important are electron spin resonance and microwave diagnostics.

12.9.1   Laser interferometry/reflectance and ellipsometry

In this technique, light reflected from an etching surface is measured. For transparent films, e.g., SiO_2, an oscillating signal is observed for the reflected laser light intensity due to interference of the reflected light from the film surface and the substrate surface. Etch rates can be determined in real time. For nontransparent films, e.g., metals, a change in reflectivity is observed upon complete removal of the metallic film.

12.9.2   Spectroscopy

In optical emission/absorption spectroscopy (OES/OAS) the change in emission from a characteristic species is monitored or observed (with bare eyes) as etching is completed. Table 12.2 lists emission lines for some important plasma species. The sensitivity of this technique depends on how much etchant is consumed or how much film material is etched per unit time. Other spectroscopic measurements are laser-induced fluorescence (LIFS), coherent anti-Stokes Raman (CARS), and mass/energy and optogalvanic spectroscopy.

Table 12.2. *Emission lines of some species.*

Species	λ (nm)	Species	λ (nm)
O	777; 843	CO	484
F	704	CN	387
N	674	HO	309
H	656	SiF	440; 777
N_2	337	SiCl	287

12.9.3 Probes

The self-bias and/or plasma potential are changing with substrate material. For example, the amount of Si etched in an F-based IBARE has a strong influence on the created self-bias. So, it can be used as an end point detector for a layer of Si on top of SiO_2 or vice versa. Other probe techniques are Langmuir probes, double probes, and emissive probes.

12.10   Current trends

Currently, a great deal of development effort has gone into producing low-pressure ($p < 10$ mTorr) single-wafer etchers with adequate throughput, such as magnetron ion etching (MIE) or electron cyclotron resonance (ECR), which would perform tasks normally accomplished in IBARE batch reactors. Anisotropic etching is easier to achieve in low-pressure reactors because of a high ion-to-neutral flux ratio and the reduced probability of ion-neutral collisions in the sheath region at low pressure.

Low-pressure reactors are much more demanding in terms of pumping equipment and wafer cooling than conventional IBARE systems. For IBARE a roots blower and a turbopump are required to maintain pressures down to 10 mTorr at adequate gas flows. The pressure for MIE processing is near 1 mTorr, and for ECR etching it can be even lower. Moderate gas flows at these low pressures demand very high pumping speeds, e.g., for a flow of 30 sccm a 1500 L/s turbopump may need to be employed. Wafer cooling is a critical issue because of the achievement of high etch rates, significant ion bombardment, and low-pressure operation. Backside helium cooling using a wafer clamp or an electrostatic chuck is necessary in order to control the etching process.

12.10.1   Magnetron ion etching (MIE)

In MIE, magnetic fields from magnets are used to enhance ionization by confining a discharge. Arrangements similar to those used for magnetron sputtering have been incorporated into etching hardware. Magnetic fields and parallel and electric field lines normal to the cathode surface (self-bias) confine electrons on cycloidal trajectories near the cathode [43]. The probability of an electron undergoing collisions with gas phase species is thus enhanced, and the ion/neutral ratio can be 50 times greater in MIE than in IBARE. The mobility of electrons toward the cathode is decreased because of this confinement, causing the self-bias to be lower than conventional IBARE. A large flux of low-energy ions is thus produced in magnetrons at low pressure, whereas in IBARE a small flux of high-energy ions is produced for the same input power.

In dynamic magnetron apparatus, a racetrack-like region of enhanced ionization is mechanically scanned over the wafer surface [47, 48]. But it is also possible to design a static magnetic field that creates an intense discharge zone without the need for mechanical motion. This type of system is related to post magnetron used for sputtering. These are two examples of commercially available etch systems that afford high enough etch rates at low pressure to compete as single-wafer systems. Both systems apply the magnetic field in the vicinity of the wafer. In an alternate approach, magnets are arranged around the periphery of the chamber to reduce electron diffusion losses similar to the arrangement used in broad-beam ion sources. Other arrangements include interactions between magnetic and microwave fields in beam-like machines [48].

12.10.2   Electron cyclotron resonance (ECR)

In ECR, a discharge is produced by microwave excitation (commonly 2.45 GHz) [44]. When a magnetic field of B = 875 Gauss is applied, resonance between the cyclotron motion of the electrons in the magnetic and microwave field occurs. Electrons at resonance efficiently convert microwave energy into dissociation of gas species. The wafer is placed below the discharge chamber and can be rf or dc driven to control the energy of impinging ions and radicals. This enables far greater control of the etching process than possible in IBARE.

12.10.3   Inductively coupled plasma (ICP)

Unlike the planar reactors, where the rf power is coupled capacitively, ICP reactors make use of an inductive coupling. In this case, the plasma is inductively heated by the electric fields generated by a coil wrapped around the discharge chamber. These systems typically operate at a frequency of 13.56 MHz, with powers of several hundreds to thousands of watts. The chamber is typically made of Pyrex or quartz. The inductive coupling has the advantage above capacitive coupling that there will be no large or high-voltage sheaths present in the chamber, and as a result no sputtering of the wall. These devices are often used for fast isotropic etching (e.g., photoresist stripping), like in barrel reactors. However, when a kind of triode arrangement is used, the extra internal electrode can be used in an IBARE mode to develop a dc bias voltage. In this case, the high-power ICP source is used to generate a high concentration of reactive species, and the low-power IBARE source develops the dc bias which accelerates the ions to enhance the etching.

12.10.4   Pulsed plasma

In some cases it is possible to switch a plasma chemistry continuously between the isotropic etching with a highly reactive gas and the directional deposition of an inhibiting layer. Figure 12.17 shows the basic idea behind this Bosch patent. During the etching of trenches, an intrinsic roughness at the sidewalls is developed. This roughness is caused by the isotropic etch and depends on the frequency in which the equipment is switched between etching and depositing. However, the etch and deposit runs are using different gases, so the switch frequency is limited by the time needed to change the reactor volume between these two gas chemistries: the residence time.

Another method to control the directionality is to switch the frequency or power and leave the gases undisturbed. This is feasible because plasma deposition rates are depending on

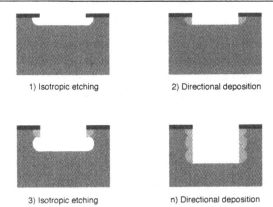

Fig. 12.17. Pulsed plasma etching. After an isotropic etching a directional deposition always follows. The sidewall roughness is in the order of the etched depth during one isotropic etch run.

the ion energy. So at high power or low excitation frequency, the reactor is in the etch mode. In contrast, at low power or high frequency, the reactor switches over to deposition. The advantage of this method is the higher switch frequency possible with respect to gas switching.

12.10.5 Cryogenic cooling

High etch rates, needed for production of single-wafer etch systems, usually require substantially higher power density than batch systems. Special techniques are necessary to handle the resultant thermal and electrical loads. Enhanced helium backside cooling may require hardware features like wafer clamping and special gas control features to permit adding high-pressure gas behind the wafer, so-called helium chucks. In all cases, coolant fluids need to be circulated to the electrodes in sufficient quantities to remove the heat created by the discharge. It is desirable to control the temperature of all electrodes, because they may affect process performance, so refrigerated cooling systems are common. Electrode temperatures from 0°C to 90°C may be needed for different processes.

Etching at extremely low substrate temperatures, down to −150°C, has gained more and more interest in the past years. Advantages seem to be a better control of the undercut during directional etching and an increased selectivity with respect to polymer mask materials. Commonly, the substrate is cooled by circulating a coolant through the platen on which the substrate rests. However, this technique is insufficient for cryogenic cooling, because the heat transfer between the wafer and the platen is unadequate at such low temperatures and at the pressures normally used for plasma etching. Therefore, helium backside cooling is generally utilized.

12.10.6 Clustering

IBARE process integration is introduced to effectively integrate IBARE into the overall fabrication sequence, e.g., by connecting deposition and etching chambers by clean, evacuated transport chambers. At the same time, real-time process monitoring equipment will detect process/equipment malfunctioning. The near future may see the utilization of computer models to scaleup the IBARE reactors and to control the etching process.

12.10.7   Others

Other interesting new techniques not treated here are microwave multipolar plasma reactors equipped with confinement magnets which surround the etching chamber and use ECR sources [45], rf driven double cathode etcher [46], hallow cathode (HC), distributed electron cyclotron resonance (DECR), and magnetically controlled reactive ion etching (MCRIE), to name only a few.

References

[1] S. M. Irving, U.S. Patent 3,615,956, assigned to Signetics (1971).
[2] A. Reinberg, Proc. Symp. Etching, Electrochem. Soc. 91 (1976).
[3] B. Chapman, *Glow Discharge Processes* (John Wiley & Sons, New York, 1980).
[4] D. L. Flamm and V. M. Donnelly, Plasma Chem. and Plasma Process. **1**, 317 (1981).
[5] J. A. Mucha and D. W. Hess, *Plasma Etching* (American Chemical Society, 1983), p. 215.
[6] R. A. Morgan, *Plasma Technology Vol. 1: Plasma Etching*, ed. L. Holland (Elsevier, Amsterdam, The Netherlands, 1985).
[7] G. S. Oehrlein, *Reactive Ion Etching: Handbook of Plasma Processing Technology*, ed. S. M. Rossnagel (Noyes Publications, Park Ridge, New Jersey, 1990), p. 196.
[8] H. W. Lehmann, *Thin Film Processes II: Plasma-Assisted Etching*, eds. J. L. Vossen and W. Kern (Academic Press, Inc., San Diego, 1991), p. 673.
[9] H. F. Winters and J. W. Coburn, Surface science aspects of etching reactions, Surface Science Rep. 14, North-Holland (1992), p. 161.
[10] M. Konuma, *Film Deposition by Plasma Techniques* (Springer-Verlag, Berlin Heidelberg, Germany, 1992).
[11] *AEDEPT (Automatic Encyclopedia of Dry Etch Process Technology)*, Du Pont Electronics, Wilmington, Delaware (1989).
[12] R. d'Agostino, *Plasma Deposition, Treatment, and Etching of Polymers* (Academic Press, Inc., San Diego, 1990).
[13] H. F. Winters, J. Appl. Phys. **49**, 5165 (1978).
[14] H. V. Jansen et al., Sensors and Act. **A41–A42**, 136 (1994).
[15] H. V. Jansen, M. J. de Boer, R. Leglenberg, and M. C. Elwenspoek, Journal of micromechanics and microengineering, nr: (5), (pp. 115–120). ISSN 0960-1317.
[16] G. C. Schwartz and P. M. Schaible, J. Vac. Sci. Tech. **16**, 410 (1979).
[17] C. J. Mogab, J. Electrochem. Soc. **124**, 1262 (1977).
[18] S. Tachi et al., Appl. Phys. Lett. **52**, 616 (1988).
[19] R. A. Haring et al., Appl. Phys. Lett. **41**, 174 (1982).
[20] M. Seel and P. S. Bagus, Phys. Rev. **B23**, 5464 (1981).
[21] G. C. Tyrrell et al., Appl. Surf. Sci. **43**, 439 (1989).
[22] S. Matsuo, Appl. Phys. Lett. **36**, 768 (1980).
[23] R. d'Agostino et al., J. Appl. Phys. **52**, 1259 (1981).
[24] J. W. Coburn and H. F. Winters, J. Vac. Sci. & Tech. **16**, 391 (1979).
[25] G. S. Oehrlein et al., Appl. Phys. Lett. **46**, 686 (1985).
[26] W. R. Harshbarger, Solid State Tech., Vol. 25. No. 4. 126 (1982).
[27] E. Gogolides et al., Microelectr. Eng. **27**, 449 (1995).
[28] G. C. Schwartz and R. A. Gottscho, The Electroc. Soc., Pennington, 201 (1987).
[29] M. de Boer, H. Jansen, and M. Elwenspoek, Proc. of Transducers '95, (Stockholm 1995, June 25), pp. 565–568.
[30] K. A. Shaw et al., Sensors and Act. **A40**, 63 (1994).
[31] Y. X. Li et al., Proc. IEEE MEMS, 398 (1995). Amsterdam Jan. 29–Feb. 2.
[32] C. J. Mogab and H. J. Levinstein, J. Vac. Sci. Tech. **17**, 721 (1980).
[33] Y. H. Lee and M. M. Chen, J. Vac. Sci. Tech. **B4**, 468 (1986).

[34] H. Jansen et al., Proc. IEEE MEMS p. 88 (1995).

[35] H. F. Winters and D. Haarer, Phys. Rev. **B36**, 6613 (1987).

[36] Z. H. Walker and E. A. Ogryzlo, J. Appl. Phys. **69**, 2635 (1991).

[37] Y. H. Lee et al., Appl. Phys. Lett. **46**, 260 (1985).

[38] H. Jansen et al., Microelectr. Eng. **27**, 475 (1995).

[39] G. K. Wehner, General Mills Report 2309, Litton Industries, Beverly Hills, CA (1962).

[40] T. H. Fedynyshyn et al., J. Elect. Soc., Solid-State Sci. Techn. **134**, 2580 (1987).

[41] D. Chin et al., J. Elect. Soc. **132**, 1705 (1985).

[42] R. Legtenberg et al., J. Elec. Soc. (1995), (pp. 2020–2028), ISSN 0013-4651.

[43] H. Okano and Y. Horiike, *Plasma Processing* (Electrochemical Society, Pennington, 1982), p. XXX.

[44] K. Suzuki et al., Vacuum **34**, 953 (1984).

[45] Y. Arnal et al., Supplement a la Revue, Le Vide, les Couches Minces **237**, 73 (1987).

[46] R. W. Boswell et al., Supplement a la Revue, Le Vide, les Couches Minces **237**, 78 (1987).

[47] Y. Horiike, H. Okano, Y. Yamazaki, and H. Horie, Jpn. J. Appl. Phys. **20**, L817 (1987).

[48] K. Suzuki, S. Okudaira, N. Sakudo, and I. Kanomada, Jpn. J. Appl. Phys. **16**, 1979 (1977).

[49] A. R. Reinberg, *Plasma Etch Equipment and Technology: Plasma Etching; An Introduction*, eds. D. M. Manos and D. L. Flamm (Academic press, Inc., San Diego, 1989), p. 339.

[50] J. L. Mauer, J. S. Logan, L. B. Zielinski, and G. C. Schwartz, J. Vac. Sci. Tech. **15**, 1734 (1978).

[51] Y.-Y. Tu, T. J. Chuang, and H. F. Winters, Phys. Rev. **B23**, 823 (1981).

[52] Y. X. LI, Thesis, Technical University of Delft, Delft, The Netherlands (1995).

[53] McFeely et al., Phys. Rev. **B30**, 764 (1984).

13

Remote plasma etching

Now that the contact plasma etchers have been treated extensively, it is instructive to show also some aspects from the remote plasma etchers. The ion milling equipment and downstream etchers are probably the most well-known members of this group. Although the ion–surface interactions are quite basic for the milling purpose, it will have almost no place in this chapter. The reason is that this subject can be found in many reviews about ion milling [50, 51]. Instead, this chapter will give some insight into a special member of this group; the thermally assisted ion beam etch (TAIBE) technique. For the etching of fluorocarbonic materials, this method is especially applicable.

The sputtering of fluorocarbon (FC) polymers is of significant industrial interest. A special member of this family is the linear, quasicrystalline polymer polytetrafluoroethylene (PTFE), better known under its trademark Teflon^{TM}. Stringent application requirements concerning friction, adhesion, dielectric properties, hydrophobicity, chemical inertness, and high-temperature survivability have fostered the widespread utilization of FC polymers. Similarly, there are many applications in which a thin coating or adhesively bonded FC laminate would be of great utility [1].

This chapter reviews FC polymer vacuum sputtering technologies from which the ion beam etching (IBE) of PTFE seems to be the most promising [1]. Although laborious work concerning the IBE of PTFE was already performed by NASA in the mid-seventies, most investigators were mainly interested in increasing grasslike roughness on the PTFE surface to be able to attach it to other materials. However, for moulding purposes and for most other microsystem uses, the grass is highly unwanted.

The review is followed by an experimental investigation of the thermally assisted ion beam etching (TAIBE) of PTFE. The etch apparatus is schematically shown and it is attempted to explain the etch mechanism of this technique. After this, the difference in etch rate with respect to other materials and the forming of grass is treated. After the experimental setup and the results, some applications will be shown.

13.1 Review of vacuum etching

Already in 1978, Banks et al. [1] reviewed several FC polymer sputtering techniques, including thermal, plasma, and ion beam sputtering. Ion beam etch rates (ER) up to 2 mm/hr for PTFE were determined and shown to depend predominantly upon the ion beam power flux:

$$ER = P^{1.4} \quad [W/cm^2], \tag{13.1}$$

where P is the power flux [1, 12, 13]. High peel strengths were measured for epoxy bonds to various ion beam sputtered FC polymers and appear to be related to the surface microstructures (grass) resulting from IBE. Fluorinated ethylene propylene (FEP), polychlorotrifluoroethylene (PCTFE), perfluoroalkoxy (PFA) Teflon^{TM}, Tefsel^{TM}, and polyoxymethylene (POM) gave identical results [1, 3, 12].

Jansen et al. [2] treated a few other FC deposition (spin coating) and sputter (focused electron beam and plasma) techniques. The thin film properties were comparable with those of commercially available bulk PTFE [4]. In order to track the chemical nature of the deposited FC thin film, in Ref. [5] the authors presented the results of X-ray photoelectron spectroscopy (XPS) analyses.

To be able to control the etched profile (especially steep walls are useful), the particles are collimated and directed to the surface to be etched. This can be achieved by parabolic mirrors (photons), wind tunnels (neutrals, radicals), or electrical fields (ions, electrons). In fact, the PBE and IBE equipment are typical examples of this technique.

During etching deep trenches into substrate materials, specific problems may occur which will hamper the aspect ratio (AR = depth/width) of the trench such as the depletion of etching particles due to diffusion lag. Until now, many techniques have been developed in order to achieve high AR structures in polymers: photolithography such as LIGA (PBM), reactive ion etching (IBARE), thermally assisted ion/electron beam etching (TAIBE or TAEBE), and others.

- *PBM*: A direct way is the use of conventional photolithography. However, because of the rather high wavelength of the exposing source and the limited thickness of the, e.g., spin-coated photosensitive polymer, the AR of the structures seldom exceed 7 and it is difficult to create submicron features [6]. To increase the aspect ratio, short wavelength directional X-rays from a cyclotron source are used. However, this so-called LIGA technique is expensive [7].

- *IBARE*: In order to extend the limit of PBM, special three-level "RIE" techniques have been developed [8]. Electron sensitive polymers are used in the first level to create submicron structures (FEBM) [9]. Nevertheless, the AR is usually not higher than 10 because of the relatively high operating pressures during copying of the pattern with IBARE into the third level. To increase the AR, low-pressure and high-density IBARE machines are developed, but these machines are costly [10].

- *TAIBE*: IBE is a simple technique to create high AR trenches into PTFE. The technique was demonstrated first by NASA in the mid-seventies [11–13]. It uses the extraordinary properties of PTFE, e.g., its extremely high melt viscosity [4]. Because of this, PTFE preserves its shape even at temperatures near its melting point. Therefore IBE can be used at elevated temperatures to achieve high etch rates up to several tens of microns per minute. The most straightforward name for this type of etching equipment is thermally assisted ion beam etching (TAIBE). The selectivity with respect to the mask typically exceeds 1000. Because of the low operating pressure during IBE, the AR is superior to IBARE.

- *TAEBE*: The use of electrons to etch a material is found in the e-beam evaporator. Energetic electrons are directed to a material what will melt and evaporate (FEBE). As for TAIBE, highly directional profiles are created in PTFE. Instead of the focused beam of an evaporator, a broad band electron source can be generated with the help of an IBE. Electrons from a current heated tungsten wire are directed to the substrate with the help of a grid at a positive potential with respect to the wire. A logical name for such a system would be electron beam etcher (EBE), and at elevated temperatures the process becomes thermally assisted electron beam etching (TAEBE). The lower operating pressure opens the possibility of achieving even higher aspect ratios, as in the case of TAIBE.

- *TAPBE*: It is straightforward to use a laser or a halogen lamp, assisted with a heat source, to etch a substrate (i.e., TAFPBE and TAPBE, respectively). A 1200-nm photon has an energy content of 1 eV, so polymers, especially PTFE, are easily sputtered at elevated temperatures. To be able to create submicron structures, it is necessary to use photons with a small wavelength like X-rays, and special X-ray-opaque materials are needed to transfer a pattern.

13.2   Etch apparatus

In Fig. 13.1 a simplified schematic of the IBE apparatus which is used in this study is shown. It consists of four independent parts: (i) the vacuum system, (ii) the ion source, (iii) the substrate holder, and (iv) the heater.

13.2.1   The vacuum system

To direct a beam of atomic particles over a large distance (>5 cm), it is necessary to create a pressure below 1 mTorr. This is achieved by mounting a turbopump in series with a rotary pump to the reactor. The turbopump also removes reaction products.

13.2.2   The ion source

The energetic particle flux is generated with the help of a remote electron bombardment ion source (Kaufman type). Inert (Ar) or reactive (O_2) gases are let into the small chamber ($l = 42$ mm and $\emptyset = 33$ mm). To create a plasma, a current is forced through a tungsten wire which gives a cloud of electrons surrounding the wire, i.e., the cathode. These electrons are forced to the walls of the chamber which have a positive potential, i.e., the anode. On their way to the walls the electrons collide with the gases creating all kinds of energetic particles, including ions and electrons. Due to the increasing amount of charged particles, a plasma is started. To increase the efficiency of the electrons in creating plasma particles, magnets are surrounding the Kaufman source in a way that the electrons are describing a spiral-like path

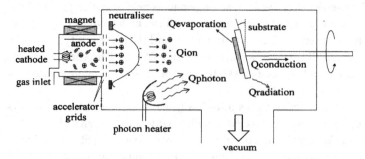

Fig. 13.1. Basic schematic for thermally assisted IBE.

to the anode. So, their lifetime is increased and therefore so is the particle generation. For the same "efficiency reason," this ion source has to be generated in the remote high-pressure chamber (high with respect to the "sample room" in the reactor), because otherwise there would be no electron–gas collisions. Beam extraction is accomplished by a flat, two-grid carbon ion optics system. A neutralizer is used to prevent (i) beam diffraction, (ii) charging of the substrate, and (iii) electrical breakdown of electronic circuits in the substrate when turning the beam off.

13.2.3   The substrate holder

To be able to control the etched profile, the system features substrate rotation with an adjustable angle and water cooling at 18°C.

13.2.4   The heater

To control the temperature of the substrate, it is mounted on top of a 150-Watt Joule heater or a 150-Watt halogen lamp with a 50 cm^2 spot size.

13.3   Review of thermally assisted ion beam etching

Heating of PTFE in the range 360–600°C caused thermal decomposition involving formation of free radicals due to random polymer chain scission. It left a carbon residue and the largest scission fragment was C_3F_8, with C_2F_4 as the primary product [14, 15]. The plasma decomposition of PTFE at 11.5 Torr resulted in a similar distribution of reaction products, probably due to the short mean free path "λ" and further fragmentation in the plasma glow [15]. However, the low-pressure environment in IBE (<0.2 mTorr) allows large scission fragments ($>C_3F_8$) to escape from the substrate with minor carbon residue. This would explain the quasi-crystallinity of the ion beam sputter deposited FC thin films. This is in contradiction to the other techniques which produce amorphous highly cross-linked FC thin films [1].

Hall and Green [16] and Rost et al. [17] measured the IBE etch rates of FEP and PTFE, respectively, and found them to be approximately two orders of magnitude higher than those of metals and glass ($\sim$100 CF_2 groups per incident ion). Rost et al. attributed the higher etch rates of PTFE to the breakup of strong bonds in molecular chains producing a wide distribution of scission fragments that leave the target surface by thermal effects where only small intermolecular forces must be overcome. Hall and Green also measured erosion rates of similar linear addition polymers, i.e., polyvinylfluoride and the linear polypropylene, but did not observe elevated erosion rates. More rapid cross-linking than depolymerization was felt to cause this result. The FEP sputter yields exhibited a significant sensitivity to ion current density. This is a unique characteristic in that most materials have sputter yields nearly independent of current density [18–20]. This last result is indicative especially for local heating as an important factor in the IBE of PTFE.

13.4   Etch mechanism

So, the etch rate of PTFE is highly temperature dependent. The sample can be heated, e.g., from the backside with the help of a Joule heater or from the frontside by way of a lamp or incoming ion flux. The ions especially carry with them a large amount of kinetic energy (typically 1000 eV per incident ion). The heat at the frontside is partly leaving the surface as evaporation/kinetic energy of etch products and black body radiation. The rest will travel through the PTFE bulk to the backside, which can be water-cooled at 18°C or Joule heated at a fixed temperature T_b. Together with the heat resistance, this flow of heat will create a temperature gradient. It is assumed that the temperature distribution in the PTFE is not changing in time: In other words, the distributed heat capacity is fully charged. In the same way, it is assumed that the surrounding area is at a stable temperature T_s. Writing the energy balance per unit area and time for the frontside of the PTFE sheet, we find at thermodynamical equilibrium

$$Q_{ion} + Q_{photon} = Q_{conduction} + Q_{radiation} + Q_{evaporation} \quad [\text{W/m}^2] \qquad (13.2a)$$

$$U_i J_i + \eta \varepsilon U_p I_p / A = \lambda (T_f - T_b)/t + \varepsilon \sigma \left(T_f^4 - T_s^4\right) + E R(\Delta p) \Delta H^\circ + Q_{kin} \quad [\text{W/m}^2], \qquad (13.2b)$$

with U_i [V] being the acceleration voltage of the ions, J_i [A/m^2] the ion current density, η the efficiency of the lamp in heating the sample with photons, ε the emissivity depending on material properties, U_p [V] the voltage over the lamp, I_p [A] the current through the lamp, A [m^2] the spot size, $\lambda = 0.24$ [W/mK] the specific heat, conductance of PTFE, T_f, T_b, and T_s the temperature [K] of the front, back, and surrounding surface, respectively, t[m] the thickness of the sample between T_f and T_b, $\sigma = 5.7 \times 10^{-8}$ [W/m^{2}K^4] the Stefan-Boltzmann constant, and $E R$ [mol/m^{2}s] the etch rate, which is a function of the pressure difference Δp between the vapor pressure p_v of the PTFE and the partial pressure p_p in the vacuum system. At high etch rates or sample temperature, $\Delta p \approx p_v$, which is a function of the temperature T_f. The molar enthalpy (of sublimation = fusion + vaporation) at 1 atm is per definition

$$\Delta H^\circ = \Delta H^\circ_{sub} = \Delta H^\circ_{fus} + \Delta H^\circ_{vap} \quad [\text{J/mol}]. \qquad (13.3)$$

It is the energy needed to bring one mole of solid PTFE into the vapor phase. This figure is assumed to be independent of temperature in the range we are working (i.e., 20–300°C). Q_{kin} [W/m^2] is the heat flux due to kinetic energy. Because the amount of particles removed per incident ion (i.e., the yield) is so high and not dependent on the ion energy or ion current flux alone but rather on the power flux, it is thought that Q_{kin} can be neglected.

13.4.1 *Photon source*

To find the efficiency of a halogen photon source of $P = 150$ W with a spot size of 50 cm^2 to heat a sample, three different materials with the same dimensions ($10 \times 10 \times 1$ cm^3) were placed into the vacuum directly in front of the photon source: aluminium, oxidized steel, and PTFE. The sample was mounted with thin plastic wires to ensure a bad thermal contact with the surroundings. Indeed, after turning off the lamp, the temperature of the samples of steel and aluminium at 45°C didn't change for more than 1° over 1 hour, indicating that there is little heat transported due to conduction or radiation. After creating a vacuum of 10^{-6} Torr the sample was heated for a period of time while monitoring the sample temperature ΔT [K] with a thermocouple. The heating is stopped when a certain temperature rise ($\Delta T = 15$°C) is achieved and the time is clocked. Now, we are able to calculate the efficiency with the

Table 13.1. *Data to find the efficiency of the photon heater.*

Material	m [g]	c [J/gK]	ε	ΔT	t [s]	η
Al	168	0.88	0.24	20	420	0.2
Steel	484	0.48	0.80	19	180	0.2
PTFE	147	1.05	0.9	15	420	0.04

help of the formula

$$\eta = cm\Delta T/\varepsilon P t \quad [-], \tag{13.4}$$

with c [J/gK] the specific heat capacity of the sample and m [g] its mass, as found in Table 13.1. For example, to increase the temperature of steel 19°, 3 minutes photon heating was needed and we calculate $\eta = 0.2$. So, only 30 Watts are used for heating the sample. Nevertheless, the incoming heat flux is still 0.6 W/cm². As can be concluded from Table 13.1, the efficiency of the lamp is 0.2 when aluminium or steel is taken as a reference, but 0.04 for the PTFE sample. This is unlikely, because the photon source was placed in the same way with respect to the sample. It is known that a difference in roughness for the same material is responsible for a major change in emissivity. So, when we assume that the emissivity $\varepsilon = 0.2$ of PTFE, then we calculate $\eta = 0.2$. It is also thinkable that the PTFE is losing material, i.e., energy during the heating of the photon source. So, little of the absorbed energy is used to increase the heat in the heat capacity.

In principle, the photon source is a black body radiator with an equivalent temperature:

$$T_p = (\eta U_p I_p/\sigma A)^{1/4} = 600 \quad [K]. \tag{13.5}$$

However, because it is driven by a remote power source, it is treated separately. Nevertheless, when the lamp is included in the radiation term, then

$$Q_{rad} = \varepsilon\sigma\left(T_f^4 - T_s^4 - T_p^4\right) \quad [W/m^2]. \tag{13.6}$$

Note that the cathode and the neutralizer are also radiating photons to the sample. This energy input should be added in the same way as already done.

13.4.2   Conduction

Let's take a typical IBE setting of 1000 eV at 1 mA/cm², i.e., $Q_{ion} = 1$ W/cm². If the only loss of heat was due to conduction and $t = 3$ mm, we are able to calculate the temperature gradient:

$$T_f - T_b = Q_{ion}t/\lambda = 120 \quad [K]. \tag{13.7}$$

This is a major temperature rise and can be even higher when the thermal contact of the PTFE sample is incorrect.

Sovey found that the etch rate increased from 180 to 640 μm/hr as the target reference temperature increased from 200°C to 280°C [12]. At 280°C conditions, photomicrographs indicated evidence of local melting and spirelike structures no longer existed. The ion etch rates tended to asymptote to approximately 100 μm/hr at target temperatures below 150°C. He concluded: "This implies that physical sputtering probably dominates over thermal processes at target reference temperatures less than 150°C." However, at the fixed parameters Sovey used (750 eV, 0.5 mA/cm^2 $\Rightarrow$ 375 mW/cm^2, and a 3-mm thick PTFE sample) we are able to calculate a possible temperature rise of 45°C because of the ion heat flux. Indeed, 280°C + 45°C = 325°C is close to the melting point of PTFE at atmospheric pressures (327°C). The local pressure might be 1 atm due to the high etch rate.

13.4.3   Radiation

To find out the importance of the black body radiation we assume $\varepsilon = 0.9$ and the sample to be at 600 K, i.e., the temperature of melting for PTFE:

$$Q_{\text{rad}} = \varepsilon\sigma\left(T_{\text{f}}^4 - T_{\text{s}}^4\right) = 0.6 \quad [\text{W/cm}^2]. \tag{13.8}$$

Again, an important drain of energy.

13.4.4   Sublimation

To calculate the heat flux due to sublimation, it would be convenient to know what the p-T phase diagram looks like. This plot contains the triple point which is characteristic for a certain material and represents the only pressure and temperature at which the vapor, liquid and solid of a specific material can coexist in equilibrium *under the pressure of the vapor alone*. For example, in the left side of Fig. 13.2 the p-T phase diagram of pure water is given. Its triple point is located at 0.0098°C and 4.5 Torr. At pressures lower than this point, only vapor and solid can coexist in equilibrium *under the pressure of the vapor alone*, as represented by the sublimation line. This line connects a certain temperature T_{f} with a single pressure: the vapor pressure p_{v}. The vapor pressure is increasing exponentially with the sublimation temperature:

$$p_{\text{v}}(T_{\text{f}}) = p^{\circ}\exp((T_{\text{f}}\Delta S^{\circ} - \Delta H^{\circ})/RT_{\text{f}}) \quad [\text{Pa}], \tag{13.9}$$

with $R = N_{\text{A}}k = 8.3$ J/Kmol being the universal gas constant, $N_{\text{A}} = 6 \times 10^{23}$ mol^{-1} the number of Avogadro, $k = 14 \times 10^{-24}$ J/K the constant of Boltzmann, ΔS° the molar entropy at 1 atm, and $p^{\circ} = 1$ atm the reference pressure where

$$\Delta G^{\circ} = \Delta H^{\circ} - T_{\text{f}}\Delta S^{\circ} = 0 \Rightarrow T_{\text{f}} = T^{\circ} = \Delta H^{\circ}/\Delta S^{\circ} \quad [\text{K}]. \tag{13.10}$$

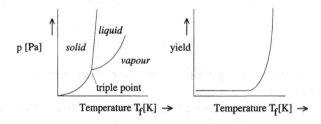

Fig. 13.2. (left) p-T diagram and (right) temperature dependency of the sputter yield.

So, by definition T° is the temperature (usually the boiling temperature T_b) where $p_v = p^\circ$. In equilibrium, the amount of solid going into the vapor phase [mol/m²s] equals the amount of vapor going into solid at a rate [47]:

$$ER(p_v) = p_v(2\pi m N_A R T_f)^{-1/2} \, [\text{mol/m}^2\text{s}], \qquad (13.11)$$

with m [kg] being the mass of the molecule. Now when a piece of ice is placed into vacuum the vapor is removed (turbo) or frozen to cold surfaces (cryo) and the backstreaming is made impossible. It is reasonable to assume that the sublimation is going on with the same rate, thus the ice is losing mass, presuming the pump can handle the amount of vapor which is produced. The rate of disappearance is low, of course, because the vapor pressure is low. To give an indication, assume the vapor pressure is 10^{-6} Torr, i.e., 133×10^{-6} Pa at a temperature of 1000 K, the interatomic distance is 3 Å, and the mass of the molecule is $33 \times 1.6 \times 10^{-27}$ kg. Then the rate of sublimation is approximately 3×10^{-6} mol/m²s $\approx$ 0.2 monolayers/s $\approx$ 0.6 Å/s. Note: Every vacuum system removes gases selectively. For example, a turbopump removes water vapor much easier than hydrogen gas. This means that the water vapor pressure is lower than indicated by the pressure sensor of the vacuum system. Nevertheless, the vapor pressure is not always low and increases exponentially with temperature. So, a different material can be removed quite fast when it is heated sufficiently. When the sample is heated and the reactor walls are at room temperature, the walls will act as an adsorption pump and the etch rate is completely determined by the vapor pressure of the sample.

Now, let's take a PTFE sample. PTFE consists of many different sized FC chains. Therefore, the p-T plot of PTFE will not have single equilibrium lines but a broad spectrum and not a triple point but a triple line. During heating PTFE, the short chains will sublimate (or melt and evaporate) first, leaving the longer chains in the solid. Increasing the temperature more and more, the longest chains will evaporate at last. It would be quite interesting to see how the perfluoroalkanes (C_nF_{2n+2}, like $C_{10}F_{22}$) behave under the same conditions as PTFE, because PTFE is believed to consist of such oligomers [48]. Because these molecules have a unique chain length, their p-T plot will be quite normal and we should be able to calculate the ER of such materials with the help of the theory presented in this chapter.

Note that, in principle, TABE opens a whole scale of possibilities to pattern an arbitrary substrate material. For example, zinc could be IBE patterned with a high selectivity if we reach a vacuum low enough for its triple point (otherwise it would first melt and then vaporize, i.e., it would become isotropic) together with substrate heating [47]. As a mask, a material with a much higher melting point or a cooled (zinc) shadow mask could serve.

The low thermal conductivity together with the comparable high vapor pressure at moderate temperatures might account for the difference in etch rate between PTFE and inorganic materials. In the right side of Fig. 13.2 a plot is shown giving the yield of an incident ion as a function of the temperature. Evidently, the yield of an incoming ion is increasing with temperature because particles are leaving the solid at a higher rate than particles coming back. At relatively low sample temperatures the etch rate is determined by the pumping speed of the vacuum system because $\Delta p = p_v - p_p \neq p_v$ and the yield of an ion will be between 1 and 4 for most materials. The energetic ion is removing just a few atoms due to its impulse. However, when the temperature of the sample is much higher than the surrounding surfaces, the IBE system will act as an adsorption pump and all the PTFE material is deposited at the reactor walls (note that this insulating layer could destroy the function of the Kaufman source). In these cases the etch rate is determined by the vapor pressure and the yield can be over 100. The TAIBE of PTFE is enormously anisotropic, which might be linked with its high melt viscosity. But why does PTFE possess such a high anisotropy/melt viscosity? It is possible that the longer chains in the PTFE, which are melting at much higher temperatures than the smaller chains, are acting as the framework of the PTFE. So, it preserves the shape of the PTFE in the regions where ion bombardment fails. In contrast, in the open regions after the smaller chains have left, the longer chains can be broken by the ions and the products can leave the surface. The only question left is why do other comparable polymers not etch at increased speed? The answer might be found in the way they degrade at elevated temperatures. For example, polypropylene would leave a carbonic residue which prevents further high-speed etching. Or, another possibility, the poly chain length responsible for the triple line makes it difficult for the PTFE to etch isotropically at elevated temperatures: There will always be a part of the PTFE having chain lengths too big to sublimate.

13.5 Grass

Large cone or grasslike surface microstructures, with diameters of a few microns and lengths up to more than 100 microns, resulting from ion or rf bombardment have been observed for FEP and PTFE substrates [16, 21]. Figure 13.3 shows a typical example. The structures are found to be always parallel to the direction of the incident ions.

IBE guarantees an excellent anisotropy because of its highly directional ion flux and low operating pressure. Due to the high selectivity of PTFE with respect to other materials, micromasking will cause the grass problems which are identical with high anisotropic IBARE trench etching [22, 44]. To overcome this problem, substrate rotation can be used. Also, the substrate normal should be tilted with respect to the ion flux direction. This solution is visualized with the help of Figs. 13.4(a) and (b).

However, care should be taken while using substrate rotation, as demonstrated with the help of Fig. 13.4c. In this figure the depth profile of a pinhole in the mask design is shown. Because of the tilted substrate rotation, a conelike structure will develop directly under the

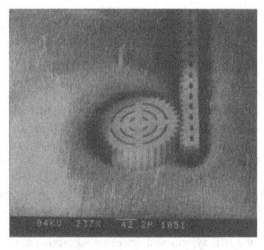

Fig. 13.3. Forming of grass during IBE of PTFE.

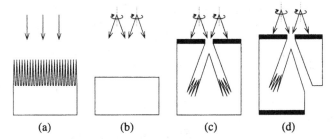

Fig. 13.4. (a) The growing and (b) preventing of grass. (c) and (d) Profile development
of a pinhole under a tilted substrate rotation.

pinhole position and, again, IBE grass in these trenches can't be prevented because of the
highly directional ion flux. Moreover, due to geometrical reasons the etch rate in depth will
decrease inversely proportional in time. It will decrease even more due to the exponential
behavior of the sublimation with temperature.

Another way to prevent the forming of grass is preventing non-PTFE particles from
depositing on top of the PTFE surface, as shown in Ref. [45]. The authors found that as
much as $5\frac{1}{2} \times 10^{-4}$ particles from miscellaneous sources per incident ion were sputtered
on the sample, i.e., a monolayer in 4 minutes.

Following are some causes of contamination.

- *Mask particles*: The first contamination of particles is due to sputtered mask material.
  This deposit can be lowered by decreasing the ion energy. However, the ion current
  density must be increased simultaneously to ensure a certain etch rate. A different
  solution is the use of a cooled PTFE shadow mask. At the same time the PTFE sample
  should be heated to increase the selectivity between mask and sample.
- *Reactor particles*: A major source of particles is due to sputtered reactor material.
  During the IBE of PTFE we found that stainless steel from the sample holder was
  deposited at the sidewalls of the PTFE trenches. This type of grass was prevented by
  using a PTFE sample holder.
- *Grid particles*: The next important particle source is the sputtering of the second grid
  of the carbon ion optics system. After passing the first grid, the ions are accelerated
  to the second grid at, e.g., -1000 V. When the grids are misaligned, the second
  grid is hit by the ions and will sputter carbon particles on top of the PTFE sample.
  Accurate alignment and making the holes in the second grid bigger than those in
  the first grid decrease this type of micromasking. Note that this solution increases
  beam diffraction. At this moment, effort is put into manufacturing a micromachined
  self-aligned double-grid ion optics system.
- *Ion particles*: Another source of particles might be imbedded ions in the PTFE subsurface from the ion source (Ar or O_2).
- *Neutralizer particles*: The next possible source of particles is the tungsten neutralizer
  which is in front of the ion beam to prevent beam diffraction. We didn't notice beam
  diffraction, so we removed the neutralizer completely.
- *Cathode particles*: A similar source of particles is the tungsten wire in the Kaufman
  source which starts the plasma. So far we have not able to remove this source without
  changing the basic idea of IBE.

- *Bulk particles*: The last source of particles could be the PTFE bulk itself. PTFE always contains a small amount of non-PTFE material, sometimes on purpose to give the PTFE a different characteristic, like sodium, which is difficult to etch. But, even when the PTFE is pure, carbon residue may cause grass to form. During ion sputtering PTFE, gases like F_2, CF_4, or C_2F_6 are easily created, leaving a carbon-rich PTFE surface behind. To remove this carbon overdose, oxygen ions can be used, i.e., RIBE.

13.6   Experimental

Before starting the experiments, the samples have to be cleaned properly, otherwise micromasking will occur. For example, salt from a fingerprint is difficult to etch and will cause irreproducible results. The surface roughness of commercially available PTFE is not good enough to pattern micron-sized structures. So, it is necessary to polish the PTFE surface. Then, in most cases, a mask has to be placed on top of the PTFE. Because of the extremely low surface free energy of PTFE, this is not straightforward. When the mask is applied on top of the clean and polished PTFE, the sample can be etched for, e.g., MEMS applications.

Sample cleaning

PTFE samples to be etched were ordinary bulk material (1–10 mm thick) as well as PTFE foils (50–500 micron thick) [23]. To clean the PTFE surface, a diluted piranha mixture at 80°C was used, i.e., 100 ml of sulphuric acid (H_2SO_4) gently poured into 500 ml of demi-water and mixed with 100 ml hydrogen peroxide (H_2O_2). After this treatment the PTFE surface becomes nicely white.

Sample polishing

To create a smooth surface, some samples were polished with silica or alumina balls ($\varnothing \sim$ 50 nm). After this treatment, the PTFE surface becomes highly reflective, i.e., mirrorlike. Eventually, the PTFE samples can be cleaned again to remove any polish residue.

Sample patterning

In order to copy a pattern into the sample, the first problem is the extremely low free energy of the PTFE surface. Therefore direct patterning with photoresist is difficult. Techniques to overcome this problem are (1) roughening the PTFE surface with a plasma before spinning resist, (2) the deposition of an intermediate layer such as evaporated aluminium or titanium, and (3) the use of a shadow mask, e.g., fabricated in silicon [24].

Roughening the surface to increase the adhesion for resist seems to be rather clumsy after the polishing treatment and will not be considered.

The deposition of an intermediate layer is an excellent way to pattern the PTFE. However, care should be taken in this step. We studied two different metals, aluminium and titanium, from which titanium has the lowest etch rate, and thus the highest selectivity in IBE. To copy the resist pattern into the metal, we found that wet etching is almost impossible because the layer is stripped. After all, the layer has a bad adhesion with the PTFE. Therefore, the whole etch process should be dry using, e.g., IBARE. Aluminium can be etched with the help of chlorine-based chemistry and titanium with fluorine-based chemistry. During deposition of titanium with the help of an evaporator, the titanium particles leaving the melt are extremely hot and will dig themselves into the PTFE subsurface [49]. Although this effect will increase the adhesion of the titanium to the PTFE, after patterning the mask with resist it can be difficult to remove the subsurface titanium particles from the spots which should be etched. In contradiction, aluminium which has a low melting point, can be evaporated at a minimum deposition rate to prevent aluminium from entering the surface of the PTFE.

The use of a shadow mask is a rather old technique, but for our purposes it has a tremendous potential. In this technique, a thin membrane with the pattern is placed in between the ion beam and the sample to be etched. If properly cooled, the mask will etch slowly, whereas the heated sample is etched at high speed. Therefore many samples can be etched with the help of the same shadow mask. Even better, the mask can be "refreshed" many times just by depositing a new layer in, e.g., an e-beam evaporator. An advantage of this technique is the lag of subsurface particle contamination, such as titanium, because the surfaces to be etched are 100% clean.

Evidently, it is possible to create a shadow mask with holes but impossible to create a mask with "open" structures such as pillars. Therefore, in order to create any kind of open structure, (i) two different shadow masks are needed, (ii) the same mask is used twice but the second time it is rotated by a certain angle, or (iii) it has to be used indirectly: A metal layer is evaporated through the shadow mask on top of the PTFE substrate and then the PTFE is etched with this metal as a mask. However, it should be remembered that the metal film is fading near the pattern etches. Information about this subject is found in Ref. [24].

Sample etching

The prepared PTFE samples were placed downstream from the ion source on top of the substrate holder and bombarded with, e.g., 1000 eV and 1 mA/cm^2 argon ions and examined with SEM afterward. Vacuum grease was used to provide a better thermal contact between the sample and the sample holder. Ion incidence was normal to the target surface except when substrate rotation was used. The vacuum facility pressure was in the 0.1 mTorr range during ion beam sputtering. However, because of the high etch rate of the FC polymer, the pressure near the etching sample can be much higher. To increase the selectivity between a metal mask and the PTFE even further, an oxygen source is used (i.e., TARIBE) which also should remove carbon residue. During etching PTFE samples with argon and oxygen at similar conditions, there is no difference in etch rate found. However, oxygen should be able to transform, e.g., titanium into the much harder to etch titaniumoxide.

13.7   Results and applications

Work pioneered by NASA Lewis Research Center has led to the nonpropulsive use of ion thrusters in the area of FC surface modification for biomedical applications including sputter texturing for soft tissue implants [27, 28], cardiovascular prostheses [29], percutaneous connectors [30, 31], and hydrocephalic shunts [32]. Thin polymer sheets may be ion beam textured and subsequently epoxy bonded to surfaces for moisture protection, load bearing surfaces, or chemically inert low-friction surfaces [33–35]. The IBE-textured FC polymers enable the capability of writing or printing on the PTFE surface, decal applications, bonding using adhesive tapes, or as encapsulants for solar cells using silicone adhesives [36]. In literature, additional information is found concerning the roughening and subsequently bonding of non-PTFE surfaces like graphite, surgical stainless steel, titanium, titanium 6–4 nickel, Co-Cr 46–20, CTFE, PFA, FEP, polyurethane, and polyolefins [36, 43]. These materials are found in the field of biomedicine and implantology, e.g., dental implants, orthopaedic and vascular prosthesis, and artificial heart components.

Adhesion

Many methods have been studied to improve the strength of adhesive bonds to fluoropolymers, including chemical treatment [37], rf sputtering processes [21, 25, 38], glow discharge [39], metal evaporation [40], and others [41]. Among these methods, the sodium etching method has been adapted industrially. The extremely rough surface morphology of sputter-textured FC polymers enables strong mechanical attachment to high modules adhesives. Sputter-etched PTFE has demonstrated superior adhesiveness to epoxy resins in comparison to conventional sodium/naphthalene (Matheson's Poly Etch) treated PTFE [1, 21, 25, 36]. The bond can be even stronger when using both techniques subsequently. The use of the intermediate adhesive layer technique to transfer a resist pattern is shown in Fig. 13.5. After evaporating a thin titanium layer on top of a PTFE sheet, resist is spun on top of this layer and IBARE patterned with a line pattern. After this, the PTFE is etched in the IBE.

Biomaterial

The etched PTFE sheet can be used directly as a biocompatible filter for, e.g., blood filtering [26]. Such filters are strong because of the high plasticity of PTFE. In Fig. 13.6 an example of a 5-micron filter is given.

Direct moulding

An even more exciting application is the use of an etched PTFE substrate for direct moulding [11]. Due to the low adhesion property and remarkably good thermal stability (up to 300°C), a patterned Teflon sheath could be used directly as a stamp for the replication of other

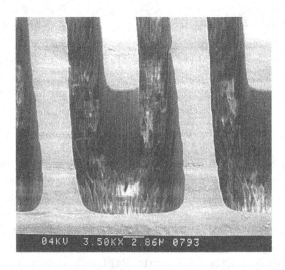

Fig. 13.5. The use of a titanium intermediate layer to transfer a resist pattern into PTFE.

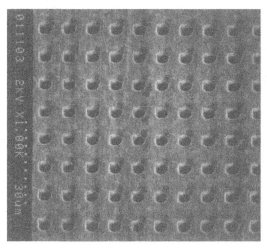

Fig. 13.6. A Teflon sieve with holes of 5 microns etched with the help of a 1-micron-thick nitride shadow mask sieve.

polymers. In this way, typical moulding problems such as the difference in thermal expansion between a nickel mould and the polymer to be moulded, the brittleness of a silicon mould, or the embossing of the mould insert are minimized. Because of the low surface free energy of PTFE, it should be easy to remove an electrochemically grown structure from the PTFE. However, presently we are not able to prevent the micrograss completely ($\Rightarrow$adhesion), because the IBE system features carbon grids for ion extraction. Figure 13.7 shows a 55-micron-deep structure into a PTFE sheet. The structure is etched with the help of a shadow mask within 30 minutes TAIBE. At this moment the use of electrons instead of ions to etch PTFE is studied, i.e., TAEBE, because in this mechanism it is much easier to control and minimize the particle contamination. Moreover, the TAEBE equipment is straightforward and astonishingly simple. First results are quite promising, because the grass problem disappeared, and will be presented in the future.

Microvalve

If substrate rotation is used under a tilted angle, a unidirectional valve will be etched directly under a pinhole, as shown in Fig. 13.4d. A flow coming from the pinhole side may easily pass the microvalve, whereas a flow from the opposite direction will be blocked.

3-D structuring

Another interesting possibility is to use a PTFE sheet to structure surfaces which are not flat, because the PTFE is easily folded around the structure. To give another fantastic example, it is possible to IBE holes ($\varnothing = 18\ \mu$m) into a PTFE tube ($\varnothing = 375$–$625\ \mu$m) using electroformed nickel mesh masks for hydrocephalic shunts, as demonstrated by Garner et al. [13]. Holes with the smoothest walls and the sharpest definition were obtained by using low beam power densities and a tubing target temperature of less than 50°C. The perforated microtubules have an important application in medicine for sufferers of hydrocephalus, a malady which results in the buildup of cerebrospinal fluid in the brain. The tubing is inserted into the ventricle and serves as a shunt by draining off the excess cerebrospinal fluid into

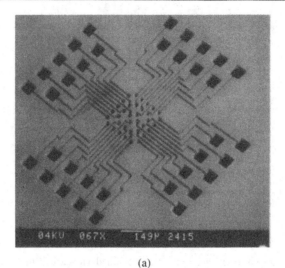

(a)

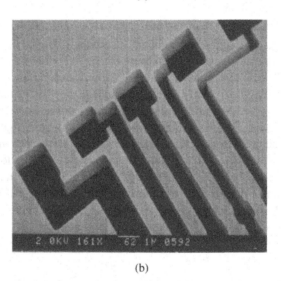

(b)

Fig. 13.7. (a) A 50-micron-thick silicon shadow mask with 10 micron resolution. (b) A Teflon sheet 55 microns thick etched with the help of a shadow mask.

another part of the brain, where the fluid is absorbed by normal processes. Figure 13.8 shows an example of the indirect technique to construct 3-D structures. It shows PTFE pillars 3 microns in diameter and 55 microns high after 30 minutes IBE. The evaporated mask is 200 nm of titanium.

Bonding

FC polymers can be first textured and then bonded to themselves using epoxy resin [36, 42]. We experienced that a PTFE sheet with an evaporated titanium/aluminium sandwich layer can be bonded to a glass or prepared silicon wafer. This technique opens a complete new set of possibilities.

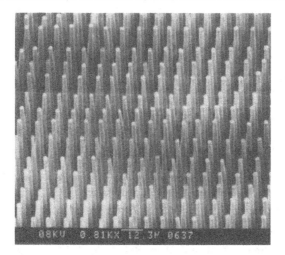

Fig. 13.8. Pillars 55 micron high, 3 micron in width etched with the reverse shadow mask technique.

MEMS

Etched PTFE structures with deposited metal layers can be used to construct MEMS. Many techniques are available, as found in de Boer et al. [46].

References

[1] B. A. Banks, J. S. Sovey, T. B. Miller, and K. S. Crandall, NASA TM-78888, 1978, p. 2, Eight International Conference on Electron and Ion Beam Science and Technology, Seattle, Washington, USA, May 21–26, 1978.

[2] H. Jansen, H. Gardeniers, M. de Boer, M. Elwenspoek, and J. Fluitman, Journal of micromechanics and microengineering, 6 (pp. 14–28), ISSN 0960-1317.

[3] A. J. Weigand, NASA TM-78851, 1978.

[4] H. V. Jansen, J. G. E. Gardeniers, J. Elders, H. A. C. Tilmans, and M. Elwenspoek., Sensors and Act. **A41–42**, 1994, pp. 136–140.

[5] J. Elders, H. V. Jansen, and M. Elwenspoek, proc. MEMS, Oiso Japan, (1994, Jan. 25), pp. 170–175.

[6] W. M. Moreau, *Semiconductor Lithography* (New York, 1991). Plenum Press.

[7] H. Lehr and W. Ehrfeld, Aix-en-Provence, France, 1994.

[8] J. Moran and D. Maydan, J. Vac. Sci. Tech. **19** (1981), p. 872.

[9] L. Fried , J. Lechaton, P. Totta, J. Logan, J. Havas, and G. Paal, IBM J. Res. Dev. **26** (1982), p. 362.

[10] J. W. Bartha, J. Greschner, M. Puech, and P. Maquin, Microelectr. Eng. **27**, 1994, p. 453.

[11] O. Auciello, J. Vac. Sci. Tech. **19** (1981), pp. 841–867.

[12] J. S. Sovey, J. Vac. Sci. Tech. **16** (1979), pp. 813–816.

[13] C. E. Garner, S. B. Gabriel, and Y. S. Kuo, Thin Solid Films, **95** (1982), pp. 351–362.

[14] M. White, Thin Solid films **18**, 157 (1973).

[15] E. Mathias and G. H. Miller, J. Phys. Chem. **71**, 2671 (1967).

[16] D. F. Hall and H. E. Green, AIAA paper No. 72-446, April (1972).

[17] M. Rost, H. J. Erler, H. Giegengack, O. Fieder, C. Weissmontel, Thin Solid Films **20** (1974). p. S15.

[18] G. K. Wehner, Phys. Rev. **102** (1956), p. 690.

[19] P. K. Rol, J. M. Fluit, and J. Kistemaker, Physica **26** (1960), p. 1000.

[20] N. Laegreid and G. K. Wehner, J. Appl. Phys. **32** (1961), p. 365.

[21] T. Moriuchi, S. Yamamoto, M. Ezoe, H. Tabata, K. Uemori, F. Shigeta, Y. Ohya, A. Tsumuro, and J. Nakai, 7th Intern. Vac. Congr. and 3rd Intern. Conf. Solid Surfaces, Vienna, 1977, p. 1501.

[22] H. Jansen, M. de Boer, R. Legtenberg, and M. Elwenspoek, J. Micromech. Microeng. **5** (1995), pp. 115–120.

[23] Goodfellow Cambridge Ltd., Cambridge CB4 4DJ England.

[24] G. J. Burger et al., Transducers '95, Stockholm, Sweden, June 25–29, 1995, p. 144.

[25] J. Nakai and K. Fukunaga, US patent No. 4,064,030, 1977. Dec. 20.

[26] C. J. M. van Rijn and M. Elwenspoek, IEEE MEMS 199XIII., Amsterdam, p. 83, Proc. MEMS '95, January 29–February 2, 1995. Amsterdam, The Netherlands.

[27] D. F. Gibbons, NASA contract Rep. CR-159358, 1980, p. 2.

[28] D. F. Gibbons, NASA contract NAS3-22443, 1980, p. 2.

[29] B. A. Banks, A. J. Weigand, C. A. Babbush, and C. L. van Kampen, NASA Tech. Memo TMX-73512, (1976), p. 1.

[30] J. Miller and C. E. Brooks, J. Biomed. Mater. Res., Symp. **2** (1971), p. 251.

[31] G. J. Picha and D. F. Gibbons, NASA contract Rep. CR-165255, 1981, p. 2.

[32] E. Foltz, NASA contract NAS3-21963, 1980, p. 1.

[33] J. R. Hollahan, T. Wydeven, and C. C. Johnson, Appl. Opt. **13** (1974), p. 1844.

[34] D. T. Morrison and T. Robertson, Thin Solid Films **15** (1973), p. 87.

[35] R. Harrop and P. J. Harrop, Thin Solid Films **3** (1969), p. 109.

[36] M. J. Mirtich and J. S. Sovey, J. Vac. Sci. Tech. **16** (1979), p. 809.

[37] A. A. Benderly, J. Appl. Polym. Sci. **6**, 221 (1962).

[38] D. M. Matton and J. E. McDonald, J. Appl. Phys. **34** (1963).

[39] H. Schonhorn and R. H. Hansen, J. Appl. Polym. Sci. **11** (1967), p. 1461.

[40] R. F. Roberts and F. W. Ryan, H. Schonhorn, G. M. Sessler, and J. E. West, J. Appl. Polym. Sci. **20** (1976), p. 255.

[41] G. C. S. Collins, A. C. Lowe, and D. Nicholas, Eur. Pol. J. **9** (1973), p. 173.

[42] TRA-CON, Inc. TRA-CAST BA-2114 epoxy resin.

[43] A. J. Weigand and B. A. Banks, J. Vac. Sci. Tech. **14** (1977), pp. 326–331.

[44] H. Jansen, M. de Boer, and M. Elwenspoek, (1996, February 11), Proc. of IEEE MEMS '96, pp. (250–257). San Diego, California, USA.

[45] S. M. Rossnagel, J. J. Cuomo, and H. R. Kaufman, *Handbook of Ion Beam Processing Technology*, (Noyes publications, Park Ridge, New Jersey, USA, 1989).

[46] M. de Boer, H. Jansen, and M. Elwenspoek, (1995, June 25), Proc. of Transducers '95. pp. (565–568). Stockholm, Norway.

[47] R. Kelly, Surf. Sci. **90** (1979), p. 280.

[48] H. W. Starkweather, Jr., Macromolecules **19** (1986), pp. 1131–1134.

[49] C. Chang, Y. Kim, and A. G. Schrott, J. Vac. Sci. Tech. **A8** (1990), pp. 3304–3309.

[50] James M. E. Harper, *Plasma Etching; An Introduction: Ion Beam Etching*, eds. D. M. Manos and D. L. Flamm (Academic Press, Inc., San Diego, 1989), p. 391.

[51] P. Reese Puckett, Stephen L. Michel, and William E. Hughes, *Thin Film Processes II: Ion Beam Etching*, eds. J. L. Vossen and W. Kern (Academic Press, Inc., San Diego, 1991), p. 673.

14

High aspect ratio trench etching

In the first years after the introduction of dry plasma etching in microelectronics, this technique was mainly used for the ashing of photoresist because of its cleanliness and high selectivity [1]. In the beginning, the isotropic nature of these so-called radical etchers (RE = PE) was not a problem, but the ever decreasing dimensions in the integrated circuits forced the research institutes to develop dry plasma systems for anisotropic etching, and the ion beam etching (IBE) was born [2]. However, the etch selectivity between mask and substrate was rather poor and special IBE reactors, e.g., chemical assisted ion beam etching (CAIBE), were designed [3].

A major step in the direction of large-scale integration was taken after the ion beam assisted radical etching (IBARE = RIE) became available [4]. In IBARE it is possible to get a very high selectivity with a perfect anisotropy [5]. Nowadays, to increase the electronic circuit density, the use of the surface area of the silicon wafers only is not sufficient and the research is focused on using the third dimension: depth. So, there is an increasing demand for processes which are able to create not only a high selectivity and anisotropy but also high aspect ratio trenches (HARTs). The aspect ratio (AR) is defined as the maximum depth divided by the maximum width. Such HARTs, e.g. a quarter micron in width and three micron in depth, are needed for, e.g., transistor trench isolation and trench capacitors. HARTs are not only important for silicon etching. The etching of HARTs in polymers is a prime technology [6].

In micromechanics, as in microelectronics, profile control of HARTs is increasingly important. These trenches give rise to specific problems: the aspect ratio dependent etching (ARDE) effects and tilting [7–9]. In Fig. 14.1 these effects are shown: RIE lag, bowing, and trench area dependent tapering of profiles (TADTOP, Fig. 14.1(a)), bottling (Fig. 14.1(b)), micrograss (Fig. 14.1(c)), and tilting (Fig. 14.1(d)). The main subject of this chapter is to explain and control these effects. Therefore, the IBARE process is analyzed with the help of four sets of variables: the IBARE setting, the equipment parameters, the plasma characteristics, and the trench-forming mechanisms. These variables control the final HART profile. The study continues with information concerning the equipment: a conventional high-pressure IBARE and a cryogenically cooled low-pressure IBARE with a high density source. After information about the sample preparation, results and discussions are given for the HARTs which are etched. Because the tapering of the trenches is changing with IBARE setting, the black silicon method (BSM) is used to be able to change the setting while keeping the profile anisotropic [10]. The influence of ion energy and trajectory is found to be most critical.

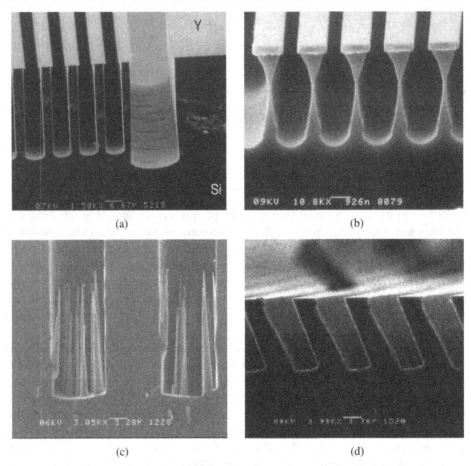

(a) (b)

(c) (d)

Fig. 14.1. (a) Important ARDE effects: (1) RIE lag is associated to the effect that smaller trenches are etched at a different speed than wider trenches. It is explained in considering the depletion of reactive particles during their trajectory in the trench. (2) Bowing is the observation of a parabolic curvature of etched sidewalls as found for the wider opening. (3) Trench area dependent tapering of profiles (TADTOP) is the effect that wider trenches have different slopes/tapering with respect to the smaller trenches. (b) Bottling shows up as bottle-shaped trenches and is caused by ion shadowing and sharpening of the collimated ion beam incoming from the glow zone. Generally, it is pronounced only at a higher pressure. (c) Micrograss is the existence of long-tailed spikes at the trench bottom. Generally, the appearance of such rough surfaces is grey or even completely black. (d) Tilting is observed near geometrical obstacles and is due to boundary distortion or local differences in radical density. In this case, the wafer edge is responsible for the tilting.

14.1 Qualitative analysis

There are two fluxes of particles of major concern in IBARE: ions and radicals. These particles are the main source for etching. Together with the neutral density they determine how the final profile will evolve in time during etching, i.e., they determine the HART effects. To get an idea of the behavior of the particle fluxes and energies, the etch process is split into four sets of variables, as found in Fig. 14.2: the (i) IBARE setting (computer control), (ii) equipment parameters (reactor dimensions, pressure regime, or maximum

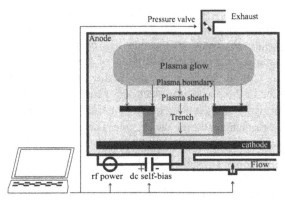

Fig. 14.2. Basic IBARE system.

power), (iii) plasma characteristics (glow, boundary, and sheath), and (iv) trench-forming mechanisms. The HART effects are explained while using these variables. The IBARE setting is controlled by the operator, whereas the equipment parameters are determined by the manufacturer. The plasma characteristics are dependent on the IBARE setting and the equipment. Together with the trench-forming mechanisms these figures are responsible for the final result of the trench profile. In this section these sets of variables are qualitatively highlighted.

14.1.1 IBARE setting

This set is directly controlled by the operator. It consists of the familiar pressure, flow, power, and temperature. Sometimes it is possible to create a dc self-bias which is not directly controlled by the rest of the parameters. For example, when the reactor geometry is changed (showerhead) or when an extra plasma source (ICP) is used, the dc self-bias will be an independent variable. Another parameter controlled by the operator is the sample to be etched. This can be a conducting material like silicon but also an insulating polymer. There are some variables depending on the other settings. The residence time t_r [s] is one of them and expressed in the others as $t_r \sim pV/Q$, with p the operation pressure, V the reactor volume, and Q the total flow.

14.1.2 Equipment parameters

The manufacturer of an IBARE system determines most boundary conditions for the IBARE setting. For example, when a high flow is needed at a low operation pressure, the manufacturer is forced to imbed a vacuum system able to handle such flows, e.g., a turbopump with a high throughput. Other examples are the anode/cathode ratio and the rf frequency which are responsible for the magnitude of the dc self-bias created by the plasma.

14.1.3 Plasma characteristics

These variables are a function of the IBARE setting and the equipment parameters. In this section a closer look at the plasma is given. Most plasma reactors consist of two electrodes connected to an rf voltage source which encloses a low-pressure gas as a dielectric.

The rf electrical field will force electrons to the positive electrode. In their way they will collide with the feed gas generating the gas phase etching environment which consists of neutrals (N), radicals (R), electrons (E), ions (I), photons (P), and phonons (T). The photons are responsible for the characteristic glow of the plasma. Since the electron mobility is much greater than the ion mobility, the electrons are able to track the rf electrical field. So, after ignition of the plasma the electrodes acquire a negative charge whereas the plasma becomes positively charged. Of course, only electrons in the direct neighborhood of the electrodes will reach them during the rf cycle. Therefore, a thin region depleted from electrons will be developed close to the electrodes: the plasma sheath. Because there are no electrons in this region to generate photons, the region is dark and the plasma sheath is also known as the dark space. The rest of the plasma is called the glow region. Both regions are separated clearly by a boundary layer. Shortly, a plasma can be divided into three components: (i) the glow region full of electrons, (ii) the sheath region depleted from electrons, and (iii) the boundary layer separating both regions. Due to the charging of the glow region, a dc electrical field will exist in the sheath region forcing positive ions to the electrode. This phenomena is typical for IBARE: a continuous flow of directed ions together with an isotropic flux of radicals. In most cases, the area of the two electrodes facing the plasma is not the same. Therefore, a different negative charging of the electrodes is a result and the dc voltage between the glow and one of the electrodes will differ from the other one. This difference in dc voltage between the two electrodes is called the dc self-bias which is a measure of extreme importance in IBARE. The electrode having the biggest area facing the plasma is always at a higher dc potential than the smaller one and is, therefore, called the anode. The smaller electrode is then, of course, the cathode.

The glow region

This region is characterized by a set of glow variables: the particle densities (ρN, ρR, ρI, ρE, ρP, and ρT) and the energy of these particles (expressed in eV or K, where 1 eV = 8,000 K). Other examples are the power and energy density and the floating potential, i.e., the potential of a floating sample in the glow region with respect to the anode potential. The electrons are the only particles able to gain energy from the rf power source. Therefore, an important variable is the electron energy distribution function (EEDF, Fig. 14.3(a)), which is a measure of the energy of the electrons. Generally, the electrons will collect energies between 1 and 10 eV before colliding with another particle, which is strongly dependent on the operation pressure of the IBARE. A typical particle generated by electron impact is the photon. The energy of the photon (expressed in eV or nm, where 1 eV = 1200 nm) is dependent on the electronic configuration of the bombarded particle and therefore the photon energy distribution function (PEDF, Fig. 14.3(b)) is related to the feed gas. The total spectrum of photons is giving the plasma a typical color. For example, the color of a nitrogen plasma is pinkish whereas an oxygen plasma generally is greyish-white, although this color is strongly dependent on the plasma gas pressure.

The boundary layer

The energetic particles from the plasma glow should be transported to the sample surface. This is accomplished by a flux of particles through the plasma boundary layer. The radicals are having only thermal energy and are leaving the surface in all directions, i.e.,

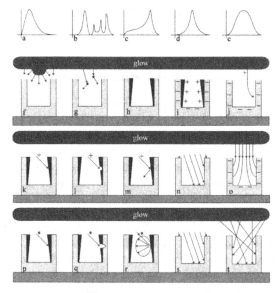

Fig. 14.3. Distribution function and trench-forming mechanisms: (a) EEDF, (b) PEDF, (c) IEDF, (d) IADF, (e) RADF, (f) boundary distortion, (g) particle collision, (h) sidewall passivation, (i) sidewall charging, (j) ion deflection, (k) ion capturing, (l) ion etching, (m) ion reflection, (n) ion shadowing, (o) ion depletion, (p) radical capturing, (q) radical etching, (r) radical reflection, (s) radical shadowing, and (t) radical depletion.

isotropically. However, because the glow is a conductor, the electrical field is pointing exactly from the boundary surface and the ions will leave the boundary layer under an angle of 90°.

The sheath region

Both radicals and ions will collide with gas particles while passing the sheath. The effect of the pressure on the ions especially is crucial [17]. Because of the collisions of ions with other particles, ion dispersion will occur, i.e., their direction will not exactly correspond with the normal of the boundary layer anymore. This effect is expressed with the help of the ion angular distribution function (IADF, Fig. 14.3(d)). At the same time, the energy of the ions is exchanged with the particles and this effect is found in the ion energy distribution function (IEDF, Fig. 14.3(c)). The rf frequency is responsible for a varying plasma potential. So, the energy of an ion is depending on the time that the ion enters and leaves the boundary layer. This effect should be incorporated in the IEDF also. Saying it differently, the IADF is a measure for the degree of collimation or dispersion of the flux of ions. A sharp IADF means that most ions travel in the same direction. The IEDF is a measure for the energy content of the ions. A sharp IEDF means that most ions arrive with the same energy at the sample surface. Generally, the IADF and IEDF are strongly correlated, i.e., they cannot be varied independently.

In most cases, the collision of radicals with other particles is not important because, generally, this flux is already isotropic. However, during extensive etching of silicon there will be a continuous net flow of radicals from the nonetching surrounding to the etching areas. Of course, this flow will be directional and the radical angular distribution function

(RADF) will become a meaningful expression (Fig. 14.3(e)). When there is a directional radical flux, the RADF is broadened due to radical collisions.

14.1.4   Trench-forming mechanisms

The flux of particles from the plasma glow region is used to etch a sample. For example, this paper discusses the SF_6/O_2-Si system, which is an ion-inhibitor process, to explain the mechanisms during etching HARTs. In this system the oxygen is passivating the silicon surface with siliconoxide and the SF_5^+ ions are etching the passivator making it possible for the F radicals to etch the silicon. During etching, HARTs specific mechanisms which control the profile are important, such as ion deflection, radical depletion, and wall passivation. These mechanisms are a function of the other variables and of the AR of the HARTs.

- *Boundary distortion* (Fig. 14.3(f)) is found when a sample is placed on top of the cathode. This sample will distort the boundary layer which will conform to the sample geometry. Therefore, ions will leave the boundary with an off-normal angle with respect to the cathode surface. This effect is pronounced when the sample/trench geometry is greater than the thickness of the sheath region.
- *Particle collision* (Fig. 14.3(g)) is caused by gas molecules in the sheath region. The mean free path of particles is ca. 50 $\mu m * Torr$, i.e., 5 cm at 1 mTorr or 0.5 mm at 100 mTorr. These collisions are the main reason that the RADF, IADF, and IEDF are necessary functions to know in the IBARE process. Particle collisions with gas molecules in the trench could be found when the pressure is high due to reaction products. However, the reaction product is SiF_4, thus completely dependent on the radical flux from the glow region. This flux is always much lower than the neutral flux. Therefore, the pressure in the trench always equals the operation pressure and we only have to consider the RADF, IADF, and IEDF.
- *Sidewall passivation* (Fig. 14.3(h)) is necessary in ion inhibitor IBARE processes to achieve anisotropy. For anisotropic etching, it is necessary that the horizontal etching (undercut) is as small as possible whereas the vertical etching, i.e., the etch rate should be large. For the SF_6/O_2-Si system, the horizontal etching is dependent on the thickness of the passivating layer and the F-atom density, which is trying to etch the silicon by penetrating this layer. The thickness of this layer is a function of, e.g., the O atom density, the ion impact, and the local temperature. The F-atom density is a function of, e.g., the SF_6 flow, power, and (micro)loading.
- *Sidewall charging* (Fig. 14.3(i)) will occur when the sidewall inhibitor is an insulator. Ions colliding with the sidewall will leave their charge, which is difficult to compensate with electrons from the silicon behind the inhibitor. This charge will create an electrostatic field which repels a next ion.
- *Ion deflection* (Fig. 14.3(j)) is mainly caused by the diffraction of ions while entering a trench (due to the image force) but also by the negative potential of conducting trench walls with respect to the plasma glow resulting in an electrostatic deflection of these ions to the walls [11, 12]. It is concluded in this chapter that ion deflection is the driving force behind most ARDE effects.
- *Ion capturing* (Fig. 14.3(k)) is found when the created dangling bond isn't used for a fluor atom and an oxygen radical is connected to the free bond. In this case the product isn't volatile and there is no net etching. The battle between the fluor and

oxygen radicals is dependent on how strong the passivation can be. For example, the surface mobility of F atoms is decreasing with temperature, although its lifetime at the surface increases. Therefore, when an ion has created a dangling bond, the F atom might be too late to fill the vacant place. Instead, oxygen gas will take the place.

- *Ion etching* (Fig. 14.3(l)) happens when the ion isn't reflecting. In most cases, the highly energetic ion will remove the inhibitor and create dangling bonds for the Si atoms. These bonds can be connected to fluor radicals, which will remove the Si, because SiF_4 is volatile.

- *Ion reflection* (Fig. 14.3(m)) will occur when ions collide with a surface under a glancing angle. In principle, the chance reflection that takes place will follow a cosine rule, as found in the literature about ion beam etching. When the incoming ion is kinetically highly energetic, the reflection will be specular. Clearly, ion reflection is broadening the IADF in the trench.

- *Ion shadowing* (Fig. 14.3(n)) is caused by the top side of the trench. When ions arrive under an angle, the top side of the trench will block ions from etching a part of the trench sidewall. In other words, the amount of ions arriving at the sidewall is dependent on its position: In the higher regions ions from approximately half the IADF will be captured by the wall. In contradiction, in the lower regions the IADF will be sharpened more and more, because ions arriving at a high angle with respect to the sample surface normal are blocked by the trench top side. Thus, the ion shadowing is responsible for the sharpening of the IADF in the trench.

- *Ion depletion* (Fig. 14.3(o)) due to ion etching or capturing is an important parameter to achieve HARTs. Ions collected by the sidewall of a HART can't be used for the vertical etching. The relative amount of ions arriving at the trench bottom with respect to the total incoming flux depends on the AR only. Therefore, the etch rate is decreasing with the AR of a HART and thought to be responsible for most ARDE effects found in this study, but in particular for RIE lag.

- *Radical capturing* (Fig. 14.3(p)) is found when the radical hits a wall. This radical will spend a certain time at the surface before it is connected to a dangling bond (etching) or leaves again (reflection). The time a radical gets to find such a place before it will desorb is $\tau = \tau_o \exp(E_{ads}/kT)$, with E_{ads} the energy for adsorption and $\tau_o \sim 10^{-13}$ s the characteristic atomic thermal vibration time of a solid. In other words, the time that a nonetching radical will spend on a surface is decreasing when the temperature is increasing. Meanwhile, it is possible for the radical to move along the surface. This surface mobility depends on the surface temperature. The higher the temperature, the higher the surface mobility.

- *Radical etching* (Fig. 14.3(q)) is possible due to penetration of the inhibitor or when the inhibitor is removed by way of ion etching, leaving the silicon unprotected. When four fluor atoms are connected to a silicon atom, this molecule is able to desorb from the surface because it is volatile.

- *Radical reflection* (Fig. 14.3(r)) is found when the radical wasn't able to find a dangling silicon bond in time. This radical will desorb from the surface in a random direction independent of its direction before collision. In most cases during IBARE trench etching, the mean free path of particles is higher than the dimensions of the trench. Therefore, gas collisions in the trench are unlikely and we have to consider collisions with the sidewall only, i.e., molecular flow. The flow resistance of a trench is caused by the random reflection of a low energetic particle, such as radicals, against the sidewall.

Therefore there is only a small chance for a particle to enter the trench. This chance is known as the Clausing factor C and depends on the AR: $C(AR) = (1 + 0.4AR)^{-1}$ for a tube. Clearly, this effect may cause RIE lag. For the highly energetic ions the reflection is specular and therefore the Clausing factor is 1, i.e., the ions which enter a trench will not be backscattered.

- *Radical shadowing* (Fig. 14.3(s)) is identical to ion shadowing and is important because the radical flux is almost isotropic, although random reflections balance this effect.

- *Radical depletion* (Fig. 14.3(t)) is found as the well-known microloading effect. It causes HARTs of Si in open areas to etch at a slower speed than HARTs with the same dimensions located in areas where most Si surfaces are shielded from the plasma with the help of a mask. The same mechanisms causing ion depletion, except for ion deflection, are causing radical depletion. So, only radical reflection or a great surface mobility may transport radicals into the lower trench region, and radical depletion might cause RIE lag problems because radicals are consumed in the higher regions of the trench.

14.1.5   HART effects

The given variables are controlling how the final profile will evolve. However, because the trench-forming mechanisms are a function of the other parameters, it is necessary and enough to consider these mechanisms only.

14.2   Equipment and experimental

In order to track and categorize important ARDE effects, two different IBARE reactors were used. The first is a conventional IBARE used for the high pressures [13]. For the low-pressure experiments a dedicated high-density IBARE is used [14].

14.2.1   Plasmafab 310/340

This reactor has been used for the high-pressure experiments. Information about the equipment can be found in Ref. [15]. This IBARE was modified with a showerhead placed in the reactor. The main purpose of this is to minimize the area ratio between the anode and the cathode surfaces. This change decreases the dc self-bias, and thus the ion energy, which is characteristic for IBARE equipment. Although the showerhead will affect the other plasma parameters also (e.g., the flow resistance of the showerhead increases the operating pressure because the pressure sensor for the automatic pressure control (APC) is located downstream), it is assumed that these changes are minor in comparison to the change in ion energy.

14.2.2   Plasmalab 100

This system features a strong 1000 L/s turbopump connected to the reactor with huge 20 cm pipelines to ensure a low operation pressure (10 mTorr). An inductively coupled plasma (ICP) source is used to create enough plasma particles at such low pressures. An extra matching unit is used to drive the reactor in the IBARE mode creating a dc self-bias. Therefore it is possible to control independently the particle flux coming from the ICP source and the ion energy due to driving the reactor in the IBARE mode. To increase the passivation capability of the oxygen at the trench sidewalls and to ensure a stable wafer temperature, wafers are floating on a thin film of helium cooled down to $-150°C$ and up to $+200°C$.

14.2.3   Experimental

To study the profile development in HARTs, submicron and supermicron patterns were deposited on top of silicon wafers. Silicon wafers, $4''$ in diameter, were processed with submicron 250-nm thick aluminum patterns (down to a quarter micron in width). The supermicron patterns were gratings with a period of 1, 2, 5, and 10 micron. As a mask, a 20-nm chromium layer was applied with the help of the lift-off technique. Both batches didn't get any special treatment to remove dirt or native oxide. Samples to be etched with an SF_6/O_2 plasma were carefully attached to the target platen.

14.3   HARTs

In Fig. 14.4 a typical cross section of some trenches is shown. It can be seen that the shape is identical, not depending on the width. (Note that this statement is not always true, as observed in Fig. 14.1(a) where the TADTOP effect is seen.) This observation enables us to use submicron patterns instead of the larger-sized patterns. After optimizing HARTs with the help of the BSM, the parameter setting will be used to etch the supermicron gratings. This paragraph will treat tilting and the ARDE effects (i.e., RIE lag due to ions or radicals, bowing, TADTOP, bottling, and micrograss) one by one and will explain them with the help of the analysis already given. Every section ends with a conclusion on how the effect is caused and, if possible, solutions are given to prevent these HART effects.

14.3.1   Tilting

This effect is observed for extremely wide trenches and at the wafer corners and is due to boundary distortion and local differences in radical density. Boundary distortion is found when the sample, trench, or wafer-clamping geometry is bigger than the thickness of the

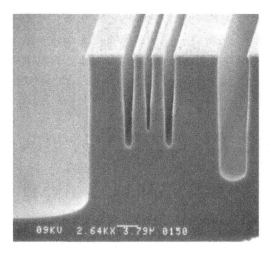

Fig. 14.4. Identical profiles for different trench openings.

sheath region. It produces off-normal ions with respect to the cathode surface area. A mechanical wafer-clamping (obstacle) produces the same effect. In Fig. 14.1(d) a typical example is shown caused by boundary distortion at the wafer edge. Similarly, a difference in radical concentration might cause tilting because this concentration is controlling the radical flux coming from a certain direction. For an Si wafer placed on top of a nonetching target platen, a directional radical flux is flowing from the direct surroundings of the Si wafer to the wafer center. Therefore, tilting is possible even when there is no boundary distortion.

So, tilting is induced by boundary distortion and local differences in radical density. The effect of boundary distortion may be minimized by increasing the thickness of the sheath region. This can be accomplished by lowering the pressure. The local differences in radical densities can be minimized by making the mask layout more uniform.

14.3.2 Bowing

In Fig. 14.1(a) a typical example of bowing is observed. In the wider trench a parabolic curvature of the etched wall (i.e., negatively tapered) is observed. This effect is explained with the help of ion deflection due to electrostatic (image) forces and can be minimized by increasing the sidewall passivation, the wall charging, the energy of the ions before entering the trench, or preventing these forces, as in ion beam etching. This can be achieved by cryogenically cooling the sample or by letting an extra passivation gas like CHF_3 or O_2 into the plasma. For nonconducting samples like Teflon or siliconoxide this effect should not be observed.

The effect of bowing caused by ions can be controlled with the help of the BSM. In Fig. 14.5 the bowing has been reduced by making the silicon trench more insulating-like, thus increasing the passivator in the plasma by way of extra oxygen. To ensure a certain profile, the ion flux has to be increased at the same time by way of extra CHF_3 conformed to the BSM. The $SF_6/O_2/CHF_3$ chemistry enables us to passivate the sidewall with siliconoxide or fluorocarbon (FC). The siliconoxide is much stronger but also thinner

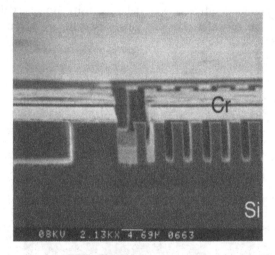

Fig. 14.5. Decreasing the ARDE effect due to a stronger passivation. The depletion of passivating particles is balancing the depletion of ions during their passage down into the trench opening.

than the FC layer. Therefore, ion deflection is more pronounced for SF_6/O_2 etching, although the sidewalls are smoother, as in the case of SF_6/CHF_3 etching, for FC is easily etched with ions [6].

14.3.3   TADTOP

The trench area dependent tapering of profiles is found again in Fig. 14.1(a) where the wider trench is more negatively tapered than the smaller trenches. Ions which enter a small cavity will be less attracted by its nearest wall because its opposite wall is closer than in the case of a wide opening. This wall is neutralizing the closest wall and therefore the smaller opening is more positively tapered. In other words, ion deflection is responsible for this effect and the solutions for this effect are identical with those for the bowing effect (see Fig. 14.5).

14.3.4   RIE lag due to ions

RIE lag is associated with the effect that smaller trenches are etched faster (negative lag) or slower (positive lag) than wider trenches, as found in Fig. 14.1(a). Positive lag may be explained in considering the amount of ions which exist during their trajectory in the trench. During this travel the outer ions will be collected by the negative (conducting) walls due to ion deflection. This process will go on until all the ions are captured by the walls. So, at last there will be no ions left to etch the bottom and etching stops in this direction. (In fact this is not completely true: during trench etching, the higher region of the trench is becoming more and more isolated due to oxidation (Fig. 14.3(i)). Therefore, wall charging will increase ion reflection, i.e., decrease ion depletion.) It is obvious that for the small trenches this ion depletion is reached sooner than for the wider, because the flux/wall area ratio is smaller after a certain etch depth (Fig. 14.1(a)).

Ion deflection should not change the profile of insulating trenches because this effect would soon be overruled by the positive charging of the walls. Therefore, we investigated the IBE of Teflon and the IBARE of SiOxNy layers [6]. During etching these layers, bowing, TADTOP, and RIE lag were barely observed, indicating that the conducting wall can explain these ARDE effects during etching HARTs in silicon [16]. However, in literature, RIE lag is without doubt observed for isolating substrate materials [7]. The reason for this discrepancy could be found in the much lower pressures at which the samples were etched. For example, Teflon is etched with the help of IBE, which operates at an extremely low pressure (<0.2 mTorr). The siliconoxynitride was etched at approximately 30 mTorr, which should make aspect ratios of over 10 possible. The samples were not etched that long to create an aspect ratio exceeding 10, so radical reflection as well as ion and radical depletion are probably not yet important. Anyway, the deflection of ions can be reduced by increasing the passivation. Again, in Fig. 14.5 it is demonstrated that the RIE lag is decreased successfully.

More evidence for the depletion of ions due to ion deflection is found in Fig. 14.1(c). In this figure a 10-micron-wide trench opening is observed with spikes in it. Quite clearly, the spikes are pointing to the center of the trench. Because the direction of the spikes is an indication for the direction of the ions, it is concluded that the ions are forced to the sidewall by the electrical fields in the trench. Of course, ion shadowing may also be responsible for this effect (see also Figs. 14.10(a) and (b)). Note that the sidewall is straight, although the ions collide under an angle.

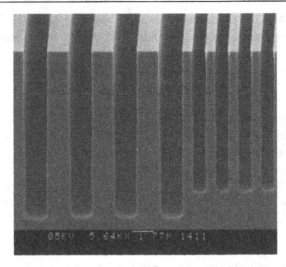

Fig. 14.6. RIE lag in a low-pressure ICP-IBARE with an SF_6/O_2 chemistry.

In Fig. 14.6 a cross section of a trench etched with the help of the ICP-IBARE operating at low pressure is shown. The effect of RIE lag is clearly visible. The phenomena can be explained by the deflection, capturing, and subsequent depletion of ions. Due to the strong wall passivation, ions are not able to etch the inhibitor and their journey is ended at the spot where they collide with the sidewall.

14.3.5   RIE lag due to radical depletion or reflection

In Fig. 14.7, three cross sections of the same channel are shown. The trenches are etched under three different circumstances. Figures 14.7(a) and (b) are etched at relatively high pressure and substrate temperature (100 mTorr, 10°C), and Fig. 14.7(c) is etched at low pressure and substrate temperature (8 mTorr, −100°C). In Fig. 14.7(a) the profile of a trench etched in the middle of the wafer is shown, and Fig. 14.7(b) shows the same channel but now located at the wafer edge. In other words, picture 14.7(a) is taken after etching and breaking the sample whereas 14.7(b) was already broken before etching. It can be observed that trenches at the edge are having an aspect ratio of approximately 20 whereas the trenches in the middle are not exceeding 10. At the edge the etching is not hampered because radicals and ions will easily arrive from the surroundings. The same pattern was also etched at low pressure and substrate temperature. Figure 14.7(c) shows a trench located in the middle of the wafer after breaking the sample. The aspect ratio is 20 and it is found that trenches at the wafer edge are identical in depth (i.e., no RIE lag). Shortly, high AR radical consumption may cause depletion responsible for RIE lag. Radical reflection isn't found, although this effect should be quite pronounced for AR > 10. Maybe, the transport of radicals along the trench sidewall is sufficient to supply enough radicals. Notice that the effect of RIE lag is still present in Fig. 14.7(c). However, this is probably due to ion and not radical depletion, or reflection. However, in order to make a more solid distinction between radical or ion depletion, a subtle approach is necessary. Therefore, in the second part of this chapter, special experiments are carried out to do so.

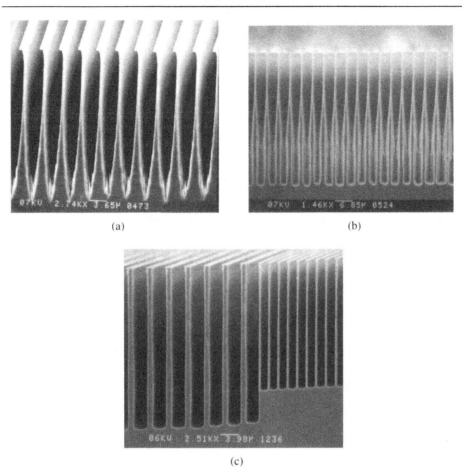

(a)   (b)

(c)

Fig. 14.7. RIE lag due to ion or radical depletion at high pressure and temperature
(100 mTorr, 10°C) in a conventional IBARE sytem using an SF_6/O_2 chemistry (a) in
the wafer center, and (b) at the wafer edge. (c) RIE lag due to ion or radical depletion
at low pressure and temperature (8 mTorr, -100°C) in an ICP-IBARE system using an
SF_6/O_2 chemistry and in the wafer center.

14.3.6 Micrograss

In Figs. 14.8(a) and (b), two trenches etched at a relatively high pressure for a different time are shown. It is observed that the trench etched for a short time (Fig. 14.8(a))
is free from micrograss whereas the deeper trench (Fig. 14.8(b)) is not. In both situations, open silicon area in the direct neighborhood stayed smooth. The reason for this
difference might be the sharpening of the IADF due to the higher AR of the deeper
trench opening. So, in the beginning the ion dispersion is high enough to prevent grass,
but during etching the AR increases and the IADF sharpens, enabling grass to form at
the bottom. Thus, grass is formed when the flux of incoming ions is highly collimated,
i.e., the IADF is sharp. However, when this effect would be the cause, then the grass
should be less pronounced in the very middle of the hole. This is not true, as found in
Fig. 14.8(b), and therefore the sharpening of the IADF is not the problem we are looking for. Another explanation is that particles sputtered from the mask edge are deposited

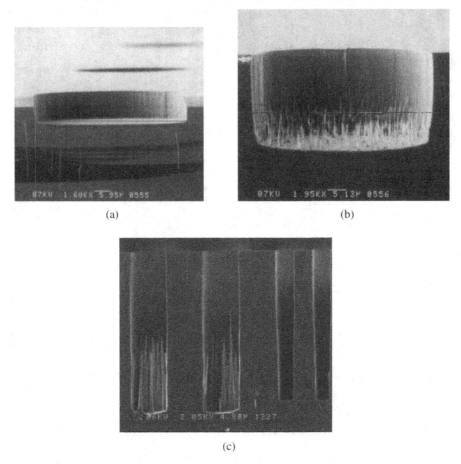

(a)   (b)

(c)

Fig. 14.8. Micrograss as an ARDE effect. (a) The picture is taken after 10 minutes of
etching. The roughness is not yet pronounced. (b) After 20 minutes of etching. Now, the
roughness is clearly observed. (c) Differences in the existence of micrograss in trenches
with different openings. The micrograss is thought to be prevented due to ion reflection
at the sidewalls in case of the smaller openings.

in the trench or hole. This effect accumulates during time, so in the beginning there
will be almost no grass but in time it will "grow." The big open structures are stay-
ing clean, because at these spots there is no sputtered material in the direct neighborhood.

The opposite effect happens when ion reflection is made possible. In Fig. 14.8(c) the
smaller trench is free from grass. Now the sharper IADF could be the cause of the grass
problem. The reflection of ions against the sidewall, which broadens the IADF and is much
more pronounced in the smaller trenches, will prevent the forming of grass. For the wider
trenches this effect is not yet enough.

So, micrograss is formed when the flux of incoming ions is highly collimated, i.e., the
IADF is sharp. The grass is prevented when the IADF is broadened, e.g., by way of a higher
pressure or due to ion reflection. In other words, a perfectly collimated ion flux is not always
preferable; a little dispersion will prevent grass.

14.3.7 Bottling

In Fig. 14.1(b) the effect of bottling is found. Such profiles are quite common in deep trench etching and are often confused with bowing. Bowing is caused by ion deflection and is responsible for the negative slope of trenches. In contrast, the bottle-shaped trenches are caused by ion shadowing and sharpening of the IADF when ions are traveling down the trench. In other words, the effect is caused by off-normal ions due to a relatively high operating pressure. In the higher regions of the profile, ions are coming in at different angles and there will be a strong undercut depending on the ion energy. However, the ions which are hitting and etching the sidewall are used up, and due to the ion shadowing (i.e., blocking of incoming ions due to the opposite sidewall) the angular distribution function is sharpened. At a certain critical angle the ions are not etching the sidewall anymore but they will be captured or reflected, i.e., bounce until they hit and etch the bottom of the trench. In other words, in the higher regions of the trench, ions will etch the sidewall, whereas in the lower regions this etching stops, because these ions will fail due to ion shadowing and ion depletion. Therefore, ion shadowing and ion depletion are thought to be responsible for the bottling effect.

To prevent the bottling effect it is possible (i) to sharpen the IADF by way of, e.g., decreasing the pressure or (ii) to decrease the IEDF by way of, e.g., the dc self-bias. In Fig. 14.9 an example of this last technique is shown. The parameter setting for both pictures is the same except that in Fig. 14.9(b) a showerhead is mounted into the conventional IBARE reactor to lower the dc self-bias, and thus the ion energy. It is observed that the amount of bottling is less pronounced for the trenches etched with the help of a showerhead. It could be that low energetic ions are more easily reflected or captured when hitting a wall under a certain angle. So, although the angular distribution of ions coming from the plasma boundary might be the same for both experiments, the energy is too small to etch the sidewall in the case of the experiments performed with showerhead. However, note that another explanation of this result is that, due to the showerhead, the thickness of the sheath might decrease. So, the plasma boundary is much closer to the wafer surface and the ions passing the sheath will encounter less collisions than in case when there would be no showerhead. Therefore, the effect of the showerhead is to sharpen the IADF.

The same effect is observed for the high-density source at low pressures during changing the dc bias from 60 to 200 volts, thus indeed the ion energy is controlling bottling. When the ion energy is low enough (<10 eV), even a very broad (i.e., high-pressure) ion distribution function is completely captured by the wall or reflecting and only etching the bottom. Figure 14.9(c) shows a HART etched at relatively high pressure (100 mTorr) with a conventional IBARE.

14.4 Quantitative analysis of RIE lag

As stated in the preceding sections, etching high aspect ratio trenches into silicon and other materials is becoming increasingly important for microengineering and nanoengineering, as in the case of, e.g., comb-driven structures, trench capacitors, and trench isolation for vertical transistors [18–20]. Currently, the most important technique to etch HARTs is IBARE. In the first part of this chapter it was observed that, while etching HARTs with IBARE in an SF_6/O_2 chemistry, the etch rate depended on time and the mask opening. In general, smaller trench openings are etched slower than those which are wider (Fig. 14.10(a)). This effect is

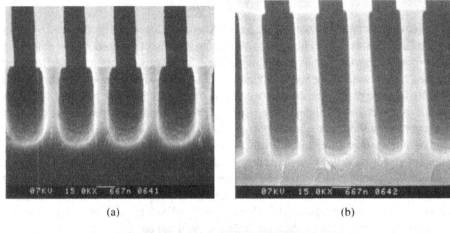

(a)   (b)

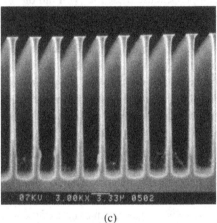

(c)

Fig. 14.9. The influence of the ion energy and collimation on bottling. (a) When a showerhead facility is not installed, the incoming ion beam is quite broad at a relatively high
pressure of 100 mTorr resulting in a bottle-shaped profile. (b) When the showerhead is
installed, the incoming ion beam has a lower kinetic energy and might be narrowed due
to the decreased sheath thickness, resulting in a more straight-down profile, notwithstanding the relatively high pressure of 100 mTorr. (c) The influence of the ion energy
on bottling. A HART etched with a conventional IBARE system at 100 mTorr and 10°C
while using the showerhead.

known as RIE lag and seems to depend on the AR of the trench (i.e., AR scaling) rather than
on the depth or width of the trench (i.e., feature size scaling). RIE lag is not only found in
the conventional plasma reactors; dedicated machines equipped with high density sources
and cryogenic cooling are facing identical problems. Moreover, the chemistry used to etch
the sample and even the sample material is relatively unimportant with respect to RIE lag
[21]. When the effect which is causing RIE lag is known, it is possible to minimize the
decrease in etch rate with increasing AR or decreasing feature size.

    In the previous sections, the RIE lag was analyzed qualitatively. In this section the RIE
lag will be analyzed quantitatively. First, a short overview of experiments carried out by
other researchers will be given. Then, the inhibitor and radical depletion will be estimated as

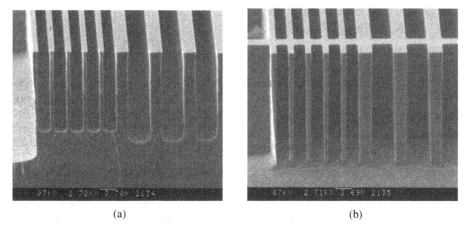

(a)   (b)

Fig. 14.10. (a) RIE lag midways a grating pattern and (b) RIE lag free trenches etched at the grating edge. Both etched with an SF_6/O_2 chemistry in an ICP-IBARE system.

well as the apparently more pronounced ion depletion. Finally, the experimental techniques to find the mechanism(s) causing the lag effect will be treated.

14.4.1 Problem description

RIE lag is clearly visible in the cross section of the grating shown in Fig. 14.10(a). However, the RIE lag is nearly observed at the edge of the grating of Fig. 14.10(b). It is observed in Fig. 14.10(b), when we look into the grating, that the RIE lag is increasing fast. So, RIE lag is not only formed top-downward the grating but also from sideways the grating.

There are three sources which may cause RIE lag: the flux of the etching ions, of radicals, and of the inhibiting neutrals which form a passivating layer. When the flux of one of these streams of particles changes during its passage through the trench, RIE lag will be the result. It is difficult to decide which mechanism is responsible for the RIE lag observed in Fig. 14.10. Radical as well as ion depletion are candidates.

14.4.2 Ion beam assisted radical etching

Reactive ion etching (RIE), or better, ion beam assisted radical etching (IBARE), enables the achievement of profile control due to the synergetic combination of physical sputtering with chemical activity of reactive species with high etch rate and high selectivity. Ion-enhanced etching can be divided into two main groups: ion-induced IBARE (reaction-controlled etching) and ion-inhibitor IBARE (desorption-controlled etching).

Ion-induced IBARE

This technique is used when the substrate is not etched spontaneously, such as in the Cl_2/Si or O_2/polymer system at room temperature. In ion-induced IBARE a directional flux of ions coming from the plasma boundary region assists a random stream of radicals. In such processes, the yield per ion and the reaction probability of radicals depend on both the radical and ion flux. This synergetic mechanism is expressed with a model proposed by Mayer and Barker based on Langmuir adsorption kinetics [23]. They assume that the etch rate R [m/s] is proportional to the surface coverage χ_R of the radicals times the ion energy E_I [eV] and the ion flux ϕ_I [mol/m^{2}s]:

$$R = k_I \phi_I E_I * \chi_R \quad \text{[m/s]}, \tag{14.1}$$

where k_I [m^3/eVmol] is the volume removed per mol-unit ion bombardment energy. Another assumption is that the etch rate is proportional to the number of bare sites $(1 - \chi_R)$ on the surface times the radical flux ϕ_R, that is,

$$R = k_R \phi_R S_R * (1 - \chi_R) \quad \text{[m/s]}, \tag{14.2}$$

where k_R [m^3/mol] is the volume removed per mol-unit radicals, and S_R the sticking probability of the radical on a bare surface. We see that for zero ion energy flux ($\phi_I E_I = 0$), the etch rate is zero (Eq. (14.1)) and therefore the radical surface coverage becomes one, $\chi_R = 1$ (Eq. (14.2)). We will come back to this subject later on. Extracting χ_R from both equations results in an expression for the etch rate R as a function of the ion energy flux and radical sticking flux:

$$\frac{1}{R} = \frac{1}{k_R \phi_R S_R} + \frac{1}{k_I \phi_I E_I} \quad \text{[s/m]}. \tag{14.3}$$

Note that the ion-radical synergism is clearly evident in this expression: In both cases when the ion or radical flux is small, the etch rate becomes vanishingly small. This implies that this model isn't able to handle purely chemical (i.e., spontaneous) or physical (i.e., sputter) etching. The equation predicts that at a constant ion energy flux, the etching rate will initially increase in proportion to the radical flux and then saturates as the radical flux continues to increase. This behavior has been observed for a number of material/ion/radical combinations [23–28]. So, we come to the conclusion that ions as well as radicals are important in the ion-induced IBARE of HARTs.

Ion-inhibitor IBARE

In ion-inhibitor IBARE the substrate is etched spontaneously. To achieve directionality, an inhibiting layer is needed, such as in the SF$_6$/O$_2$/Si system. Sidewalls of trenches are not exposed to ion bombardment and will be covered by this layer. However, the bottom of the trench is exposed to ion bombardment, which increases the desorption probability. Thus the bottom is free from this deposit and etching can proceed. Therefore, the model of Mayer and Barker has to be modified to incorporate the inhibitor flux.

To find the etch rate as a function of the ion, radical, and inhibitor flux we start with the calculation of the surface coverage of the inhibitor χ_P ($p =$ passivator). Per unit of time the number of inhibiting particles N_P [mol/m^2] arriving from the gas phase and subsequently adsorbed is

$$(\partial N_P / \partial t)_{\text{ads.}} = S_P \phi_P * (1 - \chi_P - \chi_R) \quad \text{[mol/m^{2}s]}, \tag{14.4}$$

with ϕ_P [mol/m^{2}s] the flux of inhibitors in the gas phase, S_P the probability of sticking of the inhibitor to the wafer surface, and $(1 - \chi_P - \chi_R)$ the number of bare sites on the surface. These particles will prevent the substrate from etching unless ions remove them. Thus, the desorption of the inhibitors per unit of time is proportional to the ion flux:

$$(\partial N_P / \partial t)_{\text{des.}} = \gamma_I \phi_I E_I * \chi_P \quad \text{[mol/m^{2}s]}, \tag{14.5}$$

where γ_I [1/eV] is the yield of removed inhibitors per unit ion bombardment energy. In equilibrium there is no net flux due to adsorption and desorption. So, Eqs. (14.4) and (14.5) should match, giving

$$\gamma_I \phi_I E_I * \chi_P = S_P \phi_P * (1 - \chi_P - \chi_R) \quad [\text{mol/m}^2\text{s}]. \tag{14.6a}$$

Identically, per unit of time the number of radicals N_R arriving from the gas phase at bare sides and sticking is

$$(\partial N_R / \partial t)_{\text{ads.}} = S_R \phi_R * (1 - \chi_P - \chi_R) \quad [\text{mol/m}^2\text{s}]. \tag{14.7}$$

These radicals will react with the surface, giving rise to the etch rate

$$(\partial N_R / \partial t)_{\text{des.}} = R/k_R = (R_S/k_R) * \chi_R \quad [\text{mol/m}^2\text{s}], \tag{14.8}$$

where R_S [m/s] is related to the surface reaction probability. The etch rate $R = R_S$ when the supply of radicals from the plasma phase is high enough to sustain a surface coverage of one. Again, in equilibrium Eqs. (14.7) and (14.8) are identical, so

$$(R_S/k_R) * \chi_R = S_R \phi_R * (1 - \chi_P - \chi_R) \quad [\text{mol/m}^2\text{s}]. \tag{14.9a}$$

Finally, Eqs. (14.6a) and (14.9a) are put into a matrix equation, resulting in

$$\begin{bmatrix} S_P \phi_P \\ S_R \phi_R \end{bmatrix} = \begin{bmatrix} (\gamma_I \phi_I E_I + S_P \phi_P) & S_P \phi_P \\ S_R \phi_R & (\rho R_S + S_R \phi_R) \end{bmatrix} * \begin{bmatrix} \chi_P \\ \chi_R \end{bmatrix} \begin{bmatrix} \text{mol/m}^2\text{s} \\ \text{mol/m}^2\text{s} \end{bmatrix}, \tag{14.10a}$$

and therefore:

$$\frac{1}{R} = \frac{1}{k_R \phi_R S_R} + \frac{1}{k_I \phi_I E_I} * \frac{\phi_P S_P}{\phi_R S_R} + \frac{1}{R_S} \quad [\text{s/m}]. \tag{14.10b}$$

The ion-inhibitor mechanism can be found from this expression by taking the following situations:

(1) When there is no flux of inhibitors ($\phi_P = 0$), the etch rate becomes $1/R = 1/k_R S_R \phi_R + 1/R_S$, which is indeed not depending on the ion flux. For a low radical flux ($\phi_R \Rightarrow 0$), the etch rate approximates $R = k_R S_R \phi_R$, i.e., it is directly related to the radical flux (i.e., diffusion controlled). At a high radical flux the etch rate will be $R = R_S$, i.e., it is depending on the surface reaction mechanism only, and the supply of adsorbing radicals is sufficient (i.e., reaction controlled).

(2) When the ion flux is vanishingly small ($\phi_I \Rightarrow 0$), the etch rate is found to be $R = k_I \phi_I E_I * (S_R \phi_R / S_P \phi_P)$. So, the etch rate decreases when the flux of inhibitors increases with respect to the flux of radicals.

The surface coverage ratio between inhibitors and radicals can be found by dividing Eqs. (14.6a) and (14.9a), giving

$$\frac{\chi_P}{\chi_R} = \frac{S_P \phi_P}{S_R \phi_R} * \frac{R_S}{k_I \phi_I E_I} \quad [-]. \tag{14.10c}$$

So, the ratio is directly related to the sticking-flux ratio. When the ion energy flux is increased or the surface reaction probability is decreased, then the surface coverage by the radicals will increase at the expense of the inhibitor coverage.

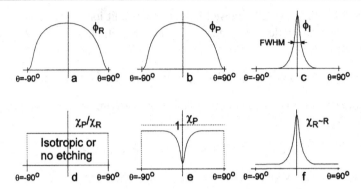

Fig. 14.11. Schematic presentation of the directional ion-inhibitor etch process.

We will now carry on with a schematic representation of the ion-inhibitor process to clarify the directional etching even more. In Fig. 14.11(a), (b), and (c) the fluxes of the radicals, inhibitors, and ions are shown coming from the plasma glow region through the dark space or sheath region. The fluxes of the radicals and inhibitors are almost isotropic, whereas the flux of ions is directional because of the directed electrical field in the sheath region. Generally, the degree of collimation of the ion flux is expressed with the help of the so-called full-width of half-maximum (FWHM, Fig. 14.11(c)). When there is isotropic or no etching due to a lack of ions or a small surface reaction probability, then the surface coverage ratio of inhibitors to radicals is nearly constant for all angles of the substrate surface normal with respect to the plasma boundary normal, as shown in Fig. 14.11(d). The directed ion flux will remove the inhibitors around angles which are facing the plasma glow. So, the surface coverage of the inhibitors will be almost zero at normal incident angles, as found in Fig. 14.11(e). These bare sides will be directly taken by the radicals and therefore the etch rate is much higher for these angles, as depicted in Fig. 14.11(f).

Note that in both equations with respect to the etch rate (Eqs. (14.3) and (14.10(b))) we didn't take thermal desorption into account, i.e., a particle has only a finite lifetime at a surface. This means that the surface coverage of the particle is never completely 1 (at least not for the Langmuir adsorption theory). It is straightforward to incorporate the thermal desorption process into the reaction mechanism of Eqs. (14.6a) and (14.9):

$$(\gamma_I \phi_I E_I + N_P/\tau_P) * \chi_P = S_P \phi_P * (1 - \chi_P - \chi_R) \quad [\text{mol/m}^2\text{s}] \qquad (14.6\text{b})$$

$$(R_S/k_R + N_R/\tau_R) * \chi_R = S_R \phi_R * (1 - \chi_P - \chi_R) \quad [\text{mol/m}^2\text{s}], \qquad (14.9\text{b})$$

where N/τ accounts for the thermal desorption, but we will revert to this subject later on. With the help of these equations it is easy to arrive at formulas identical to those in Eqs. (14.10).

14.4.3   Literature

In this section a comprehensive overview of recent literature on RIE lag will be given. In the excellent review from Gottscho et al., eight mechanisms are considered to explain RIE lag. The first four are responsible for ion depletion and the last four for radical or inhibitor depletion [21]:

(1)   differential charging of insulators [30, 31],

(2)   field curvature near conductors [31–33, 39],

(3)   image force deflection [34],

(4)   ion shadowing with the ion angular distribution [35],

(5)   radical/inhibitor shadowing [36],

(6)   molecular flow [37],

(7)   bulk diffusion, and

(8)   surface diffusion [38].

Gottscho et al. [21] argue: *"Three mechanisms, surface/bulk diffusion and field curvature, predict that for features with a given aspect ratio, RIE lag will increase with absolute feature size. This is not consistent with the observed scaling [39].... In addition, bulk diffusion is not relevant for vacuum processing of micron-sized features because feature dimensions are much less than the gas-phase mean-free path. The image force mechanism predicts that RIE lag will increase as feature size decreases.... Nevertheless, it is expected to contribute to RIE lag in high aspect ratio, subquarter-micron trenches [34]. The relative importance of the remaining four mechanisms, molecular flow, ion/neutral shadowing, and differential insulator charging, consistent with aspect ratio scaling can be inferred by comparing observed trends with pressure, voltage, and ion flux to the scaling predictions for each mechanism."*

In this paragraph we will reexamine these eight mechanisms and add some new ones. To two mechanisms, the image force and ion shadowing in corporation with the ion angular distribution function, extra attention have been paid, since these two mechanisms are believed to remain the most important mechanisms which cause RIE lag into Si using SF_6/O_2 IBARE.

Jansen et al. recently showed that RIE lag during Si etching was decreased by increasing the amount of CHF_3 in an $SF_6/O_2/CHF_3$ gas mixture [40]. At the same time they increased the amount of oxygen to ensure a certain anisotropic profile. They suggested that the higher concentration of O_2 and CHF_3 in the feed gas might increase the thickness of the sidewall passivation. This in turn would reduce the deflection of ions due to image forces because of insulator charging. Vitkarage et al. [41], and others, were able to reduce and could even produce a negative lag by increasing the pressure during oxide etching using a mixture of $CF_4/Ar/CHF_3$ [41–43]. They suggested that at higher pressures, the amount of inhibitor precursors from the CHF_3 feed would increase and the transport of these inhibitor species would be more difficult into HARTs. At this moment, it is difficult to decide which mechanism is causing the reduction of RIE lag for conducting materials; the increased thickness of the inhibitor, the shadowing of the inhibitor, or maybe another mechanism are all able to explain the improved uniformity.

In another paper, Jansen and coworkers found that the influence of the bottling effect and the lag effect could be decreased by installing a showerhead into a conventional IBARE machine without changing the pressure [22]. They suggested that the lower ion energy caused by the showerhead construction would enable ions, coming in at a glancing angle, to reflect from the sidewalls more easily. As a result the effect of ion depletion is less pronounced.

14.4.4 Inhibitor depletion in a trench

In Eq. (14.10b) we arrived at the synergetic etch mechanism of radicals, ions, and inhibitors in ion-inhibitor IBARE. From this equation we see that the depletion of inhibitors will result

in a higher etch rate. So, in small trenches where the supply of inhibitors is restricted due
to inhibitor shadowing, or whatever, inverse RIE lag should be the result.

14.4.5 Radical depletion in a trench

Depletion of radicals may be caused by transport problems of radicals in a trench. They
enter the trench due to gas/volume transport or surface migration (= diffusion). For example,
when the cathode of the reactor is completely covered by a nonmasked silicon wafer, the
etch rate in the center of the wafer will be determined by volume transport. Surface transport
is not possible because the silicon surrounding is consuming all the radicals. So, the etch rate
in the center of the wafer depends on the flux of radicals coming directly from the plasma
glow region. However, when the silicon wafer is covered by a nonetching material with a
little hole in the center, radicals may enter the silicon hole by way of surface diffusion from
the nonetching surrounding. This effect directly accounts for the so-called (micro)loading
effect. Loading occurs whenever the reactant density is depleted due to an excessive substrate
load. As a result, the etch rate will decrease inversely proportionally to the silicon area which
is exposed to the plasma glow giving rise to inverse RIE lag [40]. Moreover, the etch rate
or etched volume depends on the shape of the mask on top of the silicon. For example, a
long small silicon structure etches faster than a square, both with the same area.

So, during the IBARE of silicon HARTs it is important to know if the transport is caused
by the surface transport of radicals from nonetching areas and/or by volume transport
directly from the glow region, because this is determining the lag effect. Therefore, first
the surface coverage of radicals will be estimated. Subsequently, the volume and surface
transport mechanisms are treated as well as how they influence the lag effect.

Surface coverage of radicals

To calculate the radical surface coverage χ when there is no ion bombardment or radical
etching, again adsorption kinetics is used. In this theory, Langmuir [44] stated

(1)   There is no difference between the energy of adsorption and desorption, i.e.,
      physisorption.
(2)   The energy of adsorption is not dependent on the other adsorbed molecules.
(3)   Only one monolayer of the foreign molecules is adsorbed.

With these assumptions, the desorption of N radicals per unit of time is (compare with
Eqs. (14.5) and (14.8) where we did not take thermal desorption into account)

$$(\delta N/\delta t)_{\text{des}} = N_s(\delta \chi/\delta t)_{\text{des}} = N_s \tau^{-1} \chi \quad [\text{mol/m}^2\text{s}], \qquad (14.11)$$

with N_s [mol/m^2] (s = saturation) the amount of radicals necessary to form one monolayer
on top of a clean surface, χ the surface coverage of the radical, and τ the average residence
time on the surface. Per unit of time, the number of radicals arriving from the gas phase and
sticking is (see Eq. (14.2))

$$(\delta N/\delta t)_{\text{ads}} = N_s(\delta \chi/\delta t)_{\text{ads}} = \phi_R S_R(1 - \chi) \quad [\text{mol/m}^2\text{s}], \qquad (14.12)$$

with ϕ_R [mol/m^{2}s] the flux of radicals in the gas phase, and $S_R(1 - \chi)$ the probability
that a radical will stick to the surface. In equilibrium the expressions ruling the adsorption

and desorption (Eqs. (14.11) and (14.12)) are identical, therefore the surface coverage of radicals χ is found to be

$$\chi = \beta/(1+\beta) \quad \text{with} \quad \beta = \tau S_R \phi_R / N_s \quad [-]. \tag{14.13}$$

This expression is known as the Langmuir isotherm. To find the radical flux in equilibrium, kinetic gas theory can be utilized [44]. In equilibrium, the flux of particles leaving a gas volume equals the amount of particles arriving in the volume at a rate

$$\phi = p(2\pi m N_A R T)^{-1/2} \quad [\text{mol/m}^2\text{s}], \tag{14.14}$$

with p [Pa] the partial pressure of the particle, $N_A = 6 \times 10^{23}$ mol^{-1} the number of Avogadro, $m = 146 * 1.6 \times 10^{-27}$ kg the mass of an SF$_6$ molecule (i.e., 1 mole SF$_6$ gas = 0.146 kg), and $R = kN_A = 8.3$ J/molK the universal gas constant ($k = 14 \times 10^{-24}$ J/K the constant of Boltzmann). At $T = 300$ K and $p = 1$ Pa we calculate $\phi = 2 \times 10^{-2}$ mol/m^{2}s. To estimate the net radical flux ϕ_R, it is important to know the etch rate, because the Si is consuming radicals and therefore radicals will move to the Si surface. Assume an Si wafer, covering the complete cathode area, is etching with 10 Å/s at 20 W input power and the reactor in the RE mode, i.e., minimal ion bombardment. The Si density is ca. $\rho_{Si} = 2330$ kg/m^3, i.e., 0.05 atom/Å^3 or 5×10^{18} at/Åm^2 or 8×10^{-6} mol/Åm^2; therefore 8×10^{-5} mol/m^{2}s Si is removed. The reaction end product of Si etching is SiF$_4$, so 4 radicals are needed to remove 1 Si atom, $\phi_R \geq 3 \times 10^{-4}$ mol/m^{2}s, and the partial pressure of the F radicals $p_R \geq 0.005$ Pa, i.e., 1/2% of the partial pressure of the SF$_6$ neutrals.

Lundström wrote [55]: "*Both S_R and τ may be temperature activated with*"

$$S_R = S_O \exp(-E_s/kT) \, [-] \tag{14.15a}$$

$$\tau = \tau_O \exp(E_\tau/kT) \, [\text{s}]. \tag{14.15b}$$

With $S_O = 1$, $S_R\tau$ is therefore equal to $S_R\tau = \tau_O \, exp(E_{ads}/kT)$, where $E_{ads} = E_\tau - E_s$ is the heat of adsorption per molecule. τ_O is determined by the "escape" frequency which is of the order of the molecular vibration frequency, i.e., $\tau_O \sim 10^{-12}$ [s].

This expression shows that the ratio of the adsorption energy E_{ads} and the thermal energy kT determines how long a particle will spend at a surface. The adsorption energy may vary between 10 eV for N—N and 0.001 eV for He—He. A strong bond of several eV is called a chemical bond. For the noble gases and molecules with saturated bonds, only the much weaker van der Waals force will form a bond (order 0.01 eV). Such gases will only become a liquid when $kT < 0.01$ eV. Molecules with a polar moment like HCl and H$_2$O are able to create a strong bond (order 0.1 eV) due to image forces at solid surfaces and therefore they will be in the liquid phase at much higher temperatures.

Now, let us try to estimate the surface coverage of the radicals (F) and inhibitors (O) at a nonetching surface. When the mean interatomic distance is 2.7 Å, i.e., $N_s = 14 \times 10^{18}$ m^{-2} = 2.2×10^{-5} mol/m^2, S_o is taken as 1, and $\phi_R = 10^{-4}$ mol/m^{2}s, then $\beta = 4\frac{1}{2}\tau$. So, for $\beta = 1$ we need $E_{ads} = 0.68$ eV at room temperature (and $E_{ads} = 0.34$ eV at a cryogenic temperature of 150 K). So, $\chi = \beta/(1+\beta) \sim \beta = 4\frac{1}{2}\tau$ when $\beta < 0.1$, i.e., $E_{ads} < 0.62$ eV and $\chi = 1$ when $\beta > 10$, i.e., $E_{ads} > 0.74$ eV. The problem is now to estimate E_{ads}. However, because the foreign molecule is a radical, it is assumed that E_{ads} is high (>1 eV) and the surface is highly fluorinated/oxidized even at partial pressures as low as $p_R = 0.01$ Pa. So, the assumption that the effect of thermal desorption is negligible seems to be correct.

This figure indicates that radical transport along the wafer surface might have a major contribution to the reaction mechanism.

Volume transport

The description of gas flow in vacuum systems is generally divided into three parts, the division being specified by three ranges of values of a dimensionless parameter called the Knudsen number. The Knudsen number is defined as the ratio of the mean free path of a molecule λ to a characteristic dimension d of the channel through which the gas is flowing, for example, the radius in the case of a cylindrical tube.

Viscous flow

For flows at small Knudsen numbers (i.e., high pressures), collisions between molecules occur more frequently than collisions of molecules with the walls. Consequently, inter-molecular collisions predominate in determining the characteristics of the flow. In this range the gas can be considered to be a continuous medium, that is, a viscous flow. The coefficient of viscosity, which appears in all viscous flow equations (e.g., Poiseuille or Navier-Stokes), reflects the influence of intermolecular interactions.

Molecular flow

At large Knudsen numbers (i.e., low pressures), the flow of gas is limited by molecular collisions with the walls. The analysis of the flow is primarily a geometrical problem of determining the restrictive effect of the walls on the free flight of a molecule. Since there are comparatively few intermolecular collisions, each molecule acts independently of the others. This flow is therefore called molecular flow. Molecules do not usually bounce off a surface in the way that light is reflected, but are instead adsorbed, if only for a brief period. After this period, the molecule is again liberated from the surface (i.e., desorbed). However, it has lost knowledge of whence it came and the direction is completely random and follows the cosine distribution with an energy depicted by the reactor walls. The cooperative flow behavior of the viscous regime is thus lost and is replaced by molecular flow, where there is virtually no interaction.

Transitional flow

The transition from viscous flow to molecular flow occurs at intermediate values of the Knudsen number where both types of collisions are influential in determining the flow characteristics. There are no general derivations of flow equations in this range which are based on first principles. It is therefore necessary to describe the flow by semiempirical equations.

A practical assignment of limits to the Knudsen numbers which delineate the ranges of the high- and low-pressure descriptions is as follows:

- When $\lambda/d < 0.01$, the flow is viscous.
- When $0.01 < \lambda/d < 1$, the flow is transitional.
- When $\lambda/d > 1$, the flow is molecular.

The mean free path λ [cm] for air at 25°C is related to the pressure p [mTorr] by $\lambda \sim 5/p$. Typical pressures in IBARE are in the range of 10–100 mTorr, so 0.5 mm $< \lambda <$ 5 mm. For a trench smaller than 0.1 mm in width, the Knudsen number is 5 and the flow is molecular. Of course, the flow in the reactor is still viscous ($d \sim 100$ mm $\Rightarrow \lambda/d = 0.005$), because otherwise the plasma (conventional) wouldn't start.

So, for dimensions characteristic of MNE applications the flow is molecular. What is the molecular flow through long tubes and channels? The manner of attacking the problem of flow in this range is due to Knudsen [45]. The flow resistance of a trench is caused by the random reflection of a low energetic particle against the sidewall. Therefore there is only a small chance for a particle to enter the trench. This chance is known as the Clausing factor C, and is dependent on the AR. For a tube [46],

$$C(AR) = (1 + 0.4AR)^{-1} \quad [-]. \tag{14.16}$$

Clearly, this effect may cause RIE lag: When the $AR = 5$, the radical flux is decreased by almost $1 - 1/3 \sim 67\%$.

Despite the enormous success of the Knudsen theory to describe molecular flow through channels, there are some points which should be noticed. Lundström recently showed some remarkable wall-induced effects in molecular gas transport through extremely smooth micromachined channels in silicon around 100 nm in width [55]. They found an unexpected decrease in diffusion lag time with increasing pressure and decreasing temperature, and an influence of one species on the lag time of another species in a gas mixture. This could be explained by using a model based on the adsorption of gas molecules onto the channel walls. In the Knudsen theory the diffusion constant D_K [m^2/s] is derived with the assumption of a completely diffuse molecular scattering at the channel walls. However, the influence of partly specular reflection (i.e., elastic collisions) of radicals due to already adsorbed radicals at the channel walls and diffusion along the surface could be enhanced in ultrasmooth micromachined channels because of the high surface coverage of radicals. So, the constant D_K should be replaced by a diffusion constant which depends on the roughness and surface coverage, $D(\chi)$. In the higher regions of a trench, radicals will completely cover the flat surface, and therefore incoming radicals cannot absorb (Langmuir's third statement) and will reflect specularly deeper into the trench. Lundström et al. [55] wrote: "*A possible contribution to the difference between scattering of molecules on the covered and uncovered surfaces may be due to the delay of molecules caused by the adsorption which may decrease the size of the apparent diffusion constant for the uncovered surfaces. In addition to this delay it is likely that adsorbed molecules desorb with zero average velocity along the surface. A molecule which collides with an already adsorbed molecule has, however, a larger probability of being elastically scattered.*"

In conclusion, because of the extraordinarily high surface coverage of radicals, even at partial pressures as low as 0.01 Pa, the influence of the partial specular reflection could be rather high and the Knudsen diffusion model would not be applicable anymore. In other words, the Clausing factor is not as strong as predicted and could be even close to 1.

Surface transport

As estimated in the previous section, the transport of radicals may take place along the substrate surface. Under these circumstances, the supply of radicals depends on the geometry of the trench opening. Consider two trenches with identical areas, one small and long (9×1) and the other being a square hole (3×3, Fig. 14.12). The radical transport along the surface

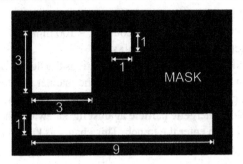

Fig. 14.12. Surface transport causing RIE lag.

is $(9 + 1)/(3 + 3) = 10/6 = 5/3$ times higher for the long-shaped trench opening with respect to the square. If we take two square holes, one is 1×1 micron and the other measures 3×3 microns, the etch rate ratio due to surface transport is expected to be $([3 + 3]/3 * 3)/([1 + 1]/1 * 1) = 1/3$. So, the smaller the hole, the faster it will etch, i.e., inverse RIE lag.

14.4.6   Ion depletion in a trench

Ion depletion may be caused by the angular distribution of incoming ions into the trench opening (which strongly depends on the pressure) and the deflection of ions in the trench due to electrostatic fields (which depends on the ion velocity). In this chapter, it will be assumed that there is no ion reflection, so ions which are captured by the sidewalls due to one of these mechanisms will fail to etch the trench bottom and this in turn will decrease the etch rate. We do not mean to imply that ion reflection is never important, but merely unnecessary for our discussions here. Ion deflection is caused by electrostatic fields in the trench, such as interactive forces due to differential charging at insulating surfaces, attractive image forces due to influencing fields in solids, distorted external fields such as, e.g., the sheath potential near (conducting) surfaces, repulsive Coulomb forces between free-moving ions, and attractive Lorentz forces due to the velocity of the ions (i.e., currents). In short, this can be written as

$$F_{total} = F_{charging} + F_{image} + F_{distortion} + F_{Coulomb} + F_{Lorentz} \quad [N]. \qquad (14.17)$$

In this section we try to find which mechanism has the most important influence and is therefore responsible for the depletion of ions in a trench.

Differential charging

Since the work of Bruce and Reinberg, there has been considerable speculation on the effects of mask charging on ion trajectories [30, 31, 33, 43, 52]. We won't go into detail about this subject, because silicon is a conductor and therefore it is believed that surface charging is of no concern to us. Although the inhibitor which is covering the sidewalls is probably a good insulator, it is believed to be too thin to prevent the tunneling or conduction of electrons, and thus there is no surface charging effect from this mechanism.

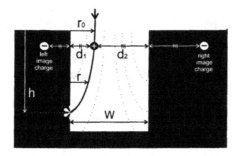

Fig. 14.13. Ion trajectory due to image forces in a trench.

Image forces

Davis was probably the first one who addressed the image force as a mechanism to explain RIE lag in submicron HARTs for low-energy ions [34]: When a charged particle is close to a solid material, it is attracted due to influencing fields. This force is inversely proportional to the square of the distance, so the acceleration toward the solid medium is ever increasing until the particle collapses with the surface of the solid. In a trench we can imagine two such forces, one from the left and the other from the right trench sidewall, trying to capture the particle (Fig. 14.13). (In fact, we should take an infinite number of image charges as Davis did, but it can be reasoned that the "higher order" image charges have a quickly decreasing influence.) Ions captured by the sidewalls are depleting the ion flux to the bottom of the trench, so the etch rate is decreasing in time. Both the acceleration force to the sidewalls and the calculation of the etch depth ask for a nonlinear analysis. Davis integrated the equation of motion numerically. We prefer to calculate the lag effect analytically and found a close match between both approaches. First, the ion deflection to a conductor for a constant acceleration is calculated, i.e., the linear analysis. To find the nonlinear equation of motion, the variation principle can be utilized. When the etch rate is found, the etch depth is derived by integrating the etch rate with respect to aspect ratio or time. After that, the influence of the opposite wall and the conductance of the sidewall are taken into account.

Linear analysis

A charged particle in the neighborhood of a conductor will be forced to the surface in order to preserve the boundary conditions of the electrical **E** and **D** field. Therefore, the particle will move toward the conductor. To calculate this force, the concept of "image charging" is generally used [47]. In this concept an imaginary particle of the opposite sign has been thought to exist under the surface of the conductor at the same distance from the surface as the real charged particle, i.e., a mirror image of the particle. The attractive force is then calculated with the help of Coulomb's formula:

$$F = q^2/4\pi\varepsilon(2r_0)^2 \quad [\text{N}], \tag{14.18a}$$

with $q = 1.6 \times 10^{-19}$ C the charge of the particle, $\varepsilon = 8.9 \times 10^{-12}$ F/m the permittivity of vacuum, and r_0 [m] the shortest distance of the particle to the surface of the conductor.

Assume an ion is entering a trench $w = 1$ μm in width at a distance of $r_0 = 50$ nm from one of the sidewalls of the trench (Fig. 14.13). The question is: What is the distance, h [μm], the ion will travel before it will collide with the sidewall? First, it can be recognized that the opposite wall is far enough away to be neglected. The force due to the influencing field is then calculated to be $F \sim 2 \times 10^{-14}$ N. For the SF_5^+ ion, $m \sim 2 \times 10^{-25}$ kg, the acceleration $a = F/m \sim 10^{11}$ m/s^2, and after 10^{-9} s the ion is displaced by $y = \frac{1}{2}at^2 = 50$ nm, i.e., it has just landed on the sidewall. Meanwhile, an ion with a kinetic energy of $E_{kin} = \frac{1}{2}mv^2/q = 10$ eV (i.e., the velocity $v = 4000$ m/s) is able to travel 4 μm down the trench. Mathematically this is

$$h^2 = 64\pi \varepsilon E_{kin} r_0^3/q \quad [\text{m}^2], \tag{14.19}$$

with E_{kin} in [eV]. If we assume the etch rate to be directly proportional to the ion flux, then the etch rate R [μm/min] of a trench of 1 μm is decreased by 10% when it has reached a depth of 4 μm (i.e., $AR = 4$), and it is decreased by 40% at $AR = 32$. In formula,

$$R = \delta h/\delta t = R_{max}[1 - 2(r_0/w)] \quad [\text{m/s}]. \tag{14.20}$$

In case the trench has an opening of 4 μm, we calculate this decrease of 10% at $AR = 8$ and 40% at $AR = 64$. In case the kinetic energy $E_{kin} = 40$ eV, the AR – in all cases – is two times higher. In other words, the wider the trench opening and the higher the kinetic energy of incoming ions, the smaller the influence of the image force on the lag effect.

With the help of the two preceding equations it is now possible to arrive at the differential equation governing the etch rate. Extracting r_0 from both equations gives

$$\frac{\partial h}{\partial t} = R_{max}\left[1 - \left(\frac{kh^2}{E_{kin}w^3}\right)^{1/3}\right] \quad [\text{m/s}], \tag{14.21}$$

with $k = q/8\pi\varepsilon$ [Vm] ~ 7ÅV. The relative etch rate $RelR$ is therefore

$$RelR = \frac{\partial h/\partial t}{R_{max}} = 1 - \left(\frac{kAR^2}{E_{kin}w}\right)^{1/3} \quad [-], \tag{14.22}$$

with $AR = h/w$. The aspect ratio corresponding to the depth at which half of the ion flux into the trench has collided with an etched wall ($RelR = 1/2$) is denoted by $AR_{1/2}$:

$$AR_{1/2} = \sqrt{\frac{E_{kin}w}{8k}} \quad [-]. \tag{14.23}$$

The solution of the differential Eq. (14.21) is found to be

$$\text{etch time} = \frac{3 * (\arctan h(g) - g)}{R_{max}} * \sqrt{\frac{E_{kin}w^3}{k}} \quad [\text{s}], \tag{14.24a}$$

with $g = (kh^2/E_{kin}w^3)^{1/6}$. Up to the second-order Taylor this is

$$\text{etch time} \approx \frac{h}{R_{max}}\left[1 + \frac{3}{5w}\left(\frac{kh^2}{E_{kin}}\right)^{1/3}\right] \quad [\text{s}], \tag{14.24b}$$

and thus

$$\text{timelag} = \frac{3}{5w}\left(\frac{kh^2}{E_{kin}}\right)^{1/3} \quad [\text{s}]. \tag{14.25}$$

In the graphs shown in Fig. 14.14 the maximum etch rate is taken as $R_{max} = 1$ μm/min, the ion kinetic energy $= 10$ eV, and $\alpha = 1$ (see later on). For a trench of 1 μm in width (i.e.,

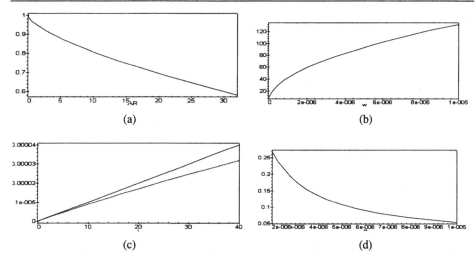

Fig. 14.14. Output of the mathematical program Maple at $E_{kin} = 10$ eV, $R_{max} = 1\,\mu$m/min, $w = 1\,\mu$m, $h = 100\,\mu$m, and, $\alpha = 1$. (a) $RelR = RelR(AR)$, (b) $AR_{1/2} = AR_{1/2}(w)$, (c) etch depth = etch depth(t), and (d) timelag = timelag(w).

the most curved line in Fig. 14.14(c)) it is found that the etch depth is approximately 30 μm after 40 min of etching. So, the RIE lag is ca. $(40 - 30)/40 = 25\%$. Indeed, as calculated before, the etch rate R is decreased 40% at that moment (Fig. 14.14(a)).

Nonlinear analysis

To find the nonlinear ion trajectory, we may use the balance of energy which gives us the differential equation:

$$\Delta E_{pot} = (q^2/4\pi\varepsilon)\left(\frac{1}{4}r - \frac{1}{4}r_o\right) = \frac{1}{2}mv^2 = \Delta E_{kin} \quad [\text{J}] \tag{14.26a}$$

$$2r(\delta r/\delta t)^2 + cr/r_o = c \quad [\text{m}^3/\text{s}^2], \tag{14.26b}$$

with $c = q^2/4\pi\varepsilon m$. Identically, the balance of forces results in

$$a = \delta^2 r/\delta t^2 = F/m = q^2/4\pi\varepsilon m(2r)^2 = c/4r^2 \quad [\text{m/s}^2] \tag{14.27a}$$

$$4(\delta^2 r/\delta t^2)r^2 = c \quad [\text{m}^3/\text{s}^2]. \tag{14.27b}$$

Indeed, differentiating the equation from the balance of energy with respect to time gives the equation for the balance of forces. In order to solve the last nonlinear equation, the variation principle can be used. In this principle a complete and independent set of functions is used to describe the movement of the ion $r = r(t)$. In this case it is straightforward to take the polynomials as a set:

$$r(t) = r_o - v_o t - \frac{1}{2}a_o t^2 - O(t^3) \quad [\text{m}]. \tag{14.28}$$

When this set is put into the nonlinear equation together with the boundary conditions

$$r(0) = r_o,$$
$$\delta r(0)/\delta t = v_o = 0,$$
$$\delta^2 r(0)/\delta t^2 = a_o = c/4r_o^2,$$

we find

$$r(t) = r_o(1 - \alpha - \alpha^2/3 - 11\alpha^3/45 - \cdots) \quad [\text{m}], \tag{14.29}$$

with $\alpha = ct^2/8r_o^3$. The particle collides with the sidewall at $t = t_c$ where $r(t) = r(t_c) = 0$, i.e., $\alpha \sim 0.7$. Before the ion hits the sidewall, it is able to travel a distance

$$h^2 = 64\alpha\pi\varepsilon E_{\text{kin}} r_o^3/q \quad [\text{m}^2]. \tag{14.30}$$

So, because the acceleration force is increasing while approaching the sidewall, the ion will collapse a little sooner, as in the case of linear approximation. The difference is expressed with the help of the factor α.

Higher-order image forces

Equation (14.30) becomes even more complicated when we take the opposite sidewall into account. If we take $\beta = d_1/w$ and the balance of forces, the next expression can be found:

$$h^2 = 64\alpha f(\beta)\pi\varepsilon E_{\text{kin}} w^3/q \quad [\text{m}^2], \tag{14.31a}$$

with

$$f(\beta) = \beta^3(1 - \beta)^2/(1 - 2\beta) \quad [-], \tag{14.31b}$$

a function which only depends on the relative distance of the ion with respect to the opposite sidewall. Again, for $\alpha = 1$ and small values of β, thus $f(\beta) \sim \beta^3$, the expression turns into the linear equation of movement. For $\beta = 1/20$, $f(\beta) = (361/360)\beta^3$, and for $\beta = 1/4$, $f(\beta) = (9/8)\beta^3$, so the effect of the attraction of the opposite sidewall is rather small. Whereas the factor α always increases the lag effect with respect to the linear approximation, the function $f(\beta)$ decreases the lag effect.

Dielectrica

The solid dielectricum does not have to be conducting in order to attract a charged particle. In general, the attractive force will be [47]

$$F_{\text{solid}} = F_{\text{conductor}}(\varepsilon_r - 1)/(\varepsilon_r + 1) \quad [\text{N}]. \tag{14.18b}$$

For example, glass with a dielectric constant of 4 has an image force 3/5 times the force imaged in a conductor. Of course, an insulator is able to "trap" an incoming ion for a while. This charge will repel a next ion, and the image force is more or less compensated for. Therefore, an insulator is acting completely different under ion bombardment as a conductor. This effect is even more accentuated because the flux of ions is directional, whereas the flux of electrons is isotropic [30].

In total, it is concluded that the influence of the image force on the lag effect is pronounced for small micron-sized trenches only. The effect can be minimized by increasing the bias voltage, that is, the ion velocity. However, as already said, a higher ion energy might increase the attack of the sidewalls due to the decrease of ion reflections. Another disadvantage is the decrease in mask selectivity.

Coulomb and Lorentz forces

To find out the importance of forces between charged particles traveling at the same time in one trench, it is calculated first if this happens often. Let's take a volume element of 1 cubic micron, i.e., a typical trench of our interest. The flux of ions during conventional RIE is typically $\phi_I = 10^{-6}$–10^{-5} mol/m^{2}s $= 10^7$ ions/μm^{2}s, i.e., approximately 1% of the radical flux [29]. The ionic pressure in the plasma glow is therefore around 100 μPa. The density is at its maximum for ions traveling with low kinetic energy and therefore 10 eV is taken for the minimal kinetic energy. When the mass of the ion is 127 amu, i.e., SF$_5^+$, we are able to calculate with $E_{kin} = \frac{1}{2}mv^2$ the velocity of the ions: $v \sim 10^3$–10^4 m/s. So, an ion needs 10^{-9}–10^{-10} seconds to travel a distance of 1 micron. Therefore, 10^{-2}–10^{-3} ions are at the same time in a volume of 1 cubic micron. In other words, the chance that two ions are at the same time in one trench is small and thus the Coulomb and Lorentz forces are of no concern to us. However, for RIE equipment with a high-density source outfit the ion density is roughly 100 times higher and these forces could become influential.

Distorted external fields

Trenches will distort external electrical fields like the one from the dark space region, and this effect is also known as "boundary distortion" or the "diverging field effect" [22, 39]. However, Ingram points out that the fraction of the total voltage drop associated with the distorted part of the field will scale with the feature width relative to the sheath thickness, thus, RIE lag from this mechanism would scale with absolute feature size [31]. Liu et al. calculated that this effect is unimportant for micron-sized features [53].

Nevertheless, the concentrated surface charge is deflecting incoming ions in the trench opening and thus the ions are spread out and giving rise to inverse RIE lag. Moreover, in case the dark space region is close to the dimension of the trench opening, the boundary layer will adapt its form to follow the surface structure. Therefore, the etch rate will have a minimum as shown schematically in Fig. 14.15.

Ion energy and angular distribution

Besides the electrostatic forces which deflect ions to the sidewalls of a trench, ion collisions in the plasma sheath and plasma glow (i.e., the thermal energy of the ions) may

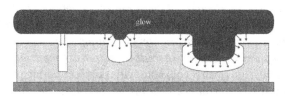

Fig. 14.15. Distorted boundary layer.

direct them to the sidewalls, leading to ion depletion too. So, because of the collisions of ions with other particles, ion dispersion will occur, i.e., their direction will not exactly correspond with the normal of the boundary layer anymore. This effect is expressed with the help of the ion angular distribution function (IADF). At the same time, the energy of the ions is exchanged with the particles, and this effect is found in the ion energy distribution function (IEDF). The rf frequency is responsible for a varying plasma potential. So, the energy of an ion depends on the time that the ion enters and leaves the boundary layer. This effect should be incorporated in the IEDF. At the same time, the thermal vibration energy of the ions in the plasma glow alters both distribution functions as well. Generally, the IADF and IEDF are strongly correlated, i.e., they can't be varied independently.

In Ref. [53], Liu et al. review: *"In a plasma, the sheath electric field accelerates ions along the macroscopic surface normal, creating a directed flux of energetic particles which induces directional etching. Ion scattering in the sheath produces a distribution of bombardment energies and angles at the surface. Therefore, to model directional plasma etching processes, the phenomena associated with ion transport through the plasma sheath must be quantitatively understood. . . . The energy and angle of an ion incident on a surface are determined by the electric field in the sheath, the sheath thickness to mean-free path ratio, and the number of rf cycles required for an ion to cross the sheath. These parameters are determined by operating conditions such as power, frequency, electrode spacing, and pressure. If pressure increases and the sheath thickness remains about the same, the increased number of collisions lowers the ion energies and randomises their direction."* For example, the measured sheath thickness in an argon plasma is typically $d \sim 10$ mm at 10 mTorr, 5 mm at 50 mTorr, and 4 mm at 500 mTorr. Moreover, Kuypers and Hopman reported a nearly constant sheath thickness as a function of power [58]. When the mean free path is calculated with the help of the relation $\lambda * p \sim 50$ mm $*$ mTorr, we are able to find the ratio of the sheath thickness to the mean free path for an argon plasma: i.e., $d/\lambda \sim 2$ at 10 mTorr, $d/\lambda \sim 5$ at 50 mTorr, $d/\lambda \sim 40$ at 500 mTorr. Liu et al. continue: *"The IEDs and IADs are said to be fully developed when the shape of the distribution functions no longer change with additional collisions. . . . At 500 mTorr, the ions go through a sufficient number of collisions to form a fully developed IED. Note that the IED is similar to a Boltzmann distribution. The ions have reached a balance between energy lost through collisions and energy gained by acceleration in the electrical field."*

Ion energy distribution

With respect to the IEDF, Liu et al. note that there are no collisions necessary for the IEDF to be broadened: *"The ion transit time, the number of rf cycles an ion takes to cross the sheath, defines the width of the ion energy distribution. In a collisionless sheath, if the ion transit time is a fraction of an rf cycle, the ion's energy will depend on the phase of the rf cycle when it enters the sheath. The maximum and minimum energy occur when the ion crosses the sheath during the maximum and minimum sheath potentials. If an ion takes many rf cycles to cross the sheath, its energy will depend on the average sheath potential. Modulation of ion energy is still apparent in cases where the ion transit time is about five rf periods."* Nevertheless, generally, ion-neutral collisions will take place during passing the sheath and this will broaden the IEDF even further: *"The two major collisional processes between argon ions and neutrals are charge-exchange* (energy-transfer) *and elastic collisions* (momentum-transfer) *which have comparable cross sections. . . . Multiple peaks in the IEDs can occur in rf modulated sheath fields for ions that undergo charge-exchange collisions. In a charge-exchange collision process, an ion transfers charge and a small amount of momentum to a neutral, forming a new ion that has little directed energy perpendicular to the electrode and an energetic neutral from the original ion. For plasmas where the probability for charge-exchange collisions is high, these energetic neutrals can contribute a significant portion of the total energy deposited on the surface.*" It can be predicted that the number of fast neutrals generated per colliding ion is almost equal to the sheath thickness to ion mean free path ratio [51]. So, at a pressure of 5 mTorr where $d/\lambda \sim 1$ the flux of ions passing the sheath is close to the flux of fast neutrals. Moreover, the shape of the fast neutral distribution functions will closely match those of the ions. Therefore, with respect to the influence of the ion distribution on the profile development of a trench, we won't have to incorporate the fast neutral distribution functions. Nevertheless, for micron-sized HARTs where the image force is pronounced, the flux of fast neutrals should be identified, because they will decrease the effect of RIE lag due to the absence of the image force mechanism.

Ion angular distribution

Regarding the IAD, Liu et al. notice the following: "*The average ion incident angle (i.e., the angle measured from the surface normal) increases with the number of collisions in the sheath until the distribution becomes fully developed. The thermal energy of the ion at the bulk-sheath interface is usually small compared to the sheath potential. If ions have no collisions in the sheath, they will have their velocity directed completely perpendicular to the surface. In situations where there are only charge-exchange collisions and the sheath potential is small (~30 eV), the random velocity component could produce ions with significant incident angles [57].*" The thermal energy to the ion will result in broadening of the IAD by an increased dispersion which can be as large as $\theta = \tan^{-1}\sqrt{(kT/E)} \sim \tan^{-1}\sqrt{(0.05/10)} = 4°$ for ions gaining 10 eV in the sheath region at a plasma ion temperature of 550 K. This mechanism could be especially influential for electron cyclotron resonance (ECR) discharges, where ions are thought to be appreciably heated. Normally, for capacitive plasmas, the initial ion thermal energy is negligible and momentum transfer scattering is necessary to produce significant ion incident angles. It can be predicted that the average ion bombardment energy at small incident angles is greater than at larger incident angles.

In summary, the IADF is a measure for the degree of collimation or dispersion of the flux of ions. The IADF depends on the ion thermal energy in the plasma glow region and collisions in the sheath region. A sharp IADF means that most ions travel in the same direction. The IEDF is a measure for the energy content of the ions. The IEDF is controlled by rf variations of the sheath thickness and collisions in the sheath region. A sharp IEDF means that most ions arrive with the same energy at the sample surface.

Finite element analysis

Gottscho et al. [21] review: "*Many experimental results suggest that lower pressure reduces the magnitude of RIE lag [35, 48, 49]. Jurgensen has assumed that etching anisotropy and RIE lag are both determined by the anisotropy of the bombarding particle angular distribution; this in turn scales with the dimensionless sheath thickness in units of the mean-free path for momentum transfer collisions [50]. Shaqfeh and Jurgensen later showed*

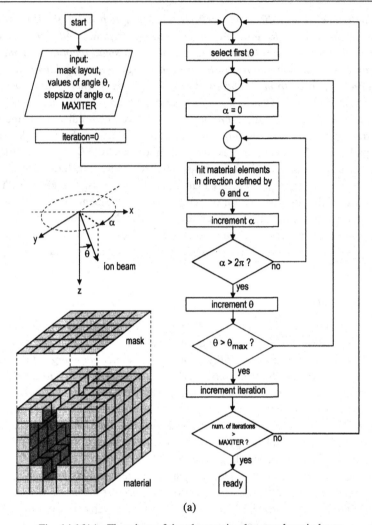

(a)

Fig. 14.16(a). Flowchart of the plasma simulator under windows.

that the predicted shape of etching profiles, magnitude of the lag effect, and their scaling with pressure were all in good agreement with experimental results for the transfer step in trilayer lithography [35]."

So, there are strong clues that the IAD has an important contribution to RIE lag. To estimate this effect, we've written a simulation program. In this section the result of this simulation is given. The program flowchart, shown in Fig. 14.16(a), simulates the influence of the angular and energy distribution of the ion flux on the lag effect and on the development of the profile. To simulate this, the substrate material is devised into many cubes (3-D matrix elements) on top of the mask. The mask layout resembles the places of the substrate material under direct ion bombardment. After starting the program, every "open" spot in the mask will fire a flux of ions with a predetermined energy (bullets) in a well-defined direction (α, θ). When a matrix element is hit by too many bullets, it will vanish and the next element in line is under direct attack. If necessary, after one

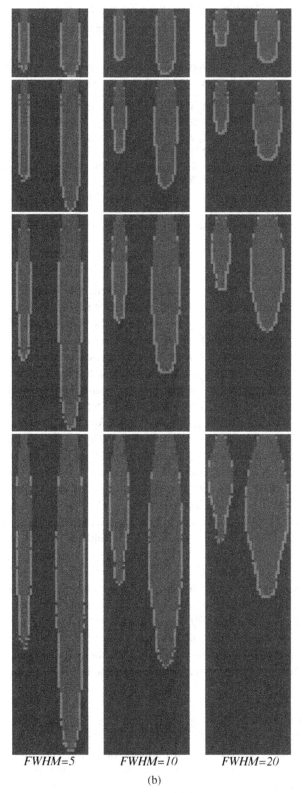

FWHM=5   FWHM=10   FWHM=20

(b)

Fig. 14.16(b). Simulator output for different (vertical) etch times and (horizontal)
FWHM of the triangle-shaped IAD.

Table 14.1. *Etch depth results from the simulator and theory.*

Etch depth (h1/h2) for holes with diameter $(w1/w2) = (3\ \mu m/7\ \mu m)$				
		FWHM $= 5°$ $\Rightarrow AR_c = 5.7$	FWHM $= 10°$ $\Rightarrow AR_c = 2.8$	FWHM $= 20°$ $\Rightarrow AR_c = 1.4$
$t = 1 * \Delta T$	simul	$26/028 = 0.93$	$21/27 = 0.78$	$15/22 = 0.68$
	theory	$27/028 = 0.96$	$21/28 = 0.75$	$15/22 = 0.68$
$t = 2 * \Delta T$	simul	$42/055 = 0.76$	$30/45 = 0.67$	$22/33 = 0.67$
	theory	$42/056 = 0.75$	$30/45 = 0.67$	$22/33 = 0.67$
$t = 4 * \Delta T$	simul	$61/090 = 0.68$	$45/66 = 0.68$	$31/49 = 0.63$
	theory	$62/089 = 0.69$	$44/66 = 0.67$	$32/48 = 0.67$
$t = 8 * \Delta T$	simul	$90/133 = 0.68$	$63/98 = 0.64$	$46/68 = 0.68$
	theory	$89/133 = 0.67$	$63/95 = 0.66$	$45/68 = 0.66$

rotation ($\alpha = 0 \ldots 2\pi$) the tilt angle θ can be changed, which accounts for the angular distribution of the incoming ion flux. At the same time it is possible to change the number of bullets per shot, which simulates the ion energy flux distribution. The simulation is finished when an adjustable (MaxIter) amount of rotations is fulfilled. The program is written in C++ under a windows environment. A characteristic of the program is that any effect, may it be a radical transport problem or an ion deflection, can be incorporated into the program without difficulties. Even better, the program is able to simulate not only RIE profile development but IBE ($\theta =$ constant) and RE trench processes also.

In Fig. 14.16(b) the output of the program is shown for two square holes in a mask: one measures 3 μm × 3 μm and the other 7 μm × 7 μm. The ion energy flux distribution of an RIE process is simulated by rotating (α) and tilting (θ) an IBE source. In the horizontal direction, the simulation for three different IADs is shown with a full-width half-maximum (FWHM) of 5, 10, and 20°, respectively. In the vertical direction the profile development in time is shown for 1, 2, 4, and 8 time intervals, respectively. The ion energy flux per incident angle θ or coneshell was kept constant from normal incident angle up to its maximum incident angle. For example, in the case of the IADF with an FWHM of 5°, 1 bullet was fired at a time in the (α, θ) direction while pointing the gun in evenly distributed directions per θ shell $\alpha = n * 8°$ and $\theta = m * \frac{1}{4}°$ ($n = 0.45$ and $m = 0.20$; m, n positive integers). The simulation was finished after 4 full (α, θ) rotations, i.e., MaxIter $= 4$. The etch time was simulated by changing the number of bullets one cube element could take before vanishing; i.e., $128(t = \Delta t)$, $64(t = 2\Delta t)$, $32(t = 4\Delta t)$, and $16(t = 8\Delta t)$. So, for all IADs at a certain etch time, the total number of ions was the same. Clearly, the effect of RIE lag and the bottle-shaped profiles are observed. Note that both the absolute feature size and the absolute etch time are not necessary for the simulation: The effect is dependent on the aspect ratio alone.

So we observe that the lag effect is strongly correlated to the FWHM of the IAD. The etch depth output of the simulator is given in Table 14.1. For a relatively sharp IAD of 5° FWHM, the RIE lag of a 3-μm hole is almost $(28 - 26)/28 \sim 7\%$ with respect to the 7-μm hole after the small hole has reached an aspect ratio of $26/3 \sim 9$. With increasing time or FWHM of the IAD, this situation becomes even worse. A quite remarkable phenomenon

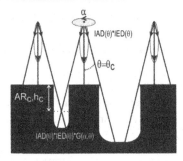

Fig. 14.17. The effect of the IAD on RIE lag.

observed when studying the output of Table 14.1 is that after longer etch times or increasing FWHM the etch depth ratio between the two holes becomes constant. In this case the ratio is approximately $h_1/h_2 \sim \sqrt{[w_1/w_2]} = 0.66$. Moreover, at higher aspect ratios the etch depth increases in proportion to the square root of time: $h \sim \sqrt{t}$. Besides, the etch depth is inversely proportional to the FWHM of the IAD: $h \sim 1/\sqrt{FWHM}$. So, it can be expected that a theoretical investigation would give something like

$$h \sim \sqrt{\frac{w*t}{FWHM}} \quad [\text{m}]. \qquad (14.38a)$$

Enough reason to look at this result in a more theoretically way.

Theoretical analysis

Now, after some simulation results have been shown with the help of a finite element program, it is useful to give some theoretical analysis for a simple ion distribution function and trench geometry. For this treatment, we make use of the schematic representation of the angle-dependent ion flux bombardment of a structure, as shown in Fig. 14.17. In this figure, the cone of angles from which ions arrive is shown schematically.

For big open structures, the ion energy flux from the whole ID will arrive at the surface and the etch rate becomes

$$R_{\max} = k_I \int_{\alpha=0}^{\theta=2\pi} \int_{\theta=0}^{\theta=\pi/2} IAD(\theta)*IED(\theta)\,\partial\theta\,\partial\alpha \quad [\text{m/s}], \qquad (14.32)$$

where k_I is the volume removed per unit ion bombardment energy. (Note: The IAD found experimentally is sometimes given as a function of both α and θ, as in the paper by Janes [59]. However, for a homogeneous plasma there is no α dependency and therefore some researchers prefer to express the IAD as a function of θ alone, as Liu et al. did [53]. So, the IAD from Liu et al. can be found by multiplying the IAD from Janes by $2\pi\theta$ and, therefore, Janes's IAD is somewhat sharper.) In first order, Eq. (14.32) closely approximates the integration of the ion energy flux over two times the FWHM of the IAD, that is,

$$R_{\max} \sim k_I \int_{\alpha=0}^{\theta=2\pi} \int_{\theta=0}^{\theta=FWHM} IAD(\theta)*IED(\theta)\,\partial\theta\,\partial\alpha \quad [\text{m/s}]. \qquad (14.33)$$

At the bottom of the structure, ions arriving at certain (α, θ) angles may fail to reach it due to ion shadowing. This effect is observed in Fig. 14.17 for the complete smaller trench and the edges of the bigger trench opening. Mathematically, this can be written as

$$R = k_I \int_{\alpha=0}^{\theta=2\pi} \int_{\theta=0}^{\theta=FWHM} IAD(\theta) * IED(\theta) * G(\alpha, \theta)\, \partial\theta\, \partial\alpha \quad [\text{m/s}], \qquad (14.34)$$

where the function $G(\alpha, \theta)$ accounts for the shadowing effect, which depends on the AR and the mask layout. For trenches, the etch rate can level with the open areas up to angles where the full IAD is still arriving, i.e., near the FWHM of the IAD. This critical angle θ_c is directly related to the critical aspect ratio AR_c of the circular hole as (Fig. 14.17)

$$1/AR_c = 2\tan(\theta_c) \sim 2\tan(FWHM) \quad [-]. \qquad (14.35)$$

Generally, for etching at low pressures the $FWHM$ is fairly small (<0.1 rad), and in such cases we may use $\tan(FWHM) \sim FWHM$ and therefore

$$AR_c \sim 1/2FWHM \quad [-]. \qquad (14.36)$$

After reaching AR_c the etch rate will decrease due to ion shadowing. If we assume the IAD to be constant per cone shell within two times its FWHM, then the etch rate will decrease inversely proportional with the AR. So, finally we are able to write down the relative etch rate of a hole as a function of the AR and the FWHM of the IAED, that is:

$$RelR = \begin{cases} 1 & \text{for } AR < AR_c \\ AR_c/AR & \text{for } AR > AR_c \end{cases} \quad [\text{m/s}]. \qquad (14.37)$$

So, for the small hole the lag effect is already pronounced, because at the mid of its bottom it is depleted from ions due to ion shadowing. For the bigger one, the etch rate in the midpoint is still at its maximum, because the ions from the whole IAD may arrive. Nevertheless, close to the sidewall the etch rate is smaller because of this shadowing. For holes etched with a steplike IAED and an FWHM of 5°, RIE lag will show up when the aspect ratio exceeds $AR = 1/2\tan^{-1}(5°) = 5.7$. For a hole of 3 μm this means a depth of 17 μm, and for the 7-μm opening the lag will start after a depth of 40 μm has been reached. With the help of Eq. (14.37) and $\delta h/\delta t = R_{max} * RelR$ it is possible to find the etch depth h as a function of time:

$$h = \begin{cases} R_{max} * t & \text{for } t < t_c \\ R_{max} * t_c\sqrt{(2t/t_c - 1)} & \text{for } t > t_c \end{cases} \quad [\text{m}], \qquad (14.38\text{b})$$

where $t_c = h_c/R_{max}$ and $h_c = w * AR_c$. So, the etch rate ratio between two holes with widths w_1 and w_2 is equal to

$$h_1/h_2 = (t_{c1}/t_{c2})\sqrt{[(2t/t_{c1} - 1)/(2t/t_{c2} - 1)]} \sim \sqrt{[t_{c1}/t_{c2}]}$$
$$= \sqrt{[h_{c1}/h_{c2}]} = \sqrt{[w_1/w_2]} \quad [-], \qquad (14.38\text{c})$$

as already predicted by the simulator. In Table 14.1 the theoretical data are given with respect to the simulator output. Mark how close both analysis techniques match.

At this moment, most effort is put into the development of high-density low-pressure sources, because this increases the ion mean free path. But, decreasing the sheath thickness prevents ion collisions in the sheath region too and this could be a much cheaper way to go. However, the minimum sheath thickness without an applied dc bias is limited by the Debye length and we should be aware of the effect of boundary distortion. The Debye length is the distance within which the ion's potential falls to $1/e$ of the value it would have without electron screening and is derived in Chapman as [61]

$$\lambda_D = \sqrt{\frac{\varepsilon * T_e}{q * n_e}} \quad [\text{m}], \tag{14.39}$$

where ε = electrical permittivity of vacuum = 8.9×10^{-12} F/m, q = electron charge = 1.6×10^{-19} C, T_e = electron T [eV], and n_e = plasma density [m^{-3}]. A typical conventional plasma having $n_e = 10^{16}$ m^{-3} and $T_e = 3$ eV has $\lambda_D = 130$ μm.

Also, the thermal ion vibration is responsible for a certain minimal FWHM for the IADF (up to 6°). Increasing the bias voltage across the sheath region will minimize this effect, but this has the disadvantage of decreasing the mask selectivity and all kinds of RIE damage.

In conclusion, if we assume that the ion enhances the etch process only at the spots where it collides with the surface, then the lag effect will be already pronounced for medium aspect ratios (5–10), even for low-pressure RIE equipment with fairly sharp IADs (*FWHM < 5°*). So, it seems to be effective to sharpen the IAD as much as possible.

14.4.7 Ion/solid interactions

Until now, we have assumed that an ion entering a trench and colliding with the wall would knock an inhibiting particle (O) away, enabling radicals (F) to etch the substrate (Si). The yield of a collapsing ion can be as large as 100; at the moment the ion has created a hole in the passivating layer a flood of radicals will attack the surface until the gap is closed by the inhibitor again. But, when the inhibitor is strong enough, it could be that the gap is closed before any radical had the chance to react with the bare surface. This delicate balance is controlled by the flux of incoming radicals and inhibitors and by the surface coverage of these particles which is strongly dependent on the temperature.

The contribution of the IAD on the lag effect would be less pronounced if the ions coming in at a glancing angle would simply reflect from the sidewall and were forced to the trench bottom. This effect is well-known for IBE researchers and it is responsible for typical IBE problems like trenching. The image force theory predicts that the lag effect will decrease when the ion velocity is increased (Eq. (14.25)). However, while changing the ion velocity, the ion impact angle is changing too. Unfortunately, it is difficult to separate both effects and to decide which phenomena has the main contribution to the lag effect. So, we have to be careful, because the faster the ions, the higher the impact of off-normal ions at the sidewalls. This higher impact energy could make it possible for ions to modify the surface and thus to increase the lag effect. In contradiction, it could also be that at higher velocities the ions rebound due to the more glancing impact angle. The interpretation of the results is even more disturbed, because changing the ion velocity by changing the rf power supply will change the sheath thickness also and therefore the FWHM of the IAD.

An ion which bounces off a solid medium will have a certain chance to lose its charge due to charge transfer processes. Such an ion would follow a straight line until the next collision with the opposite sidewall, not disturbed by any electromagnetic interaction. Ions which preserve their charge will follow a parabolic curve due to the electrostatic fields until they collapse with the same sidewall again, only somewhat deeper into the trench. So, the highly energetic ions will etch or reflect specular, and therefore the Clausing factor is 1, i.e., the ions which enter a trench will not be backscattered.

Besides the kinetic energy transferred from the electrostatic field in the dark space region, an ion which is approaching a solid surface will gain energy from the image field. This extra collision energy can be calculated with the help of Eq. (14.26). In this formula the energy transferred between the ion potential and ion kinetic energy is found. This energy can be quite large, especially at the moment the ion is very close to the solid surface. The kinetic energy gained by an ion due to the image force is calculated to be $E_{kin} = q/16\pi\varepsilon r_0 \sim 1\,eV$ at an ion-solid-distance of $r_0 = 3$ Å. This energy could be already enough to remove a sticking inhibiting particle at the surface and therefore it could promote etching. So, the image energy could cause an ion to react always with the solid surface and ion reflection is made impossible. At this moment it is unclear if there is ion reflection in the trench but, evidently, this information is extremely crucial in the battle against RIE lag.

14.4.8 *Experimental*

In order to study RIE lag, two different IBARE reactors were used, both using an SF_6/O_2 chemistry to etch Si samples. The first is a conventional IBARE used for the high-pressure experiments (>50 mTorr). For the low-pressure experiments a dedicated high-density IBARE is used.

It might be expected beforehand that the depletion of radicals and inhibitors is not important. This is because the same mechanisms which transport inhibitors in a trench control the stream of radicals. So, the depletion of inhibitors could be counterbalanced by the depletion of radicals. Nevertheless, many mechanisms cause RIE lag and the "horizontal trench" experiment should distinguish between ion and radical depletion. After that, experimental data will be compared with the image force theory and the angular distribution simulation.

Inhibitor depletion

In Eq. (14.10) we arrived at the synergetic etch mechanism of radicals, ions, and inhibitors in ion-inhibitor IBARE. From this equation we see that the depletion of inhibitors will result in a higher etch rate. So, in small trenches where the supply of inhibitors is restricted due to inhibitor shadowing, or other effects, inverse RIE lag should be the result. Because we never observed inverse RIE lag in our experiments, inhibitor depletion is excluded to be responsible for the lag effect during the IBARE of silicon in an SF_6/O_2 chemistry.

Radical depletion

Gottscho et al. used the kinetic model for ion-induced IBARE, Eq. (14.3), as a basis to describe the dependence of etching rates on aspect ratio [21]. However, the equations governing the ion-inhibitor processes are completely different (Eq. (14.10)). We assume that during the anisotropic etching of Si in an SF_6/O_2 chemistry the supply of radicals is always sufficient. So, the ions are controlling the etch process completely. The argument for this assumption is described next.

In order to decide which mechanism is the main cause for RIE lag, radical or ion depletion, we constructed special "horizontal trenches." Wafers were prepared with a thin layer of 100, 300, 400, 1000, and 2700 nm, respectively, of polysilicon sandwiched between two nonetching oxide layers. Before depositing the top oxide layer, the polysilicon was etched

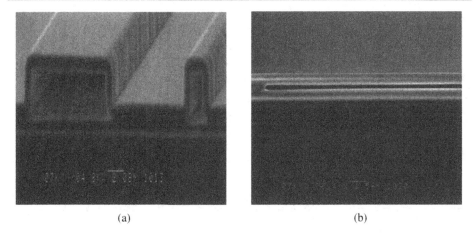

Fig. 14.18. (a) Frontside of horizontal trenches created by means of surface micromachining techniques, including sacrificial layer etching, used for the experiment to prove
that radical depletion is not important. (b) Sideview of a broken horizontal trench which
shows that radical depletion is not important. The aspect ratio of the tube is approximately $20/0.1 = 200$.

into a grating pattern with trench width of 2, 4, 6, 8, 10, and 1000 μm, respectively. In this
way, an array of extremely small prismatic tubes were fabricated. After breaking, the sample
of the polysilicon was etched in a high-pressure (200 mTorr) SF_6 plasma in the PE mode
(Fig. 14.18(a)). Under these circumstances, ions and inhibitors are not important and only
radicals are controlling the etching. We found the etch rate to be constant in time after 5, 10,
15, and 20 μm of etching for all the tubes. For example, the smallest tube (100 nm $* 2$ μm)
was etching at the same speed of ca.1 μm/min as the biggest tube (2700 nm $* 1000$ μm) up
to aspect ratios over 100. Figure 14.18(b) shows that radicals were able to underetch the
100 nm $* 1000$ μm polysilicon layer 20 μm (i.e., the aspect ratio $\sim 20/0.1 = 200$).

So, we conclude that the contribution of radical depletion on the RIE lag effect is small
in our situation.

Ion depletion

Now that the inhibitor and radical depletion has been excluded, the depletion of ions will
be considered. First, we already found that the interactions between ions traveling in the
trench at the same time were negligible and differential charging effects could be skipped
because silicon is a relatively good conductor. The thickness of the dark space region
($\sim 1–10$ mm) is much bigger than the micron-sized structures and therefore the field distortion is excluded also. So, there are only two candidates left to explain the lag effect in our
case: the image force mechanism and ion angular distribution in corporation with the ion
shadowing effect. In the next sections, first the ion distribution, then the image force, and
finally a combination of both mechanisms will be given.

Ion distribution: We will start with a simulation of the ion depletion due to the etching or
capturing of ions at the trench sidewalls caused by the IAD. After that, the analytical results
are given for a simple approximated IAD. For our experiment and simulation we used the
data given by Manenschijn, Liu et al., and Thompson et al. [51, 53, 60]. Liu et al., measured
the angular as well as the energy distribution of argon ions as a function of the pressure by

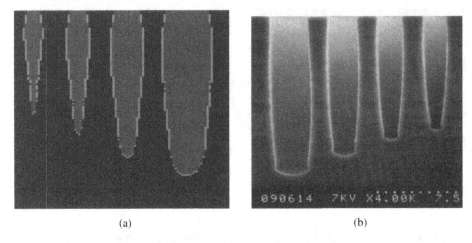

(a)   (b)

The relative ion energy flux coming in at a certain angle was settled at: {ion energy flux/incident angle} = {13/0°, 13/1°, 13/2°, 12/3°, 12/4°, 11/5°, 9/6°, 8/7°, 6/8°, 5/9°, 5/10°, 4/11°, 4/12°, 4/13°, 3/14°, 3/15°, 3/16°, 3/17°, 2/18°, 2/19°}.

Fig. 14.19. (a) Output of the simulation program which predicts RIE lag. (b) Real trenches etched with the help of a conventional RIE at 50 mTorr.

a small orifice in the cathode of an IBARE apparatus. Trenches were etched at a pressure given by Liu et al., and their data on the distribution functions were used in the trench simulation program. The data we used as input parameters in the simulator were taken from the pressure-dependent IAD argon IBARE sampling experiment Fig. 14.19 from Ref. [53] where Liu et al. measured the fraction of ions as a function of the incident angle. Of course we have to multiply the flux distribution at these angles with the average energy to find the total energy flux to the substrate as a function of the angle. We use Fig. 14.24 from Ref. [53] and we find the average ion energy to decrease with increasing incident angle. From these figures we extracted the data as found in Table 14.2.

Because Liu et al. used a set of ion collecting shells with constant width, the collecting area is increasing with incident angle. So, the real flux density as a function of α and θ should be divided by $2\pi\theta$. In contradiction, Janes used a detector system which could be tilted in the (α, θ) direction, giving the angle-dependent flux density directly [59]. Our IA/ED simulator corresponds to the data of Liu et al. and there is no need for adjusting the data given in their paper.

The result of our simulation is shown in Fig. 14.19 and it indicates that bottling is a typical result of the angular distribution of ions and predicts the lag effect. So, our results are in agreement with Jurgensen and his experiments [50]. Moreover, this time we used the measured distribution functions from literature to simulate the profile development of the trench. However, we could use the technique also backward, that is, we first examine an etched trench and after that we try to match this result with simulator output. In this way we may retrieve the IAD and IED functions in situ without extra tools like ion energy analyzers. So, we are able to characterize the apparatus with respect to beam confinement.

Image force: Let's look again at the statement Gottscho et al. gave with respect to RIE lag: "*However, further analysis of RIE lag data (Fig. 14.20(a)) shows that etching rates are not*

Table 14.2. *Experimental data for the ion angular distribution (IAD) and average ion energy distribution (IED) from Liu et al. [53].*

{ion flux fraction/incident angle} = {F} =
{91/0°, 57/4$\frac{1}{2}$°, 23/9°, 15/13$\frac{1}{2}$°, 11/18°, 10/22$\frac{1}{2}$°, 09/27°, 07/31$\frac{1}{2}$°, 05/36°} at 010 mTorr,
{59/0°, 54/4$\frac{1}{2}$°, 30/9°, 23/13$\frac{1}{2}$°, 17/18°, 15/22$\frac{1}{2}$°, 13/27°, 11/31$\frac{1}{2}$°, 09/36°} at 050 mTorr,
{16/0°, 31/4$\frac{1}{2}$°, 34/9°, 33/13$\frac{1}{2}$°, 29/18°, 26/22$\frac{1}{2}$°, 22/27°, 20/31$\frac{1}{2}$°, 16/36°} at 500 mTorr.

{average ion energy/incident angle}= {E} =
{29/0°, 26/4$\frac{1}{2}$°, 19/9°, 15/13$\frac{1}{2}$°, 13/18°, 10/22$\frac{1}{2}$°, 09/27°, 08/31$\frac{1}{2}$°, 07/36°} at 010 mTorr,
{22/0°, 20/4$\frac{1}{2}$°, 17/9°, 15/13$\frac{1}{2}$°, 13/18°, 11/22$\frac{1}{2}$°, 09/27°, 08/31$\frac{1}{2}$°, 07/36°} at 050 mTorr,
{11/0°, 11/4$\frac{1}{2}$°, 11/9°, 13/13$\frac{1}{2}$°, 13/18°, 12/22$\frac{1}{2}$°, 10/27°, 09/31$\frac{1}{2}$°, 08/36°} at 500 mTorr.

{ion energy flux/incident angle}= {$F * E/50$} =
{53/0°, 30/4$\frac{1}{2}$°, 09/9°, 05/13$\frac{1}{2}$°, 03/18°, 02/22$\frac{1}{2}$°, 02/27°, 01/31$\frac{1}{2}$°, 01/36°} at 010 mTorr,
{26/0°, 22/4$\frac{1}{2}$°, 10/9°, 07/13$\frac{1}{2}$°, 04/18°, 03/22$\frac{1}{2}$°, 02/27°, 02/31$\frac{1}{2}$°, 01/36°} at 050 mTorr,
{04/0°, 07/4$\frac{1}{2}$°, 07/9°, 09/13$\frac{1}{2}$°, 08/18°, 06/22$\frac{1}{2}$°, 04/27°, 04/31$\frac{1}{2}$°, 03/36°} at 500 mTorr.

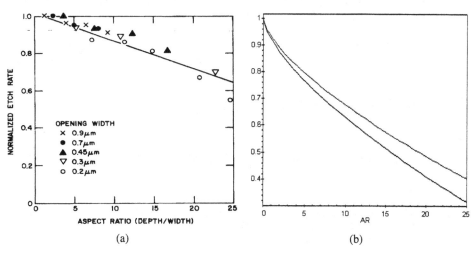

Fig. 14.20. (a) Experimental data from Chin et al. [39] showing the normalized etch rate as a function of the aspect ratio. This graph is generally used to prove the "aspect-ratio-only" dependency of RIE lag. (b) Normalized etch rate as a function of the aspect ratio calculated with the help of the image force theory. (with permission from the authors)

consistent with absolute feature-size scaling but rather aspect ratio scaling [28]. Note that because aspect ratio continually increases during an etch, etching rates are necessarily time-dependent; as the etching proceeds and the aspect ratio grows larger, the etching rate decreases (Fig. 14.21(a))." After this he continues with the image force: *"The image force mechanism [54] predicts that RIE lag will increase as feature size decreases. This mechanism cannot be responsible for the observed dependency of etching rates on aspect ratio; however, it is expected to contribute to RIE lag in high aspect ratio [28], subquarter-micron trenches [54]."* In the next discussion we will use the same data from Chin and coworkers as Gottscho et al. did and compare it with results from the image force theory.

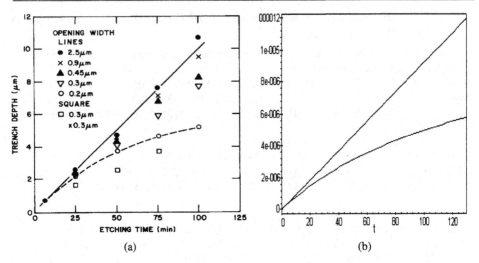

Fig. 14.21. (a) Experimental data from Chin et al. [39] showing the trench depth as a function of etch time for different trench widths. (b) Theoretical data from the image force theory showing the trench depth as a function of etch time for line opening widths of 0.2 μm and 2.5 μm. (with permission from the authors).

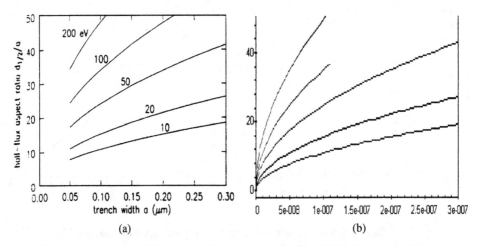

Fig. 14.22. (a) Numerical data from Davis and Vanderslice [56] showing the half-flux aspect ratio as a function of the trench width for different ion energies. (b) Data from the image force theory showing the half-flux aspect ratio as a function of the trench widths for different in energies. (with permission from the authors).

But before this, our analytical results will be compared with the numerical output found in the paper by Davis and Vanderslice [56].

Davis and Vanderslice [56] numerically computed the aspect ratio for which half of the ion flux collided with the walls due to the image force. We did exactly the same, but now the calculation is analytical with first-order assumptions. As can be observed, both graphs match perfectly (Figs. 14.22(a) and (b)). So, it is allowed to use the much easier to handle analytical expressions (Eq. (14.22)–(14.25)) instead of the numerical plots.

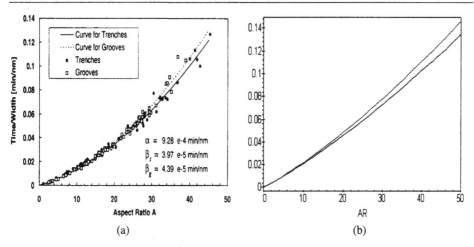

Fig. 14.23. (a) Experimental data from Müller et al. [62] showing the time to width ratio as a function of the aspect ratio. (b) Theoretical data from the image force theory showing the time to width ratio as a function of the aspect ratio (with permission from the authors).

Now, let us take the kinetic energy of the incoming ions to be $E_{kin} = 10$ eV and run the Maple program of the image theory. The graphs for all the trenches are matching extremely well, as shown in Figs. 14.21(a) and (b) for the 0.2-μm and 2.5-μm trenches. Quite surprisingly, the same data which were used to exclude the image force mechanism to be responsible for the lag effect in micron-sized trenches are now used to make it a possible mechanism. The discrepancy may be caused by the technique Chin et al. used to obtain Fig. 14.20(a) while using Fig. 14.21(a) [39]. All our attempts to reconstruct Fig. 14.20(a) from Fig. 14.21(a) failed and it is unclear to us why; a curved line seems to be convenient, also as predicted by the image force and/or ion distribution theory. As a last critical remark, it is found from the distribution theory that the relative etch rate is *inversely* dependent on the AR. So, a better plot would have been the function $RelR = RelR(AR^{-1})$.

Another experimental result also supports the image force theory. Müller and coworkers plotted the etch time t scaled by the feature width w versus the aspect ratio AR, as shown in Fig. 14.23a for differently wide grooves ($w = 0.2$–0.4 μm) and trenches ($w = 0.2$–0.8 μm). They found that the data points could be fitted by a second-order polynomial in AR [62]:

$$t/w = \alpha * AR + \beta * AR^2 \quad \text{[s/m]}. \tag{14.40a}$$

Müller et al. [62] wrote: *"The coefficient α is the inverse of the etch rate for an AR of 0.... Later on with increasing AR, the etch rate is diminished due to RIE lag quantified by the coefficient β in Eq. (14.40(a)). This diminution can be different for trenches and grooves due to their different geometrical properties. Therefore we introduced β_t for trenches and β_g for grooves. ... Trenches and grooves show essentially the same etch rate and RIE lag over the AR regime from 0 to 45. This does not confirm the data of Chin et al. [39], who have observed that grooves are etched approx. 50% faster than trenches of the same width."* However, Chin et al. were discussing square holes whereas Müller et al. used elongated trenches. This is essentially different, not only for the image force theory, but for the distribution theory as well.

From the image force theory we may rewrite Eq. (14.24b) and get

$$t/w = \frac{AR}{R_{\max}} + \frac{3}{5 R_{\max}} * \left(\frac{k}{E_{\text{kin}} w}\right)^{1/3} * AR^{5/3} \quad [\text{s/m}]. \tag{14.40b}$$

This function is plotted in Fig. 14.23b for 10 eV ions, $R_{\max} \sim 1.1$ μm/min, and $w = 0.2$ and 0.4 μm grooves. Both graphs (Figs. 14.23(a) and (b)) match reasonably, although the curvature is somewhat different. It should be noticed that the distribution theory does give the quadratic behavior, but not the linear one, as found in Eq. (14.40c). After all, rewriting Eq. (14.38) gives

$$t/w = \frac{AR_c}{2 R_{\max}} + \frac{1}{5 R_{\max} AR_c} * AR^2 \quad [\text{s/m}]. \tag{14.40c}$$

As a consequence of the image theory, the influence of the image force should decrease when the kinetic energy of incoming ions is decreased, especially for the smaller trenches. Therefore, (sub)micron-sized trenches should be etched with the help of the ICP-IBARE at different bias voltages. But, care should be taken when interpreting the results, because changing the ion energy by changing the rf power supply will change the sheath thickness also and therefore the FWHM of the IAD. We conclude that the image force cannot be excluded as a mechanism responsible for RIE lag in micron-sized HARTs.

Image force together with the ion distribution: In this last section we will combine the image force theory together with the ion distribution theory. Let's reprint the equations for the relative etch rate from the image force theory (Eq. (14.22)) and that of the ion distribution theory (Eq. (14.37)):

$$RelR_{\text{image}} = 1 - \left(\frac{k AR^2}{E_{\text{kin}} w}\right)^{1/3} \quad [-] \tag{14.22}$$

$$RelR_{\text{ID}} = \begin{cases} 1 & \text{for } AR < AR_c \\ AR_c/AR & \text{for } AR > AR_c \end{cases} \quad [-]. \tag{14.37}$$

The total relative etch rate is the multiplication of both equations, that is,

$$RelR_{\text{total}} = \begin{cases} 1 - \left(\dfrac{k AR^2}{E_{\text{kin}} w}\right)^{1/3} & \text{for } AR < AR_c \\[3mm] \dfrac{AR_c}{AR} * \left[1 - \left(\dfrac{k AR^2}{E_{\text{kin}} w}\right)^{1/3}\right] & \text{for } AR > AR_c \end{cases} \quad [-]. \tag{14.41}$$

So, for silicon structures etched with the help of the SF$_6$/O$_2$ ion-inhibitor chemistry, the etch rate will follow the feature size dependent image force theory when the aspect ratio is below the critical aspect ratio, i.e., $AR_c \sim \frac{1}{2} FWHM$. Beyond that point, the etch rate will follow more and more the aspect ratio dependency of the ion distribution theory.

In Fig. 14.24(a) the result of the theoretical analysis is shown for trenches etched with 60-eV kinetic ions. The critical aspect ratio is taken as 8, which corresponds with a $3\frac{1}{2}^{\circ}$ FWHM of the IADF. In Fig. 14.24(b), experimental data are plotted for trenches etched with the help of the ICP-RIE. Although the match is not 100%, the magnitude and the behavior are quite identical. The match could be better if we would have used a Gauss distribution

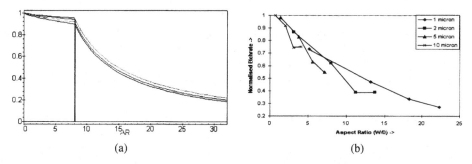

Fig. 14.24. Relative etch rate of a trench for $w = 1, 2, 5$, and $10\ \mu$m, $E_{kin} = 60$ eV, and $AR_c = 8$. (a) Theory and (b) experiment.

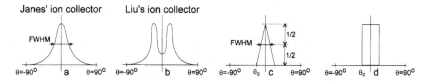

Fig. 14.25. (a) and (b) The real ion distribution and (c) and (d) the approximated triangle-shaped distribution.

(Figs. 14.25(a) and (b)) instead of a triangle-shaped distribution (Fig. 14.25(c)) to simulate the ion energy distribution. Also, the etch rate should be monitored directly with the help of a grating pattern and a laser setup. The time between two minima (or maxima) from the photo detector would then give us in situ the etch rate while running the process. Another technique is to use sufficiently big open structures to calibrate the etch time and/or etch rate between two subsequent etches. This is necessary because native oxide and the reflected power will be different between two etches.

14.5 Conclusions

A classification of the most important RIE trench effects is given. These effects can be divided into tilting and the ARDE effects, i.e., bowing, bottling, TADTOP, and RIE lag. Evidence is found that micrograss is a member of this ARDE group. It is concluded that the ion flux is the most important source of etching particles. These ions are etching the passivating layer and are controlling the etched profile by their direction. Radicals are necessary for etching the silicon, but under normal circumstances the radical density is high and they have to wait for an incoming ion before they are able to remove a silicon atom. So, to understand the RIE inhibitor process, it is necessary to follow the path of an ion coming from the plasma glow.

The ion trajectory starts its course of life at the moment a molecule is hit by a fast electron and the molecule is ionized. This ion diffuses with random thermal velocity or drifts due to small electrical fields in the plasma glow bit by bit to the plasma boundary. Arriving at the plasma boundary, the boundary distortion is influencing the electrostatic field, thus the direction of the ion which is entering the sheath region. During its passage through the plasma sheath, the ion will collide with other particles and its angle and energy will diverge in supplement to the thermal energy. This effect is expressed in the ion angular/energy distribution functions (IADF and IEDF). After this the ion is entering the trench where deflection due to electrostatic fields from the conducting sample is changing the ion's direction. The ion will end its journey at the sidewall or the bottom of the trench. Depending on the energy and collision angle of the ion, it will reflect, etch, or just end at the sidewall or the trench bottom. The ratio of the ions ending on the trench sidewall and bottom is directly related to RIE lag.

- *Tilting* (Fig. 14.1(d)) is caused by (i) boundary distortion or (ii) the local differences in radical density. Boundary distortion is found when the sample-, trench-, or wafer-clamping geometry is bigger than the thickness of the sheath region. Increasing the thickness may prevent this effect and can be accomplished by lowering the pressure. A difference in the radical density between two places causes radicals to flow into the region having the smallest density. In other words, the isotropic radical flux is becoming collimated and tilting is a result.

- *Micrograss* (Fig. 14.1(c)) is formed when the flux of incoming ions is highly collimated, i.e., the IADF is sharp. The grass is prevented when the IADF is broadened, e.g., by way of a higher pressure or due to ion reflection. In other words, a perfectly collimated ion flux is not always preferable; a little dispersion will prevent grass.

- *Bowing* (Fig. 14.1(a)) is caused by ion deflection due to electrostatic fields in the trench and the subsequent ion etching of the passivator. Increasing the passivation on the sidewall is decreasing this effect (Fig. 14.5).

- *TADTOP* (Fig. 14.1(a)), i.e., trench area dependent tapering of profiles, is closely related to bowing and is found when the opposite wall of a trench is influencing the path of an ion. Therefore, when the opposite wall is close, the ion deflection is less pronounced and the tapering will be more positive. Again, increasing the sidewall passivation is decreasing this effect (Fig. 14.5).

- *RIE lag due to ions* (Fig. 14.6) is caused by ion depletion. Ion deflection moves ions to the trench sidewall where they will be captured or will etch the passivator. The amount of ions reaching the bottom will therefore be smaller. For the smaller trenches this ion depletion is reached sooner than for the wider trenches because the flux/wall area ratio is smaller after a certain etch depth. Once again, increasing the sidewall passivation is decreasing this effect due to sidewall charging (Fig. 14.5).

- *RIE lag due to radicals* (Figs. 14.7(a) and (b)) is caused by radical depletion. The radical flux is isotropical when entering the trench, but radical etching will sharpen the RADF and lower the amount of radicals what will reach the bottom. For the smaller trenches this effect is more pronounced after a certain etch depth. To decrease this effect, radical etching of the sidewall should be prevented, i.e., ion deflection should be prevented. There is no proof found for RIE lag due to radical reflection.

- *Bottling* (Fig. 14.1(b)) is caused by ion shadowing and sharpening of the IADF when ions are traveling down a trench. The effect is caused by off-normal ions due to a relatively high operation pressure. In the higher regions of the trench, ions are coming in at different angles and there will be a strong undercut depending on the ion energy. However, the ions which are etching the sidewall are used up, and due to ion shadowing the IADF is sharpened. At a certain AR and critical angle the ions are not etching anymore but are only captured or reflected. Therefore, ion shadowing and ion depletion are thought to be responsible for the bottling effect. To prevent bottling it is possible (i) to sharpen the IADF by way of decreasing, e.g., the pressure or sheath thickness or (ii) to decrease the IEDF by way of, e.g., the dc self-bias to the point where the ions bounce off the sidewall without etching it (Fig. 14.9). Notice that the last technique forces us to use high pressures in a conventional RIE whereas the first one indicates to use a low pressure. This paradoxical nature can be overcome by using a showerhead or a dual source system able to produce highly directional low energetic ions like the ICP-RIE.

- *Quantitative analysis of RIE lag:* In our quest for the perpetrator of the lag effect into silicon high aspect ratio trenches using an SF_6/O_2 RIE we thought of some experiments which could distinguish between the candidates. There are three main groups to be considered in ion-inhibitor RIE: (1) radicals, (2) inhibitors, and (3) ions. This quantitative analysis has given arguments to believe that the image force and the IAD in conjunction with ion shadowing are the important mechanisms to consider for explaining RIE lag. The image force scales with feature size whereas the (triangle-shaped) IAD scales *inversely* with the AR. This gives us the possibility to discriminate between these two mechanisms. However, this is not straightforward. The data from Chin et al., which was used by Gottscho et al. also, with respect to the AR scaling law do not fit with our results. Nevertheless, there is another way to draw a distinction between the image force and the IAD theory. Trends may be observed with pressure and ion velocity with respect to the lag effect.

- *The IAD theory* predicts that the lag effect increases with pressure. We have written a C++ simulator in windows which had a close match with experimental data and the distribution theory for trenches where the image force is believed to be too small. The simulation shows besides RIE lag the bottling effect (i.e., the trenches are somewhat wider near the top).

The image force theory predicts that the lag effect will decrease when the ion velocity is increased. However, while changing the ion velocity, the ion impact angle is changing too. Unfortunately, it is difficult to separate both effects and to decide which phenomena have the main contribution to the lag effect. So, we have to be careful, because the faster the ions, the higher the impact of off-normal ions at the sidewalls. This higher impact energy could make it possible for ions to modify the surface and thus increase the lag effect. In contradiction, it could also be that at higher velocities the ions rebound due to the more glancing impact angle. The interpretation of the results is even more disturbed because changing the ion velocity by changing the rf power supply will change the sheath thickness also and therefore the FWHM of the IAD. Nevertheless, in comparing the image force theory with the data given by Chin et al. and Müller et al. we found a remarkably good match. So, the image force seems to be a strong candidate to explain the lag effect in micron-sized features. We have to stress that the match is in fact *too good*; we should have seen the influence of the IAD also. This can be explained because we used our only fitting parameter, the ion velocity, at a value of 10 eV to match experiment with theory. The energy used in the experiments of Chin et al. and Müller et al. was probably higher, which would decrease the influence of the image force. Both the image force and the distribution theory together would then result in a satisfying match between theory and experiment. Nevertheless, our point of view with respect to the image force is that you cannot argue with this data that, as Gottscho et al. stressed, *"This mechanism cannot be responsible for the observed dependency of etching rates on aspect ratio."*

Of course, care should be taken when using simulations. As Gottscho et al. remarked: *"In some cases, perhaps only one parameter is varied in fitting the simulation results to experiment. But such comparisons say virtually nothing about the validity of the model and many models are likely to yield comparably good 'agreement' with experiment."* However, in both our models, the image force theory as well as the ion angular distribution simulation, we did not make use of any extraordinary fitting parameter. Because of this, the validity of the models is largely enhanced.

- *In total:* The BSM is proven to be a practical tool to create highly anisotropic profiles with aspect ratios of at least 20 for etching HARTs without sidewall bowing and with smooth bottoms and sidewalls. The low-pressure cryogenically cooled ICP-RIE is an especially convenient apparatus to achieve this goal. In microengineering, the ion angular distribution is probably the most important mechanism to explain RIE lag. A lower pressure decreases this influence. In nanoengineering, the image force has to be added. Higher energetic ions decrease this effect.

References

[1] S. M. Irving, Kodak Interface Proc., 2, 1968.

[2] R. Castaing and P. Laborie, C. R. Acad. Sci. (Paris) **238** (1954).

[3] J. W. Coburn and H. F. Winters, J. Appl. Phys. **50** (1979).

[4] A. Reinberg, U. S. Patent 3,757,733 (1973).

[5] H. Jansen et al., EP appl. No. 94202519.8, Patent PCT/NL95/00221.

[6] E. Berenschot, H. Jansen, G.-J. Burger, H. Gardeniers, and M. Elwenspoek, (1996, February 11), Proc. of IEEE Micro Electro Mechanical Systems Workshop. pp. (277–284). San Diego. ISBN 1084-6999.

[7] R. A. Gottscho, C. W. Jurgensen, and D. J. Vitkavage, J. Vac. Sci. Tech. **B10** (1992).

[8] H. Jansen et al., Microelectr. Eng. **27**, 475 (1995).

[9] H. Jansen, H. Gardeniers, and J. Fluitman, Journal of micromechanics and microengineering, **6** (pp. 14–28). ISSN 0960-1317.

[10] H. Jansen, M. de Boer, B. Otter, and M. Elwenspoek, (1995, Jan. 29–Feb. 2), Proc. of IEEE Micro Electro Mechanical Systems Workshop, pp. (88–93). Amsterdam. ISBN 7803-2503.

[11] J. C. Arnold and H. H. Sawin, J. Appl. Phys. **70**, 15 (1991).

[12] S. G. Ingram, J. Appl. Phys. **68** (1990).

[13] Surface Technology Systems Limited, Prince of Wales Industrial Estate, Abercarn, Newport, Gwent.

[14] Oxford Instruments, Plasma Technology, North End, Yatton, Bristol, England.

[15] R. Legtenberg, H. Jansen, and M. Elwenspoek, J. Elec. Soc. **142** (1995).

[16] W. H. Juan, S. W. Pang, A. Selvakumar, M. W. Putty, and K. Najafi, Solid-State S&A workshop, Hilton Head, SC, 1994, p.

[17] A. Manenschijn, Thesis, Technical University of Delft, 1995.

[18] A. Hayasaka, Y. Tamaki, M. Kawamura, K. Ogiue, and S. Ohwaki, IEDM Tech. Dig. **82**, 62 (1982).

[19] H. Sunami, T. Kure, N. Hashimoto, K. Itoh, T. Toyabe, and S. Asai, IEDM Tech. Dig. **82**, 806 (1982).

[20] M. de Boer, H. Jansen, and M. Elwenspoek, Proc. Eurosensors IX and the 8th International Conference on Solid-State Sensors and Actuators, Stockholm, Sweden, 1995, p. 565.

[21] R. A. Gottscho, C. W. Jurgensen, and D. J. Vitkavage, J. Vac. Sci. Technol. **B10**, 2133 (1992).

[22] H. Jansen, M. de Boer, and M. Elwenspoek, Proc. IEEE Micro Electro Mechanical Systems '96, San Diego, CA, 1996, p. 250.

[23] T. M. Mayer and R. A. Barker, J. Vac. Sci. Technol. **21**, 757 (1982).

[24] U. Gerlach-Mayer, J. W. Coburn, and E. Kay, Surf. Sci. **103**, 177 (1981).

[25] T. M. Mayer, R. A. Barker, and L. J. Whitman, J. Vac. Sci. Technol. **18**, 349 (1981).

[26] R. A. Barker, T. M. Mayer, and W. C. Pearson, J. Vac. Sci. Technol. **B1**, 37 (1983).

[27] T. M. Mayer and R. A. Barker, J. Electrochem. Soc. **129**, 585 (1982).

[28] H. Gokan and S. Esho, J. Electrochem. Soc. **131**, 131 (1984).

[29] H. Jansen, H. Gardeniers, M. de Boer, M. Elwenspoek, and J. Fluitman, J. Micromech. Microeng. **6**, 14 (1996).

[30] J. C. Arnold and H. H. Sawin, J. Appl. Phys. **70**, 5314 (1991).

[31] S. G. Ingram, J. Appl. Phys. **68**, 500 (1990).

[32] Y. H. Lee and Z. H. Zhou, Proc. 8th Symp. Plasma **90-2**, 34 (1990).

[33] D. J. Economou and R. C. Alkire, J. Electroch. Soc. **135**, 941 (1988).

[34] R. J. Davis, Appl. Phys. Lett. **59**, 1717 (1991).

[35] E. S. G. Shaqfeh and C. W. Jurgensen, J. Appl.Phys. **66**, 4664 (1989).

[36] K. P. Giapis, G. R. Scheller, R. A. Gottscho, W. S. Hobson, and Y. H. Lee, Appl. Phys. Lett. **57**, 983 (1990).

[37] J. W. Coburn and H. F. Winters, Appl. Phys. Lett. **55**, 2730 (1989).

[38] J. Pelletier, J. Phys. **D20**, 858 (1987).

[39] D. Chin, S. H. Dhong, and G. J. Long, J. Electrochem. Soc. **132**, 1705 (1985).

[40] H. Jansen, M. de Boer, J. Burger, R. Legtenberg, and M. Elwenspoek, Microelectr. Eng. **27**, 475 (1995).

[41] D. J. Vitkavage, A. Kornblit, R. A. Nicholas, D. P. Favreau, and S. C. McNevin, 1991 Tegal Plasma Proc. Symp., San Francisco, CA, 1991.

[42] C. M. Horwitz, S. Boronkay, M. Gross, and K. Davies, J. Vac. Sci. Technol. **A6**, 1837 (1988).

[43] S. Dohmae, J. P. McVittie, J. C. Rey, E. S. G. Shaqfeh, and V. K. Singh, Proc. of the Symp. on Patterning Sci. and Technol. II **92-6**, 163 (1992).

[44] I. Langmuir, J. Am. Chem. Soc. **40**, 1361 (1918).

[45] M. Knudsen, Ann. Physik **28**, 999 (1909); **35**, 389 (1911).

[46] P. Clausing, Ann. Physik, **12**, 961 (1932).

[47] P. Lorrain, D. P. Corson, and F. Lorrain, *Electromagnetic Fields and Waves* (W. H. Freeman and Company, New York, third ed.), p. 211.

[48] H. C. Jones, R. Bennett, and J. Singh, Proc. 8th Symp. Plasma Proc. **90-2**, 45 (1990).

[49] K. Nojiri, E. Iguchi, K. Kawamura, and K. Kadota, Ext. Abstr. 21st Conf. on Sol. State Dev. and Materials, Tokyo, 153 (1989).

[50] C. W. Jurgensen, J. Appl. Phys. **64**, 590 (1988).

[51] A. Manenschijn, Thesis, Technical University of Delft, 1995.

[52] R. H. Bruce and A. R. Reinberg, J. Electrochem. Soc. **129**, 393 (1982).

[53] J. Liu, G. L. Huppert, and H. H. Sawin, J. Appl. Phys. **68**, 3916 (1990).

[54] S. Dushman, J. M. Lafferty (ed.), *Scientific Foundation of Vacuum Technique* (John Wiley & Sons, Inc., London, 2nd ed., 1962).

[55] I. Lundström, P. Norberg, and L.-G. Petersson, J. Appl. Phys. **76**, 142 (1994).

[56] W. D. Davis and T. A. Vanderslice, Phys. Rev. **131**, 219 (1963).

[57] M. J. Kusher, J. Appl. Phys. **58**, 4024 (1985).

[58] A. D. Kuypers and H. J. Hopman, J. Appl. Phys. **63**, 1894 (1988).

[59] J. Janes, J. Vac. Sci. Technol. **A12**, 97 (1994).

[60] B. E. Thompson, K. D. Allen, A. D. Richards, and H. H. Sawin, J. Appl. Phys. **59**, 1890 (1986).

[61] B. Chapman (John Wiley & Sons, New York, 1980).

[62] K. P. Müller, K. Roithner, and H.-J. Timme, Microelect. Eng. **27**, (1995).

15

Moulding of microstructures

In recent years there has been considerable interest in high aspect ratio processing. Although a variety of three-dimensional structures such as nozzles, diaphragms, and beams have been fabricated in silicon, other materials such as plastics and metals are increasingly studied as well. For certain applications the availability of such materials is a necessity, as in the case of optical devices directly fabricated in polymers. Inexpensive fabrication processes for microstructuring of different materials are desirable as well.

The LIGA technique can produce extremely high aspect ratio structures by means of irradiating thick resist layers of poly methyl meta acrylate (PMMA) with radiation of 2 Å wavelength [1] and subsequent electroplating of the mould insert for moulding or embossing. The costs, however, are high as well, making it economically less attractive. Several techniques have been proposed in order to fabricate economical high aspect ratio structures [2–4]. All of them can be used to directly fabricate the end product or to fabricate the mould for subsequent replication of the inverse structure of the mould. By using the recent innovations in IBARE plasma processes of silicon or polymers, structures with variable tapering and high aspect ratios can be fabricated. The silicon moulds can be used either directly for embossing and moulding or for electroplating and subsequent embossing, as will be explained in this chapter.

The advantage of direct embossing with silicon is that the fabrication process of a polymer structure is reduced to silicon etching and embossing, facilitating prototyping of plastic microstructures. The difference of the thermal expansion coefficient of polymer and silicon is larger with respect to, e.g., nickel. Therefore, higher shear forces are expected when releasing the structure. On the other hand, unlike in the LIGA process, the angle of the mould insert can be controlled by the IBARE process, which facilitates separating the mould from the mould insert.

When these high aspect ratio processing techniques are discussed, the following parameters are of interest: Geometrical freedom in two dimensions and in three dimensions, batch- or single-structure processing, the maximum aspect ratio (as a function of the diameter or width of the structure), the anisotropy, the roughness of the resulting structure, the resolution, and the accuracy of the desired image/structure. The term "aspect ratio" is defined as follows: The aspect ratio is the ratio of the diameter with the depth of the structure. Trenches will be defined as positive aspect ratios, while needles are denoted by means of negative aspect ratios. Typical dimensions of an example structure in microtechnology are a diameter of 5 micrometer and a depth of up to 100 micrometer.

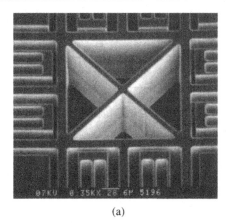 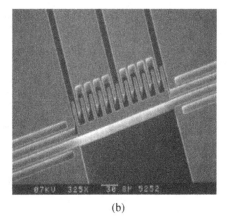

(a)   (b)

Fig. 15.1. (a) SEMs of a dry-etched silicon structure. The etching parameters were: 30 sccm SF_6, 5 sccm O_2, 75 mTorr, 75 W and the etch time was 30 min. The etch depth is over 50 μm.

15.1 Dry etching

For the experiments described here, an SF_6/O_2 plasma has been used to etch silicon. The SF_6/O_2 plasma allows elegant control of physical and chemical etching and deposition effects. Therefore, the etched profile can be controlled by adjusting the parameter setting. As has been mentioned in Section 12.3, increasing the oxygen content will decrease the chemical etching component and at high oxygen concentrations, positively tapered profiles will be obtained. It is possible that with correct control of the pressure, power, and flows, high aspect ratios and vertical or sligthly positive tapered profiles can be obtained.

In Fig. 15.1 two scanning electron microscope (SEM) photographs of trenches etched in silicon are shown. A chromium layer of about 50 nm has been evaporated onto the (100) silicon wafers as masking material. Chromium, as a mask material, has several advantages. The selectivity of a chromium mask with silicon is very high, the chromium is not sputtered up to a dc bias voltage of 200 eV, and since chromium is a conductor, the mask will not be charged during the etch process. Charging could give rise to strong local fields and influence the process in an uncontrollable manner. The silicon has been etched in a conventional plan-parallel plate reactor. The process parameters are presented in the figure heading. In this case, the roughness of the silicon is lower than 100 nm. The typical feature size of the structures varied from micrometers up to tens of micrometers.

15.2 Electroplating

The silicon trenches can be electroplated in order to fabricate a mould insert. In the case of electroplating, the silicon should be electrically conducting. For the experiments described here, a nickelsulfamate electrolyte has been used. In the electroplating process, the hydrated nickel ions diffuse toward the silicon structure, the cathode. At the silicon surface, the Ni ions and electrons combine and hydrogen is formed. There will be a concentration profile of all the reactants and products involved from the cathode (silicon) into the electroplating bath. The process parameters should be selected carefully in order to prevent too much molecular

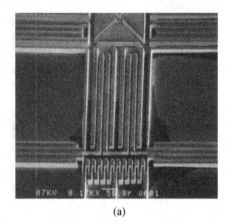
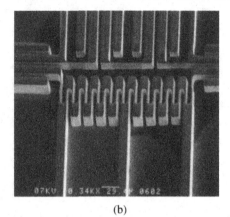

(a)   (b)

Fig. 15.2. (a) SEMs of a nickel electroplated structure, using silicon as a mould insert
for the electroplating process. The process parameters were: 101.7 g/l Ni, 2.92 g/l Br$^-$,
40 g/l H$_3$BO$_3$, releasing agent: 5 ml/l, T $= 50°$C, pH $= 3.8$, current density 1 A/dm^2,
Q $= 4.95$ A min. The silicon mould insert has been etched in KOH, which selectively
etches silicon with regard to nickel. (b) Some silicon has been left as a contrast for the
Ni, as EDX analysis demonstrated.

hydrogen evolution, which may be incorporated into the Ni structure, or too much stress
in the Ni structure or growth, induced artefacts of the solid crystal structure. The growth
should always be started very slowly, in order to ensure a homogeneous nickel seed layer.
A release agent, careful control of the pH, and careful control of the current is necessary.

In Fig. 15.2, two SEM photographs of Ni structures electroplated from a silicon structure
are shown. In this case, 50 nm Ti and 100 nm Ni have been evaporated on the silicon. The
figures demonstrate that successful electroplating on silicon is possible. As can be observed
in the figure, the structure is a fine replica of the silicon mould.

In addition to the above-mentioned experiments, direct electroplating on silicon is possible after boron doping of the silicon. Silicon wafers have been boron doped using B$_2$H$_6$
for 1 hour at 1100°C, resulting in a sheet resistance of 3.9 Ω/sq.

15.3 Embossing

The silicon structure has been used as a mould for direct embossing, as is shown in Fig. 15.3.
For the embossing process a Collin Press 200 P/M has been used, allowing a maximum
pressure of 250 bar and a maximum temperature of 300°C. One bar corresponds to a force
of 1.2 kN. The pressure on the mould insert is the ratio between the applied force and the
surface of the mould insert. During the embossing process, the surface may alter due to
flowing of the polymer.

The embossing temperature should be high enough in order to heat the polymer material above the glass-phase transition temperature and thereby decrease the viscosity of
the material substantially, allowing the material to flow. For the polycarbonate sheets of
750 micrometers, used for the silicon mould insert experiments, the glass-phase transition
temperature was 145°C. The embossing temperature has been measured in the press plates,
which overestimate the temperature of the polymer about 10°C. Typical embossing temperatures are 185°C– 200°C. When the embossing temperature is too low, the filling of the
mould will not be satisfactory and air inclusion may be observed. When the embossing

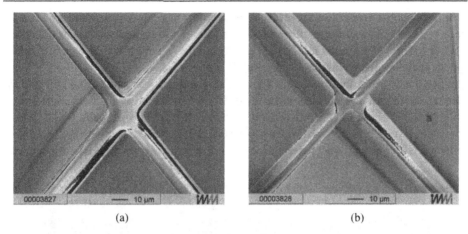

(a)   (b)

Fig. 15.3. (a) SEMs of embossed polymer structures. Silicon has been used as a mould insert and the polymer is polycarbonate. The process parameters were: $P = 4$ bar, $T = 185°C$, $t = 300$ sec. (b) The silicon surface was 1 cm^2 and the heating cycle has been performed in 300 sec. up to $185°C$, without pressure.

temperature is too high, the flowing of the material may be so abundant that no polymer is left for the microstructures to attach to.

The embossing pressure should be sufficiently high to provide sufficient filling, and a critical interplay between the temperature rising and cooling together with the pressure increase and decrease should ensure proper embossing. Generally, the pressure is applied when the temperature reaches its final value, and will be released just after the cooling started. The pressure cannot be released when the temperature is still high since the material will continue to flow and deformations will be observed. However, the pressure may not be released at too low temperatures since tension will then build up in the microstructures, leading to deformations and cracks in the structures. Process-induced stress and thermally induced stress make the releasing a tedious task. This should be done without tilting the embossed microstructures with respect to the mould insert. The structures in Fig. 15.3 have been released by hand, thereby the structures have been tilted, as careful inspection reveals.

In order to facilitate the releasing of the microstructures, the silicon has been coated with a thin fluorocarbon coating by means of a plasma deposition process. The fluorocarbon coating has been deposited in the same parallel plate reactor in which the structures have been etched [5]. Although quantitative measurements have not been performed, the fluorocarbon coating did facilitate the releasing of the microstructures.

15.4 Dry etching with photoresist as a mask

When silicon structures are etched in an IBARE system with an SF_6/O_2 chemistry and a highly selective mask such as a metal, purely straight sidewalls or a positive tapering usually correlate with grass in the valleys of the trenches. In Fig. 15.4b this grass is clearly visible. This grass can be prevented by settling the IBARE process into a chemistry where a slightly negative tapering occurs, as shown with the help of the black silicon methodology in a previous chapter. However, this tapering, like the grass, is undesirable for most moulding purposes because it is difficult to emboss such a structure without destroying (parts of) the feature. Therefore, in moulding applications, a slightly positive tapering without grass is highly preferable. In order to achieve this, it is possible to use a slowly withdrawing mask to create a positive sidewall angle, although the etch process itself is settled to create a slightly negative tapering. So, the parameter setting is forced into a point were the etch process is smooth and the withdrawing mask will result in a positive tapering.

In Fig. 15.4, spikes are etched with photoresist as a mask. The resist [Shipley 1828] is baked at 90° to prevent reflowing of the resist. Due to the almost straight sidewalls of the resist, the low selectivity of the resist does not facilitate the withdrawing of the mask. On the other hand, Fig. 15.5 shows resist which is baked at 130° in order to spread the resist a little, forming a spherical-like resist pattern. Due to the low selectivity of the resist with respect

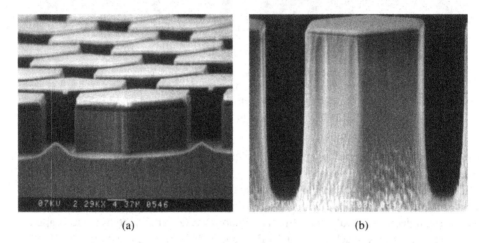

(a)   (b)

Fig. 15.4. (a) Dry etching of silicon after 10 minutes with photoresist baked at 90°C as a mask. The resist has an almost straight profile which does not facilitate a withdrawing of the mask. (b) Dry etching of silicon after 20 minutes with photoresist baked at 90°C as a mask. Because the resist is not withdrawing, grass is observed when a positive tapered profile is wanted.

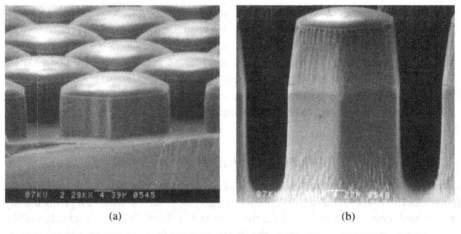

(a)   (b)

Fig. 15.5. (a) Dry etching of silicon after 10 minutes with photoresist baked at 130°C as a mask. The resist has a positive tapered profile which facilitates the withdrawing of the mask. (b) After 20 minutes. Due to the withdrawing of the mask, the profile which is transferred into the silicon is much more positively tapered while preserving the same amount of grass.

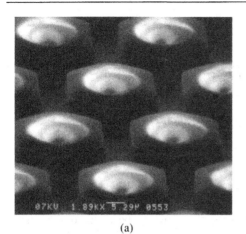

(a)   (b)

Fig. 15.6. (a) Dry etching of silicon after 10 minutes with photoresist baked at 130°C as a mask. Due to the withdrawing of the mask, the profile which is transferred into the silicon is slightly positive tapered. By settling the etch recipe for a slightly negative profile, grass is prevented. (b) After 20 minutes. Due to the withdrawing of the mask, the profile which is transferred into the silicon is slightly positive tapered. As a result, spikes with positive slopes and smooth surfaces are obtained.

to the silicon and the positive slope of this resist, the mask will repeal during etching. It is observed that the withdrawing resist has its influence halfway the structure height upward (Fig.15.5(b)). Now, when this process is optimized, by giving the resist a pre-etch in an oxygen plasma, it is possible to create smooth and positive tapered profiles, as found in the Fig. 15.6.

References

[1] H. Lehr and W. Ehrfeld, Proc. of the European Symposium on Frontiers in Science an Technology with Synchrotron Radiation, Aix-en-Provence, France, April 5–8, 1994.

[2] L. Paratte, H. Lorenz, R. Luthier, R. Clavel, and N. F. de Rooij, Conf. Proc. MEMS '94, Oiso, Japan, Jan. 25–28, 1994, p. 119.

[3] W. Ehrfeld and D. Münchmeyer, Nuclear Inst. Methods Phys. Res. **A303**, 523 (1991).

[4] T. M. Bloomstein and D. J. Ehrlich, Conf. Proc. MEMS '91, Nara, Japan, Jan. 30–Feb. 2, 1991, p. 202.

[5] J. Elders, H. V. Jansen, and M. Elwenspoek, Conf. Proc. MEMS '94, Oiso, Japan, Jan. 25–28, 1994, p. 170.

16

Fabrication of movable microstructures

After etching micromechanical silicon structures, they have to be released. This isn't straightforward and many techniques have been proposed. Frequently in surface micro-machining, a glass layer is used as a sacrificial layer which is etched using wet or vapor etchants [1–3]. They are cheap but suffer for the so-called sticking problem, due to surface tension of liquids, and many solutions have been proposed to solve this [2–9]. To avoid sticking, dry etching technologies have been presented recently such as the SCREAM, SIMPLE, and BSM-ORMS processes [10, 11, 16]. These techniques are self-aligned and are suitable for batch fabrication. The releasing of the structures takes place in the gas phase, so there is no liquid phase responsible for stiction. This chapter presents a study of plasma release techniques (bulk and surface micromachining) for the fabrication of movable silicon micromechanical structures.

16.1   SCREAM

The acronym SCREAM stands for single-crystal reactive etching and metalization [11]. In Fig. 16.1(a) the process scheme of this technique is shown, which consists of five basic steps. After

(1)   directional chlorine-based etching of silicon using oxide as a mask material,
(2)   the trenches are passivated with a PECVD oxide.
(3)   This oxide is removed at the bottom of the trench (CHF$_3$ plasma) and after that
(4)   an isotropic fluorine-based etch (SF$_6$) will release the structures.
(5)   For electrical contacting, a deposited metal layer is needed.

Notwithstanding its potential, there are some shortcomings, such as

(1)   the deposited PECVD layer should be stress-arm or tensile to avoid buckling (Fig. 16.2),
(2)   the mask layer should be stress-arm to avoid bending after the releasing (Fig. 16.3),
(3)   the released structures are hollowed out during the isotropic etch (Fig. 16.4), and
(4)   RIE lag while etching the Si directional may cause problems during the isotropic etch because the etch depth might be different at both sides of a beam (Fig. 16.4).

The last two problems make mask design rules necessary. Additionally, many process steps are needed to fabricate the MEMS (e.g., IBARE, PECVD, MIE, and sputter equipment), which are timeconsuming and might limit the yield.

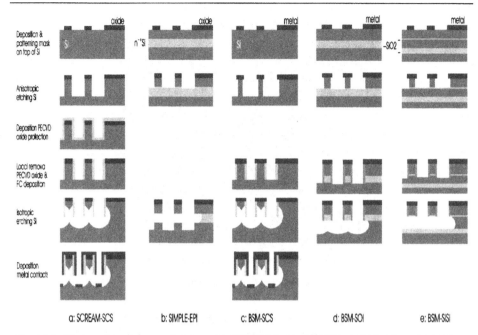

Fig. 16.1. Various dry plasma techniques: (a) SCREAM, (b) SIMPLE, and (c), (d), and (e) BSM.

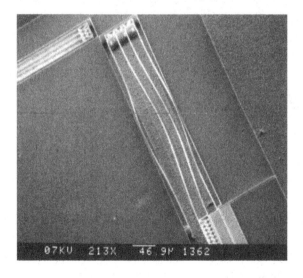

Fig. 16.2. Buckling due to compressive stress.

16.2 SIMPLE

The acronym SIMPLE stands for silicon micromachining by single step plasma etching [10]. This technique uses a Cl_2-based plasma chemistry which etches p- or lightly doped Si anisotropically but heavily n-doped Si isotropically. In such a way, movable MEMS can be patterned and released from the substrate in a single-step IBARE plasma, as shown schematically in Fig. 16.1(b). However, disadvantages of SIMPLE are the following:

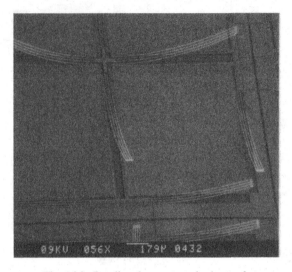

Fig. 16.3. Bending due to stress in the mask.

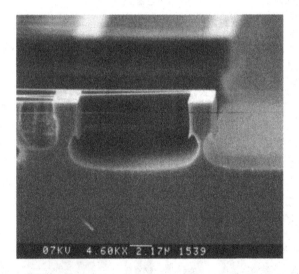

Fig. 16.4. Hollowing out of beams and asymmetrical releasing due to RIE lag.

(1)   A thick PECVD oxide mask is needed as a mask material because the Si/SiO$_2$ selec-
       tivity is rather low in a Cl$_2$ plasma.
(2)   The underetch rate is low (50 nm/min) and is a function of the doses of the buried
       layer and the spacing between the microbeams.
(3)   After etching, the bottom beneath the structures shows deep trenches, which may
       affect the moving of free-hanging structures.

16.3   BSM-ORMS

The acronym BSM-ORMS stands for black silicon method one-run multistep [16, 17]. It
solves the problems mentioned in the previous sections and, of course, it creates some

new ones. It has the ability to etch, passivate, and release MEMS in one run, and this explains its name. It is developed on a twin deposition/etch parallel-plate system operating at 13.56 MHz. There are three different process schemes developed for the BSM-ORMS process: BSM-SCS for bulk and BSM-SOI and BSM-SISI for surface micromachining.

16.3.1 BSM-SCS

This technique starts with a standard single-crystalline silicon (SCS) wafer. After the deposition of a 30-nm (lift-off) mask for the pattern definition, the movable structures can be fabricated in only one IBARE run with four individual steps (Fig. 16.1(c)):

(1)   the (an)isotropic IBARE ($SF_6/O_2/CHF_3$) of the SCS shown in Fig. 16.5,
(2)   the "IBARE-deposition" (CHF_3) of the passivation (C_xF_y film) of the sidewalls of the structures shown in Fig. 16.6,

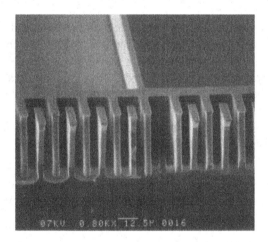

Fig. 16.5. Anisotropic etching with $SF_6/O_2/CHF_3$ plasma.

Fig. 16.6. Sidewall passivation with CHF_3 plasma.

(3)   the IBARE (SF$_6$/O$_2$/CHF$_3$) of the floor, and

(4)   the IBARE (SF$_6$) releasing of the bulk silicon shown in Fig. 16.4.

Eventually, the process can be finished with a conformal step coverage of a C$_x$F$_y$ film to protect the released structures from the environment [12]. For instance, these fluorocarbon (FC) films do have an extremely low surface tension and therefore they repel water and others.

Some remarks concerning the BSM-SCS steps are as follows:

- *Step 1*: After the anisotropic etching, the profile has to be vertical with a little underetch making it possible to deposit a FC layer where no ion bombardment occurs, i.e., under the "roof" of the mask (Fig. 16.6). The profile can be adjusted by using the BSM method [13–17]. Also, RIE lag can be suppressed by applying this method [14]. Typical parameters during anisotropic etching are SF$_6$/O$_2$/CHF$_3$ gasflow = 30/10/7 sccm, power flux = 0.3 W/cm^2, self-bias = 40 V, pressure = 75 mTorr, 3-inch silicon loading, and target temperature = 10°C. Note: The wafer has to be clamped sufficiently.

- *Step 2*: The deposition of FC is a function of, e.g., pressure and self-bias [12]. This layer is protecting the sidewalls during isotropic etching. We observed a satisfying coverage of the sidewalls directly under the mask roofs at pressure = 20 mTorr, power flux = 0.3 W/cm^2, self-bias = 600 V, and CHF$_3$ gasflow = 10 sccm. Again, in this step the wafer has to be attached carefully with the 10°C cooled target. A typical deposition rate for this process is 20 nm/min.

- *Step 3*: Before starting the isotropic etching with an SF$_6$ plasma, it is necessary to "clear" the floor of the trenches first with an oxygen-based plasma such as SF$_6$/O$_2$/CHF$_3$.

Although the bending and buckling problems are solved, BSM-SCS is not able to prevent beams from hollowing out. In the following sections, two different BSM-ORMS processes are treated which prevents this latter effect.

16.3.2   BSM-SOI

The technique starts with commercially available silicon on insulator (SOI) wafers (Fig. 16.1(d)). Now the subsequent steps are

(1)   the (an)isotropic IBARE (SF$_6$/O$_2$/CHF$_3$) of the top silicon,

(2)   the IBARE (CHF$_3$) of the insulator together with the passivation (C$_x$F$_y$ film) of the sidewalls of the structures,

(3)   the IBARE (SF$_6$/O$_2$/CHF$_3$) of the floor, and

(4)   the IBARE (SF$_6$) of the bulk silicon (Fig. 16.7).

Some remarks about BSM-SOI are as follows:

- *Step 1*: When the intermediate insulator of the SOI wafer is reached, the etch process has to be stopped to avoid unwanted underetching. This is a crucial step because when the insulator is reached, the loading is decreasing, causing a strong enhancement in lateral etching. The etching process is stopped by, e.g., visual inspection.

- *Step 2*: During this passivation step the insulator is etched at a speed of ca. 50 nm/min.

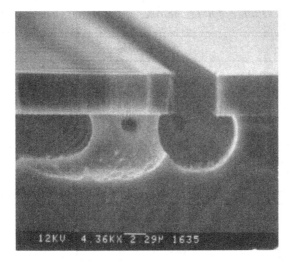

Fig. 16.7. Isotropic etch with SF$_6$ plasma.

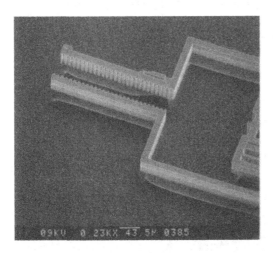

Fig. 16.8. A typical example of the BSM-SOI process scheme: a gripper.

- *Step 4*: When a relative thick intermediate layer (>0.1 μm oxide) is used, bending of the beam structure may occur due to compressive stress in the oxide layer. Another problem might be differences in stress between the two bonded silicon wafers, which may introduce bending/buckling of beams. Another disadvantage of the BSM-SOI technique is that after releasing the free-hanging structures, deep trenches are found and the underetch rate is limited due to the relatively high silicon loading.

X-y stages, microgrippers, springs, etc. were fabricated as shown in Figs. 16.8, 16.9, and 16.10 with typical dimensions listed in Table 16.1 in comparison with the SCREAM and SIMPLE processes. As can be seen, the only limiting steps are aspect ratio of trenches and beams.

Table 16.1. *Outline of BSM one-run process with typical dimensions.*

Process step	Description	Etch/deposit rate
1. Photolithography	Shipley 1805	500 nm
2. Metal deposition	E-beam: Cr 25 nm	6 nm/min
3. Lift-off	Acetone	—
4. Trench etching	$SF_6/O_2/CHF_3$	0.5–1 μm/min
5. Sidewall passivation	CHF_3	20 nm/min
Oxide etching	CHF_3	50 nm/min
6. Floor etching	$SF_6/O_2/CHF_3$	1 min
7. Releasing	SF_6	0.5–1.0 μm/min
8. FC deposition	CHF_3	20 nm/min

Beam structure	SCREAM	SIMPLE	BSM
Height (μm)	<20	<4	<400
Width (μm)	<5	<4	<50
Length (μm)	<2000	<2000	<2001
Lateral Gap (μm)	>1	>3	>1
Aspect ratio beam	<10	<10	<50
Aspect ratio trench	<7	<7	<10

Fig. 16.9. A typical example of the BSM-SOI process scheme: part of a spring.

16.3.3   BSM-SISI

To eliminate the problem of the deep trenches found in the BSM-SOI technique (and also in the BSM-SCS technique), silicon on insulator on silicon on insulator (SISI) wafers are constructed. Now a silicon layer of 1–2 μm is used as a sacrificial layer surrounded by two insulators (Figs. 16.1(e) and 16.11). The deepest lying insulator protects the Si during release resulting in a smooth/flat bottom and a small loading, thus high lateral etch rate (>1 μm/min). In Figs.16.12 and 16.13, two examples of this technique are shown.

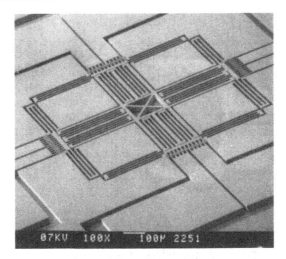

Fig. 16.10. A typical examples of the BSM-SOI process scheme: an xy stage.

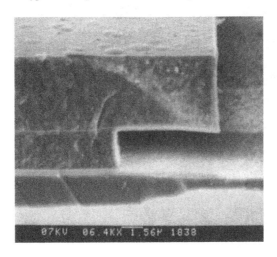

Fig. 16.11. The BSM-SISI process: profile view.

16.3.4 Conclusion

It can be stated that the BSM one-run multistep process is favorable for the releasing of MEMS with long thin beams. It includes the black silicon method as an excellent tool for profile control and to suppress RIE lag. Instead of oxide, a 30-nm thin chromium metal layer is used as a mask, which has an almost infinite selectivity with respect to silicon and creates less additional stress problems (bending). The fluorocarbon layer has a low Young's modulus which prevents stress problems in long thin beams (buckling). The intermediate layer of an SOI wafer prevents the beam from hollowing out during the isotropic etch, making an exact definition of the structure height possible. After the mask is deposited it is now possible to fabricate very quickly, accurately, and at low cost free-hanging MEMS (e.g., an accelerometer, tuneable spring/filter, AM/FM modulator, or micromechanical transistor) in one process run with an IBARE plasma without turning the plasma off.

Fig. 16.12. The BSM-SISI process: a gripper.

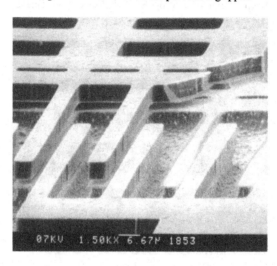

Fig. 16.13. The BSM-SISI process: a part of a combstructure.

References

[1] R. Legtenberg et al., Sens. and Act. **A43**, 230 (1994).

[2] T. Lober and R. Howe, Proc. IEEE MEMS, 59 (1988).

[3] J. Ruzyllo et al., J. Elec. Soc. **140**, L64 (1993).

[4] H. Guckel et al., Proc. IEEE MEMS, 71 (1989).

[5] R. Legtenberg and H. Tilmans, Sens. and Act. **A45**, 57 (1994).

[6] C. Mastrangelo and G. Saloka, Proc. IEEE MEMS, 77 (1993).

[7] D. Kobayashi et al., 7th Int. Conf. Solid-State S&A (1993), p. 14.

[8] A. Kovacs and A. Stoffel, Europ. Workshop on micromach (1992), p. 114.

[9] R. Alley et al., Proc. IEEE MEMS, 202 (1992).

[10] Y. Li et al., Proc. IEEE MEMS, 398 (1995).

[11] K. A. Shaw, Z. Zhang, and N. MacDonald, Sens. and Act. **A40**, 63 (1994).

[12] H. Jansen et al., Sens. and Act. **A41–A42**, 136 (1994).

[13] H. Jansen et al., EP appl. No. 94202519.8.

[14] H. Jansen et al., Microelectr. Eng. **27**, 475 (1995).

[15] H. Jansen et al., Proc. IEEE MEMS 488 (1995).

[16] M. de Boer, H. Jansen, and M. Elwenspoek, Proc. Eurosensors IX and the 8th international conference on solid-state sensors and actuators, Stockholm, Sweden, 1995, 565, 142-C3.

[17] H. V. Jansen, M. de Boer, R. Legtenberg, M. Elwenspoek, Proc. Micro Mechanics Europe (MME'94), Pisa, Italy, Sept. 1994, p. 60.

Index

Printed in the United States
By Bookmasters